Sonochemistry
Source of Clean Energy

Editors

Felipe López-Saucedo
Institute of Nuclear Sciences, Department of Radiation Chemistry
National Autonomous University of Mexico (UNAM)
Mexico City, Mexico

Amira Jalil Fragoso-Medina
Institute of Nuclear Sciences, Department of Radiation Chemistry
National Autonomous University of Mexico (UNAM)
Mexico City, Mexico

Emilio Bucio
Institute of Nuclear Sciences, Department of Radiation Chemistry
National Autonomous University of Mexico (UNAM)
Mexico City, Mexico

CRC Press
Taylor & Francis Group
Boca Raton   London   New York

CRC Press is an imprint of the
Taylor & Francis Group, an **informa** business
A SCIENCE PUBLISHERS BOOK

I0034415

Cover credit: The cover images are self-composed by Dra. Amira Jalil Fragoso-Medina with individual elements, extracted from the global information network.

First edition published 2024
by CRC Press
2385 NW Executive Center Drive, Suite 320, Boca Raton FL 33431

and by CRC Press
4 Park Square, Milton Park, Abingdon, Oxon, OX14 4RN

© 2024 Felipe López-Saucedo, Amira Jalil Fragoso-Medina, and Emilio Bucio

CRC Press is an imprint of Taylor & Francis Group, LLC

Library of Congress Cataloging-in-Publication Data (applied for)

ISBN: 978-1-032-21298-2 (hbk)
ISBN: 978-1-032-21300-2 (pbk)
ISBN: 978-1-003-26771-3 (ebk)

DOI: 10.1201/9781003267713

Typeset in Times New Roman
by Prime Publishing Services

Preface

The purpose of this book is to provide an overview of ultrasonic waves, this alternative energy source endeavors to satisfy the standards demanded by a responsible society, because Sonochemistry is a fast, simple, cost-effective, environmentally friendly, and safe method with various approaches to synthesize and treat organic and inorganic materials.

This text delves into the scientific context of Sonochemistry, emphasizing the challenges Sonochemistry faces to become a primary method of synthesis with potential applications in industry. The contents focus on synthetic developments with promising properties to achieve a sustainable society. Sonochemistry as green synthesis, reaction mechanisms, and applications in nanotechnology are highlighted topics, as well as the future perspective of Sonochemistry-based processes for water and waste treatment. In addition, the effect of acoustic waves on food and their role and application in the food industry are addressed, mentioning some guidelines on the use of ultrasound in fresh and processed meat and dairy products, where sensory quality can be improved, avoiding any detriment to the nutritional value.

The multidisciplinary view of Sonochemistry presented in this book is made possible because all the chapters were written by specialists in different areas of Sonochemistry. The fact that each chapter is developed freely and independently provides a fresh approach to Sonochemistry.

The process of compiling the necessary information for this book has been gratifying, since this text encourages the proper development and enhances the applications of Sonochemistry, providing the possibility of understanding and reasoning its importance.

Contents

CHAPTER 1

From Sustainable Development to Sonochemistry

Amira Jalil Fragoso-Medina[1,2]

1. Introduction

Human activities have deteriorated the planet, particularly the natural sources of food and water. Thus, a change in human actions is necessary to get the equilibrium back with the environment and to improve the quality of life for future generations. To reach sustainable development, the United Nations Educational, Scientific and Cultural Organization (UNESCO) has implemented the Sustainable Development Goals from 2015 to 2030 yr, in order to protect the world and provide peace and prosperity in people's lives. One of the best ways to achieve these is modification of daily activities of human beings to develop more sustainable actions. It is also important to highlight that the chemical processes used to make drugs, cosmetics, food, and materials, among othersare considered the main source of environmental pollution. To solve this problem, the concept of green chemistry is introduced to prevent environmental pollution.

Green chemistry is protocolized by 12 principles. Among these principles, principle six highlighted environmental and economic impacts, which are recognized due to the energy requirements. Also the electromagnetic spectrum is employed to identify the radiation emission-absorption region of a substance. The electromagnetic waves can be classified according to their energy, the type of radiation, the common name of the waves, the sources of radiation, wavelength, and their frequency (UTAH DEQ WM&RC 2021), for example, according to the frequency range, there exist 4 types of sound: infrasound (frequency ranges from zero to 20 Hz), sound of human auditory field (frequency ranges from 20 Hz to 20 kHz), ultrasound (frequency ranges from 20 kHz to 100 GHz), and hypersound (frequency greater than 100 GHz) (Mason and Peters 2002). The application of sound energy can alter

[1] Institute of physics, Department of Condensed Matter, National Autonomous University of Mexico (UNAM), Exterior Circuit, University City, Mexico City, 04510.
[2] Institute of Nuclear Sciences, Department of Radiation, National Autonomous University of Mexico (UNAM), Exterior Circuit, University City, Mexico City, 04510.
Email: amirakhalil2001@yahoo.com.mx

the material properties and provide suitable conditions in which molecules undergo chemical reactions. The study of this research area is called sonochemistry.

Sonochemistry has now attracted attention in diverse fields, especially in the organic chemistry and pharmaceutical industry, due to its numerous advantages, namely low-cost, high yields, clean energy source, ease of implementation, prevents or minimizes the formation of waste, use of mild conditions, better reaction times, etc. Based on these advantages, sonochemistry can reduce environmental pollution as well as improve the quality of life by developing sustainable applications. Therefore, sonochemistry is considered an alternative approach for the development of ecological techniques.

2. From Sustainable Development to Sonochemistry

2.1 Sustainable Development

2.1.1 Progress of Sustainable Development

Human activities inhabited the planet to live their way of life, which provoked significant changes to the environment such as soil desertification, exploitation of wood resources, contamination of water sources (i.e., rivers, Mediterranean Sea, etc.), heavy rainfall, and air pollution (Bollaín and Vicente 2019, Educadores por la sostenibilidad 2012, Escobar 2002, GPE 2017, Lebreton et al. 2018, Mallén 2018a, Soto de Prado y Otero 2010). These changes in the environment are triggered due to the intense human activity which is driving the degradation of the planet. Therefore, future generationsmay not get an opportunity to utilize natural resources that human beings are using now. A new strategy needs to be implemented so that people can use the natural resources and consecutively protect the environment as well as natural resources (Mallén 2018b, Mani et al. 2017, UNESCO 2018, WHO, AP 2022).

Several strategies have been established to solve the above cited problems for a sustainable future. The Kyoto Protocol was created in 1997, which discussed ideas for reducing the greenhouse gas effects (Berhe et al. 2018, MTERD 2021). The Paris Agreement, in 2015, primarily focused on the threat of climate change. The Paris Agreement also talked about reducing two degrees Celsius of the global temperature by controlling greenhouse gas emissions (UNESCO CFEAP 2021]). In addition, the World Air Quality Index Project was developed to produce an air quality map to check the air quality in real time for several world localities (TWAQI PT 2022). All of these strategies are to achieve sustainable development carried out by the United Nations to reach peace, security, economic, and social progress (UN CNU 2021).

It is essential to highlight that the concept of sustainable development was initially addressed in the Brundtland report in 1987, which was presented to the United Nations by the World Commission on Environment and Development. This report dictates "It is in the hand of humanity to make development sustainable to ensure that it meets the needs of the present without comprising the ability of future generations to meet their own". This phrase can be considered a fundamental reference for sustainable development and implies an ecological change to achieve sustainability (UN WCED 1987).

Some years later, the Earth Charter was released (in 1992). This Charter is an international declaration of principles, proposals, and deep and consistent international dialogues with the purpose of forming a just, sustainable, and peaceful society in the 21st century. This concept seeks to inspire people in a new sense of interdependence and responsibility for the good of humanity (CCT 2020). In 2000, the Millennium Goals were established at the Millennium Summit. The Millennium Goals discussed the effect of the life quality of people on the development of sustainability for global aspects. Various parameters, such as extreme poverty, hunger, universal primary education, gender equality, women's empowerment, infant mortality, maternal health, diseases, etc., can control the environmental sustainability of the planet, and development cooperation.

Subsequently, the General Assembly of the United Nations in 2002 proclaimed the Decade of Education for Sustainable Development from 2005 to 2014r. This strategy established goals to integrate the principles and practices for sustainable development in all aspects of education and learning, reorient education towards sustainability by changing people's thinking, and favor a more sustainable future in terms of environmental integrity, economic viability, and social justice (UNESCO EDS 2021). The scope of the Decade of Education for Sustainable Development is extremely broad and impactful with qualitative and quantitative data, which are important for monitoring and evaluating the decade (UNESCO DNU 2021).

In 2014, the Conference on Education for Sustainable Development was held in Aichi Nagoya, Japan. This conference addressed the previous declaration and highlighted urgent measures to strengthen and expand Education for Sustainable Development. The commitment to these strategies was made by all member countries of the United Nations (UNESCO CMEDS 2017).

During the celebration of the 70th General Assembly of the United Nations in 2015, the heads of the United Nations and civil society entities analyzed and reconsidered the fulfillment of the goals established by the previous Millennium Summit. This program introduced comprehensive action in different spheres, such as social, economic, and environmental sections, etc., to achieve sustainable development (UNESCO ODM 2017).

As a result, seventeen new objectives were reported, including prosperity and peace in this action plan, called Sustainable Development Goals (SDGs). The main goal of this program is to diminish poverty, protect the planet, and provide peace and prosperity at the end of 2030. These established strategies are called "Agenda 2030", which is designed for the following slogan "of the people, by the people, and for the people". The Scientific and Cultural Organization arranged the program with the active contribution of the United Nations Educational for the achievement of sustainable development. The division of the United Nations monitored their management, coordination, and implementation of education up to the year 2030. Since this program is made by several governments around the world, it covers a broader territory of action that influences the fulfillment of these goals (Flores et al. 2009, Girón 2016).

Thus, as mentioned earlier, sustainability is defined as meeting the needs of the present without compromising the ability of future generations to meet their own needs. However, it is important to highlight the current efforts made by various organizations are valuable but insufficient. Therefore, everyone needs to acquire their responsibility. Consequently, the development of sustainable nature must be essential and permanent for every human (UNESCO BP 2021).

2.1.2 *The Sustainable Development Goals*

The 17 SDGs (UNDP 2022), which are expected to be achieved by 2030, are listed below along with the targets associated with each of them:

One: The end of poverty; eradicate it.

Two: Zero hunger; end hunger in all its forms and malnutrition.

Three: Health and well-being; provide universal health coverage.

Four: Quality education; ensure that all children complete their free primary and secondary education.

Five: Gender equality; give women equal rights in access to economic resources.

Six: Clean water and sanitation; guarantee universal access to safe drinking.

Seven: Affordable and non-polluting energy; expand infrastructure and improve technology for clean energy in all developing countries.

Eight: Decent work and economic growth; achieve full and productive employment and decent work for all.

Nine: Industry innovation and infrastructure; promote innovation in the means we use with more than half of the world's population living in cities.

Ten: Reduction of inequalities; bridging the differences between human communities regarding mass transportation, access to renewable energies, industries, information technologies and communication.

Eleven: Sustainable cities and communities; improving the safety and sustainability of cities.

Twelve: Responsible production and consumption; achieving economic growth and sustainable development through the most efficient use of resources.

Thirteen: Climate action; acting for the climate protection and recovery.

Fourteen: Life below sea; sustainably manage and protect marine and coastal ecosystems from land pollution.

Fifteen: Life in land ecosystems; take urgent action to reduce the loss of natural habitats and biodiversity that are part of our common heritage and support global food and water security.

Sixteen: Peace, justice, and strong institutions; reduce all forms of violence and work with governments and communities to find lasting solutions to conflict and insecurity.

Seventeen: Partnerships to achieve goals; achieve global intercommunication by improving access to technology and knowledge and fostering innovation.

The action plan of the 2030 Agenda with the SDGs is shown in a circular format, Figure 1, because the world is more interconnected than ever, and encompasses the 17 goals in the following five areas.

A. Peace (Foster peaceful, just, and inclusive societies, free from fear and violence).

B. Partnership (Implement the agenda through a solid global partnership).

C. Prosperity (Ensure that everyone can enjoy a prosperous life and that economic, social, and technological progress can be achieved in harmony with nature).

D. Planet (Protect the planet from degradation for current and future generations).

E. People (End poverty and hunger in all forms and guarantee a healthy environment with dignity and equality).

The Sustainable Development Goals aim to stimulate sustainable economic growth through higher levels of productivity and technological innovation.

It is also necessary to mention that the United Nations Educational, Scientific and Cultural Organization has implemented the project to develop the SDGs for different areas, such as education, natural sciences, and social sciences (including communication and information) (UNESCO ODS 2021). It highlights the diversity in three main areas of development: poverty eradication, structural transformations,

Figure 1. Action plan of the 2030 Agenda.
Image made by Amira Jalil Fragoso-Medina, icons in the rearrangement were obtained from internet with credit to whom it corresponds.

and building resilience (UNDP LAC 2022). Additionally, valuable knowledge and interesting experiences of the SDGs have planned to fulfill their goals in 170 countries by the 2030 (UNDP 2022).

The scope of SDGs is directly involved with the Pillars of Sustainable Development such as social, economic, and environmental factors, as shown in Figure 2. Social factor is related to the value of the individual person, and their individual contribution. To accomplish a better quality of life, it is important to minimize social impacts and promote social sustainability based on the respective people. The economic factor is important to provide access to facilities and services for all people in designing and innovating new production strategies without degrading the environment. It is important that each individual must have the economic freedom to find a source of personal fulfillment instead of just surviving on the planet. Since natural resources play a fundamental role in this development, the environmental factor represents the basis for the success of the 2030 Agenda goals. It is essential to direct our efforts for the preservation of ecosystems and biodiversity, and their stability with traditional exploitation models (Zarta 2018).

The contribution of creativity, knowledge, technology, and financial resources among citizens, governments, and companies increases the possibility of achieving the Sustainable Development Goals (Silva and di Serio 2016). Thus, it is possible to construct a better planet for future generations. The citizens, societies, and

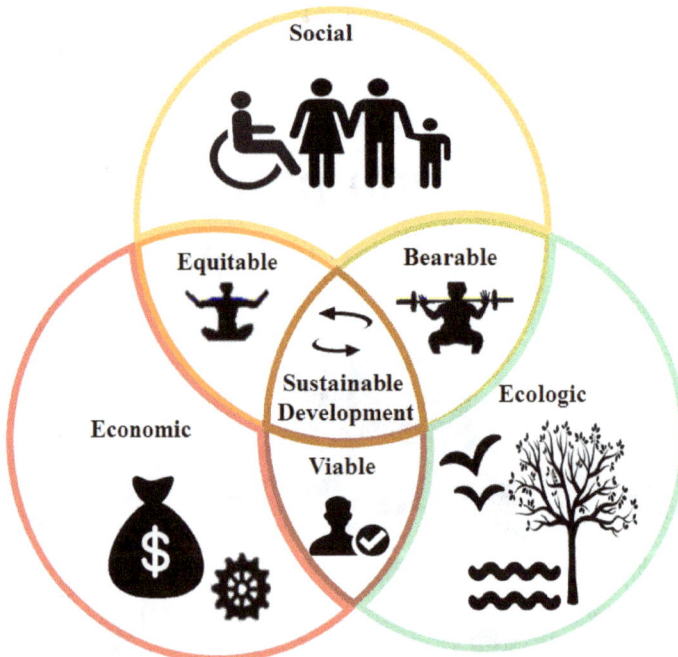

Pillars of sustainable development

Figure 2. Pillars of sustainable development.
Image made by Amira Jalil Fragoso-Medina, icons in the rearrangement were obtained from internet with credit to whom it corresponds.

governments must follow their responsibilities to comply with the visions of the SDGs. The citizens need to live their lives by activating the local economy and reducing or eliminating the use of single-use products, etc. Societies and governments can apply the circular economy to establish energy laws to make energy more efficient without pollution. Countries should develop innovative perspectives in science and technology, which can create new knowledge and appropriate designs for solving different contexts available in the respective countries. A brief description of clean energy, green chemistry, and sonochemistry are discussed in the following sections.

2.2 Clean Energy and Green Chemistry

2.2.1 From Scientific Areas to Green Chemistry

As mentioned earlier in this chapter, the knowledge achieved over time has allowed the development and production of several commodities for humanity to meet their primary needs (i.e., the provision of food and clothing), performing new and important discoveries to maintain and improve their quality of life. These commodities comes from activities as mining, mineralogy, petrography, pottery, agriculture, the extraction of chemical substances, technical chemistry, and pharmaceuticals, which is useful to improve the quantitative and qualitative analysis (Hing and Hing 2008, Reséndiz and Pastrana 2014). It is known that the "chemistry surrounds us". Although knowledge of chemistry has changed the lifestyle of individuals and society, it is also responsible for environmental pollution by producing electricity, primary metals, chemicals, paper, and waste materials (Argota et al. 2012, Bustamante-Montes et al. 2001, Falero et al. 2005, Garrido 2016). In addition, it is important to remark that this rapid advancement in chemistry is provoked by the globalization process, which improves the research and development technologies for the automation of processes in the industry (Morales R. et al. 2017, Mutobola C. et al. 2020), as well as the integration of digital strategies, principally in pharmaceutical simulation programs and virtual laboratories (García 2007, Marín et al. 2017). According to the philosophy and application of chemistry, it is convenient to mention that the area of chemistry can be classified into three important chemistry types: Green Chemistry, Sustainable Chemistry, and Environmental Chemistry.

Sustainable Chemistry is an integrated approach that includes both the application of the philosophy of green chemistry and the systems larger than basic research (i.e., process, materials, energy, and economic effects). Green engineering is an example of sustainable Chemistry. Sustainable Chemistry designs the use of technically, and economically viable processes and products while minimizing the generation of contamination at source and the risk to health and the environment. Environmental Chemistry is the chemistry of the natural environment and the polluting chemicals. This area studies the problems and the damage that occur in the global environment and the impact generated by human activities, and as a counterpart, provides a solution for the damage caused. Green chemistry is introduced to reduce the use or generation of hazardous substances, which applies across the life cycle of chemical products, manufacturing processes, design, and ultimate disposal. Therefore, it has been employed in different areas, such as organic chemistry, green analytical chemistry, green pharmacy, and others (Kar et al. 2022, Pawliszyn et al. 2022).

The Green Chemistry has been used to resolve the problems created by chemistry. Green chemistry initially concentrated on some specific areas, such as architecture, engineering, chemistry, pharmaceuticals, mathematics, agriculture-mathematics, mathematics-hydrobiology, and others (Acosta 2009, Anastas and Zimmerman 2003, Bejarano et al. 2015, De León 2021, Dunn et al. 2010, Naula 2010, Vargas and Ruiz 2007). Green Chemistry has focused on the design of preparation methods and the use of chemical products with less potential for contamination and environmental risk than conventional methods and improving the efficient use of materials and energy in the development of renewable resources. Moreover, Green Chemistry utilizes different strategies, such as the application of alternative energy sources to activate a reaction, utilization of cost-effective and eco-friendly catalysts, the use of alternative solvents, use of less toxic substances among others. This scientific discipline was protocolized in twelve principles, which was reported by Paul et al. in 1998 (Anastas and Warner 1998a). These principles of Green Chemistry are displayed in Figure 3 and these were proposed to develop a sustainable environment.

In general, all these areas ensure environmental and economic sustainability with a difference in their approaches. Sometimes redesigning basic materials for society and making them more benign for humans and the environment, is preferred for the best economic and social advantages. Thus, Green Chemistry focuses on saving and avoiding environmental deterioration that incorporates the goals treated in each type of chemistry (Sierra et al. 2014).

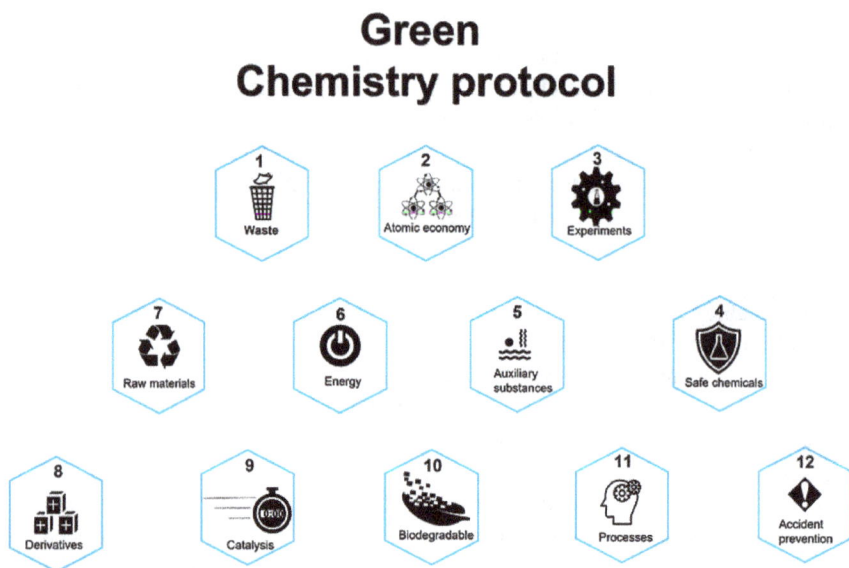

Figure 3. Green Chemistry protocol.
Image made by Amira Jalil Fragoso-Medina, icons in the rearrangement were obtained from internet with credit to whom it corresponds.

2.2.2 Tools of Green Chemistry

It difficult to create new products, processes, and services for achieving cheap and environmental benefits. It is necessary to use some new tools or approaches for minimizing the chemical processes without the dispersion of hazardous chemicals in the environment. Green Chemistry protocols are employed to obtain a sustainable environment by using renewable resources, which enhances the durability and recyclability of products. It is essential to understand the types of reactions, their components, energy activation mode, and the medium for sustainable development. Green Chemistry tools are employed to develop new materials and chemical products and include kind reactions, components of reaction, green solvents, energy activation mode, and materials. All of these are included in the Figure 4, and are described in the following section.

Tools of Green Chemistry are described as follows:

- *Kind reaction*: These reactions include biotransformation (which constitutes the modification that every xenobiotic suffers, for example, fuels of biological origin (i.e., bioethanol and biodiesel are obtained from organic remains), flow chemistry (continuous reaction process, which is the treatment of wastewater by a physicochemical system and continuous flow chemical oxidation), electrochemistry (the chemical reactions take place between electrical energy and chemical energy, for example, flow electrochemistry to produce *o*-benzoquinone

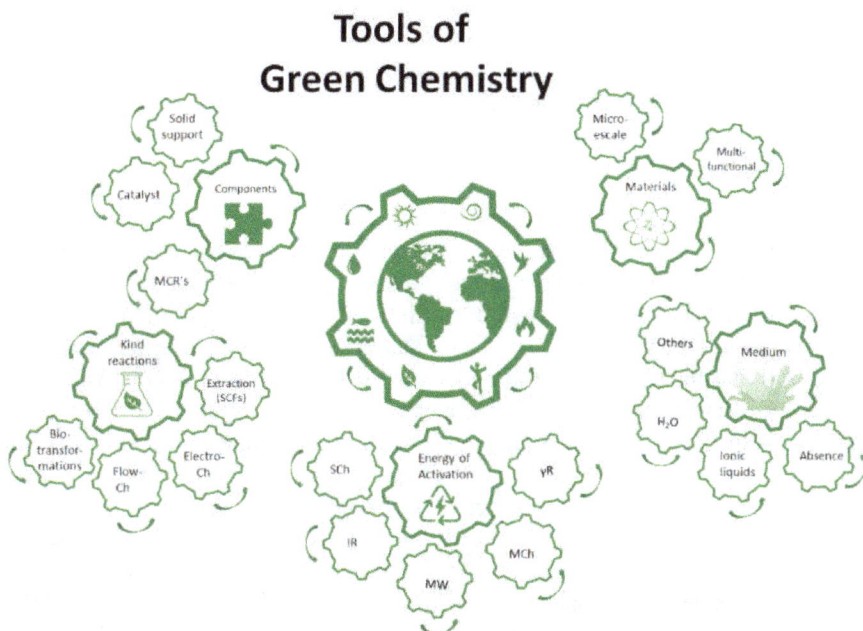

Figure 4. Green Chemistry Tools.
Figure note: Ch (chemistry), γR (gamma rays), IR (infrared), MCh (mechanochemistry), MW (microwaves), MCR's (multicomponent reactions), SCh (sonochemistry), SFs (supercritical fluids).
Image made by Amira Jalil Fragoso-Medina, icons in the rearrangement were obtained from internet with credit to whom it corresponds.

through *in situ* electro-generation), and the extraction with supercritical fluids (employing solvents above critical pressure and temperature, and coexist both phases gas and liquid). These reactions present an option when carrying out the extraction in an eco-friendly way (Bernal-Martínez et al. 2011, Herrera et al. 2007, Kashiwagi et al. 2012, Páramo et al. 2021).

• *Components of a reaction*: Reaction components have the use of solid support (which is used to immobilize the molecules to allow greater contact surface between reagents and substrates for successful reactions), the use of a catalyst (some reactions increase the chemical efficiency, in addition to reducing the cost of the process, without affecting the thermodynamic conditions involved in the process), and the multicomponent reactions (where three or more components are combined in a reaction in a single glass vessel to produce a single final product without further additions, isolations or manipulations, except for the final product isolation) (Allen and Shonnard 2001, Anastas and Williamson 1998b, Borges-Rodríguez et al. 2012, Corma and García 2003, Cuevas-García 2020, García 2009, Larhed et al. 2002, Sánchez-Rodríguez et al. 2018).

• *Green solvents*: The solvent is the medium, in general, used to carry out a reaction, but the properties of the reaction components must be considered. Water is considered an unusual organic solvent. However, water has been well-documented for biochemical reactions or in- and on-water processes. Another solvent with a wide variety of applications is the ionic liquids. The physical and chemical properties of ionic liquids allow them to respond to the demands of the chemical industry for producing materials with better environmental features and economical technologies. Also, the reactions can be performed in the absence of conventional solvents like 1,4-dihydropyridines. Therefore, sometimes it is not necessary to use solvent for the reaction process (Castillo 2015, Fuentes and Amábile-Cuevas 2013, Kowsari and Mallakmohammadi 2011, Rolando et al. 2005, Tanaka and Toda 2000, Wang et al. 2008).

• *Energy activation mode*: In the first instance, it is necessary to point out the use of conventional mantle heating to activate a reaction. This is perceived as unsustainable for various reasons, such as needing more reaction time, very low yield, high reaction byproducts, and among others. Thus, an ecological alternative for chemical synthesis is needed for the use of alternative energy to activate the reactions through sonochemistry, near-infrared and far-infrared irradiations, microwave irradiation, mechanochemistry, and gamma rays. The chemical synthesis is also carried out by the combination of two or more methods to achieve the desired product like using sonochemistry–microwaves in Williamson's ether synthesis in the absence of phase transfer catalysts, or in the generation of nanowires, which provide inherent advantages compared to the conventional mantle heating system (Chatel and Colmenares 2017, Fragoso-Medina 2017, Friščić et al. 2020, Kamble et al. 2017, Kappe 2004, López-Saucedo et al. 2020, Mazo et al. 2007, Miranda et al. 2009, Muñoz-Batista et al. 2018, Peng and Song 2002, Shen 2009).

- *Materials*: another determining tool is the materials (i.e., the multifunctional materials or the microscale) used for the chemical process. Materials integrate the most common and essential properties to obtain energy among other characteristics. The materials are necessary to be eco-friendly in the process of extraction, elaboration, and implementation for a specific use. They do not harm the environment or pollute the atmosphere. Furthermore, materials on a smaller scale can be used in the chemical reactions for the synthesis of new products. The microscale is accompanied using small vessels and a significant reduction in reagents. A difference is obtained between the waste produced by applying the microscale, which obtains the waste in the order of milliliters than the order of liters (Aponte et al. 2013, Coronado 2003, LJL 2013).

Finally, all these tools have made it possible to get the excellent current status of Green Chemistry due to the different international activities. It is necessary to highlight the education area (creation, operation of educational centers, and delivery of courses on the subject), research (contribution of workgroups of public institutions), diffusion (through institutes, dedicated to the subject and publication of literature on the subject), and activities developed around in different parts of the World (Amador 2013).

2.3 Sonochemistry

2.3.1 Development and Fundament of Sonochemistry

Sonochemistry is an example of an efficient synthesis process, which has several advantages over conventional mantle heating involved in the process design, such as energy consumption, waste reduction, atomic economy, scalability (from laboratory to industrial level), and as the most efficient and selective process to carry out a selective process for the remotion and elimination of antibiotics from water (Serna-Galvis et al. 2016).

Sonochemistry is considered a useful methodology for high yields, low energy requirements, low waste, and no use of solvents. The use of ultrasound in sonochemistry offers specific activation like acoustic cavitation. The mechanical activation in cavitation destroys the attractive forces of molecules in the liquid solution. During ultrasonication, the liquid compresses and expands, forming small bubbles through a sudden pressure drop. These bubbles change their shape and size with each cycle of the applied ultrasonic energy. Finally, these liquids collide and/or violently collapse.

Since the sustainable development is necessary, the employment of sonochemistry is used to minimize and prevent the environmental pollution.

The history of sonochemistry is briefly described here.

In 1779, biologist L. Spallanzani found the relationship between sound waves and bats where the ultrasonic signals are believed to have their origin in nature. Consequently, the theoretical basis for the application of sound waves was established in the eighteenth century. Later in 1842, the physicist A. Doppler noticed the change in the frequency emitted by a sound source. At the end of the nineteenth century, Pierre and Jacques Curie investigated the interaction between ultrasonic waves and objects with the discovery of the piezoelectric effect where the generation and detection of this type of waves was recognized. Thorneycroft and Barnaby launched the first report of the sound waves effect taking as an example the propellers of the Royal Navy ships. Lord Rayleigh explained this phenomenon, which is due to the rapid erosion of ship propellers. In 1927, Richard and Loomis studied the effect of ultrasound on chemical compounds. In 1980, the wide availability and distribution of suitable and cheap laboratory equipment (ultrasonic cleaning baths or ultrasonic probes) promoted the multiple applications of sonochemistry. The reduction of potassium iodate by treatment (with sulfurous acid and the hydrolysis of dimethyl sulfate in a basic solution) was observed in 2006 (Dency et al. 2006). Currently, the effects of ultrasonic irradiation on chemical reactions are investigated. Thus, sonochemistry represents an ecological alternative in comparison to conventional heating systems due to the broad range of applications, and design of the theoretical-experimental processes and their implementation (Fragoso-Medina et al. 2021).

In addition, it is important to comment that the energy generated during sonochemistry is sufficient to produce transformations of the matter. The sonochemistry is divided into two approaches: homogeneous sonochemistry (performed in a liquid medium) and heterogeneous sonochemistry (occurs in solid-liquid systems, solid-solid ensembles, and systems of an immiscible liquid interface) (Ayala et al. 2017, Cruz-González et al. 2017, Docampo and Pellón 2008, Pellón and Docampo 2008). Sonochemical processes are based on the same phenomenon produced by sound waves.

This phenomenon of energy transfer can be explained by the cavitation theory in three steps: (1) sound waves are emitted under mechanical stress, an electric charge, and the driving force, which modifies the microscopic structure of the receptor body through the piezoelectric effect; (2) dipoles are formed by accumulating disturbances in the medium and then identified as a compression-rarefaction process caused by forming vapor bubbles and generating energy body receptor (the bubbles have a high energy density in the order of 1,018 kW/mL), and (3) this energy is transferred through molecular movement until it is captured, subsequently, the conditions are generated in uniform physicochemical processes (Castellanos 2013, Gutiérrez-Mosquera et al. 2019, Martínez et al. 2007, Reyes-Cruz et al. 2016).

A deep knowledge of sonochemistry and its effects is necessary to discern between a reaction that occurs by this method and a transformation that is the product of mechanical stresses, considered as false sonochemistry (Arroyo and Flores 2001). Chemical reactions generally do not occur during the ultrasonic irradiation of solids or solid-gas systems because the phenomenon of cavitation takes place only in liquids. The chemical reaction occurs not only during turbulent flow but also under the high-intensity ultrasonic irradiation. However, the effects of sound waves are presented on solid materials. It is found that these effects are less perceptible than in liquids and cannot be distinguished with the naked eye.

2.3.2 Scope and Areas of Sonochemistry

The electromagnetic spectrum is employed to identify the radiation emission-absorption region of a substance. The basic information and the applications of the four types of sound waves mentioned early in the introduction of this chapter are described in the following;

A. The sound of human auditory field is used in conventional applications such as in food processing and water treatment for the degradation of organic pollutants, drugs, pesticides and herbicides, colorants, and other persistent organic compounds (González-Labrada et al. 2010, Mason 1986, 1997, Robles and Ochoa-Martínez 2012, Sillanpää et al. 2011, Vigueras-Carmona et al. 2013).

B. Ultrasound is employed in an ultrasonic treatment in a continuous flow system for the reduction of the viscosity of heavy crude oil, the experimental biodiesel reactor operation, secondary metabolites extraction, chemical compounds synthesis, measuring distances, nanosensors, synthesis of nanomaterials, biological applications and reception-emission of images, ATP-ase activity, energy, automotive and telecommunications engineering careers, radar as a teaching tool, among others (Albalawi et al. 2021, Alvarado et al. 2014, Cigales 2016, Da costa et al. 2012, Díaz et al. 2013, Filomena et al. 2012, Florez and Ortega 2019, Mansur 2010, Martínez et al. 2011, Mavares 2020, Meléndrez and Vargas-Hernández 2013, Navarro et al. 2004, Ovalle et al. 2010, Vaitsis et al. 2020).

C. Hypersounds are generally applied in the high energy astrophysics (Deaza 2005, Llamosa 2011, Mukherjee and Zanin 2022, Revueltas A et al. 2014).

D. The sounds whose frequency in the range of MHz can be studied in various applications, such as medicine (i.e., knee injury prevention and knee therapy, or in the central nervous system, differential diagnosis), meteorology, materials engineering, defense service (i.e., for naval use), etc. (Anillo et al. 2009, Franco et al. 2011, Fuentes et al. 2006, Grande and Ramírez 2015, Jiménez and Loaiza 2010, Leiva et al. 2005, Leon and Romero 2015, Ochoa-Pérez and Cardozo-Ocampo 2015, Quiroga 2018, Rauh et al. 2010, Rodríguez D. et al. 2012).

The applications of sonochemistry are increasing every day and in various areas of knowledge.

3. Conclusions and Future Perspectives

The negative results of human activities make clear the need for sustainable development. For this, various strategies were implemented to solve the problem and the ways to reduce or stop its effects. This chapter initially discussed various strategies for sustainable development. The progress in sustainable development up to so far is also included. The sustainable development goals are focused on stimulating sustainable economic growth through higher levels of productivity and technological innovation. This chapter also introduced the concept of clean energy and green chemistry to achieve sustainable development goals.

The types of green chemistry and their tools are described extensively. Finally, the development and scope of sonochemistry are described in detail for achieving sustainable development goals. Understanding all these ideas could change our practices and contribute towads improving the social, economic, and environmental development of future generations. Sonochemistry is an example of Green Chemistry, which helps in the activation of reactions by changing the shape and size of the liquid of the solution. Since the use of sonochemistry is explained with the help of the effect produced by sound waves, sonochemistry contains tremendous advantages and can be useful in diverse areas. Sonochemistry is recently being used for the synthesis of various materials and studies on the chemical process of material synthesis are very few. The degradation of pollutants through different species using a sonochemical process is also required in the future to reduce an energy-consuming and costly process. Therefore, understanding the mechanisms occurred during the sonosynthesis process can propel this technique towards the synthesis of novel materials for commercial and sustainable development.

Acknowledgements

The author of this chapter would like to thank Dr. Ashok Adhikari and Dr. Joel Omar Martínez for reviewing this text.

References

Acosta, D. 2009. Arquitectura y construcción sostenibles: Conceptos, Problemas y Estrategias. Dearq 4: 14–23. https://doi.org/10.18389/dearq4.2009.02.

Albalawi, F., M.Z. Hussein, S. Fakurazi and M.J Masarudin. 2021. Engineered nanomaterials: The challenges and opportunities for nanomedicines. International Journal of Nanomedicine 16: 161–168. https://doi.org/10.2147/IJN.S288236.

Allen, D.T. and D.R. Shonnard. 2001. Green engineering: environmentally conscious design of chemical processes and products. AIChE Journal 47(9): 1906–1910. https://doi.org/10.1002/aic.690470902.

Alvarado, R., F. Solera and J. Vega-Baudrit. 2014. Síntesis sonoquímica de nanopartículas de óxido de cinc y de plata estabilizadas con quitosano. Evaluación de su actividad antimicrobiana. Rev. Iber. Polímeros 15(2): 134–148. ISSN-e 0121–6651.

Amador, B.C. 2013. Sustentabilidad y educación química. Educación Química 24(2): 182–183. https://doi.org/10.1016/s0187-893x(13)72460-1.

Anastas, P.T. and J.C. Warner. 1998a. Green chemistry: theory and practice. Oxford University Press, New York.

Anastas, P.T. and T.C. Williamson. 1998b. Green chemistry, frontiers in benign chemical syntheses and processes. Oxford University Press, Great Britanian, UK. ISBN: 019850170697780198501701.

Anastas, P.T. and J.B. Zimmerman. 2003. Peer reviewed: design through the 12 principles of green engineering. Environmental Science and Technology 37(5): 94A–101A. https://doi.org/10.1021/es032373g.

Anillo, R., E. Villanueva, L. Dayeneri and A. Pena. 2009. Ultrasound diagnosis for preventing knee injuries in cuban high-performance athletes. MEDICC Review 11(2): 21–28. https://doi.org/10.37757/mr2009v11.n2.7.

Aponte, R.A., R. Aguilar Gonzalez and I. Austin de Sánchez. 2013. Trabajos prácticos en microescala como estrategia didáctica en cursos de química de educación media. Actualidades Investigativas en Educación 13(2): 1–19. https://doi.org/10.15517/aie.v13i2.11731.

Argota, G., Y. González, H. Argota, R. Fimia and J. Iannacone. 2012. Desarrollo y bioacumulación de metales pesados en gambusia punctata (poeciliidae) ante los efectos de la contaminación acuática. Revista Electrónica de Veterinaria 13(5): 1–12. https://www.redalyc.org/comocitar. oa?id=63624365020.

Arroyo, C.J. and V.J. Flores. 2001. Degradación ultrasónica de contaminantes orgánicos. Revista Peruana de Química e Ingeniería Química 4(2): 3–14. https://revistasinvestigacion.unmsm.edu.pe/index.php/ quim/article/view/4267.

Ayala, J.A., C.O. Castillo and R.S. Ruiz. 2017. Ultrasonic, ultraviolet and hybrid catalytic processes for the degradation of rhodamine b dye: decolorization kinetics. Revista Mexicana de Ingeniera Química 16(2): 521–529. https://doi.org/10.24275/RMIQ/CAT833.

Bejarano, F.T., H. Ramírez-León, C.R. Cuevas, M.P.T. González and M.C.V. Jaraba. 2015. Validación de un modelo hidrodinámico y calidad del agua para el río magdalena, en el tramo adyacente a barranquilla, colombia. Hidrobiologica 25: 7–23. http://www.scielo.org.mx/pdf/hbio/v25n1/ v25n1a2.pdf.

Berhe, A.A., R.T. Barnes, J. Six and E. Marín-Spiotta. 2018. Role of soil erosion in biogeochemical cycling of essential elements: carbon, nitrogen, and phosphorus. Annual Review of Earth and Planetary Sciences 46(1): 521–548. https://doi.org/10.1146/annurev-earth-082517-010018.

Bernal-Martínez, L.A., C. Solís-Morelos, I. Linares-Hernández, C. Barrera-Díaz and A. Colín-Cruz. 2011. Tratamiento de agua residual municipal por un sistema fisicoquímico y oxidación química en flujo continuo. Avances En Ciencias e Ingeniería 2(2): 69–81. https://www.redalyc.org/articulo. oa?id=323627682007.

Bollaín, P., C. and D. Vicente A. 2019. Presencia de microplásticos en aguas y su potencial impacto en la salud pública. Revista Española de Salud Pública 93: e1-10. http://scielo.isciii.es/scielo. php?script=sci_arttext&pid=S1135-57272019000100012&lng=es&tlng=es.

Borges-Rodríguez, D., A.N. San Juan-Rodríguez, A.O. Díaz-LLanes, E. Gómez-Santiesteban and R. Hernández-Sanchez. 2012. Evaluación de la zeolita como soporte sólido para la formulación del biofertilizante azospirillum. ICIDCA. Sobre Los Derivados de La Caña de Azúcar 46(2): 12–18. http://www.redalyc.org/articulo.oa?id=223124990002.

Bustamante-Montes, P., B. Lizama-Soberanis, G. Olaíz-Fernández and F. Vázquez-Moreno. 2001. Ftalatos y efectos en la salud. Revista Internacional de Contaminación Ambiental 17(4): 205–215. https://www.revistascca.unam.mx/rica/index.php/rica/article/view/25362.

[CCT] Comisión de la Carta de la Tierra, ¿Qué es la Carta de la Tierra?, International document, 2020. https://cartadelatierra.org/sobre-nosotros/

Castellanos, N. 2013. Evaluación preliminar del uso del efecto piezoeléctrico para generación de energía. INVENTUM 8(15): 35–40. https://doi.org/10.26620/uniminuto.inventum.8.15.2013.35-40.

Castillo, B.F. 2015. Líquidos iónicos: métodos de síntesis y aplicaciones. Conciencia Tecnológica 49: 52–56. http://www.redalyc.org/articulo.oa?id=94438997007.

Chatel, G. and J.C. Colmenares. 2017. Sonochemistry: from basic principles to innovative applications. Topics in Current Chemistry 375(1): 8. https://doi.org/10.1007/s41061-016-0096-1.

Cigales, C.J. 2016. Síntesis y caracterización de nanopartículas de carbono nanoluminiscentes: carbon quantum dots (CQDs). M Thesis, Universidad de Oviedo, España. https://digibuo.uniovi.es/dspace/ bitstream/handle/10651/38769/TFM_Cigales%20Canga.pdf?sequence=3.

Corma, A. and H. García. 2003. Lewis acids: from conventional homogeneous to green homogeneous and heterogeneous catalysis. Chemical Reviews 103(11): 4307–4366. https://doi.org/10.1021/ cr030680z.

Coronado, E. 2003. Materiales moleculares multifuncionales. Anales de La Real Sociedad Española de Química 2: 151–157. ISSN-e 1575–3417.

Cruz-González, G., C. Julcour and U. Jáuregui-Haza. 2017. El estado actual y perspectivas de la degradación de pesticidas por procesos avanzados de oxidación. Revista Cubana de Química 29(3): 492–516. http://www.redalyc.org/articulo.oa?id=443552968013.

Cuevas-García, R. 2020. Obtención y análisis de expresiones de cinética química. Mundo Nano. Revista Interdisciplinaria En Nanociencias y Nanotecnología 14(26): 1e-25e. https://doi.org/10.22201/ceiich.24485691e.2021.26. https://doi.org/10.22201/ceiich.24485691e.2021.26.69639.

Da Costa, A.P., E.C. Botelho, M.L. Costa, M.N. Narita and J.R. Tarpani. 2012. A review of welding technologies for thermoplastic composites in aerospace applications. Journal of Aerospace Technology and Management 4(3): 255–265. https://doi.org/10.5028/jatm.2012.04033912.

De León, E.M.D. 2021. Denitrification modeling using natural organic solid substrates as carbon sources. Tecnología y Ciencias Del Agua 12(2): 1–58. DOI: 10.24850/j-tyca-2021-02-01. https://doi.org/10.24850/J-TYCA-2021-02-07.

Deaza, R.P.I. 2005. Las primeras etapas en la formación de una galaxia. Ingeniería (Bogotá) 10(1): 16–22. https://doi.org/10.14483/23448393.1871.

Dency, J.P.E.C.A. 2006. Ejemplos selectos de sonoquímica: una alternativa ecoamigable. Nueva Época 2(1): 26–44. https://www.uniatlantico.edu.co/uatlantico/pdf/arc_12410.pdf.

Díaz, Á.J.C., R. Martínez Rey, E.J. Patiño Reyes and R. Barrero Acosta. 2013. Estudio experimental sobre la eficiencia de un tratamiento de ultrasonido en un sistema de flujo continuo para la reducción de viscosidad de crudo pesado. Revista Ion 26(2): 47–63. http://www.scielo.org.co/scielo.php?script=sci_arttext&pid=S0120-100X2013000200006&lng=en&nrm=iso>. ISSN 0120-100X.

Docampo, P., M.L. and R.F. Pellón C. 2008. Nuevo método para la arilación de aminas aromáticas con ácido 2-clorobenzoico y derivados, asistido por ultrasonido en dimetilformamida como disolvente. Revista CENIC. Ciencias Químicas 39(1): 29–33. http://www.redalyc.org/articulo.oa?id=181615032002.

Dunn, P. J., A.S. Wells and M.T. Williams. 2010. Green chemistry in the pharmaceutical industry wiley. VCH Verlag GmbH and Co. KGaA. https://doi.org/10.1002/9783527629688.

Educadores por la sostenibilidad. 2012. De la "Primavera silenciosa" al clamor por RIO+20. Revista Eureka sobre Enseñanza y Divulgación de Las Ciencias 9(2): 309–310. http://dx.doi.org/10.25267/Rev_Eureka_ensen_divulg_cienc.2012.v9.i2.13http://reuredc.uca.es/.

Escobar, J. 2002. La contaminación de los ríos y sus efectos en las áreas costeras y el mar. Serie Recursos Naturales e Infraestructura 50: 1680–9017. http://hdl.handle.net/11362/6411.

Falero, M.A., C. Pérez, B. Luna and M. Fonseca. 2005. Impacto de los disruptores endocrinos en la salud y el medio ambiente. Revista CENIC. Ciencias Biológicas 36 http://www.redalyc.org/articulo.oa?id=181220525013.

Filomena, A., L.E. Díaz, A. Puig and I. Sotelo. 2012. Efecto de ultrasonido sobre la actividad atp-asa y propiedades funcionales en surimi de tilapia (oreochromis niloticus). Vitae 19(1): S379-S381. http://www.redalyc.org/articulo.oa?id=169823914119.

Flores, B., C. Chacón and B. Galia. 2009. El desarrollo sostenible y la agenda 21 sustainable development, agenda 21. Agenda 11(2): 164–181. http://www.redalyc.org/articulo.oa?id=99312517003.

Florez, M.J.F. and D.R. Ortega A. 2019. Design and manufacturing of an ultrasonic reactor for biodiesel obtaining by transesterification. DYNA 86(211): 75–83. https://doi.org/10.15446/dyna.v86n211.78518.

Fragoso-Medina, A.J., R.G. Escobedo-González, M.I. Nicolás-Vázquez, G.A. Arroyo-Razo, M.O. Noguez-Córdova and R. Miranda-Ruvalcaba. 2017. A DFT study of the geometrical, spectroscopical and reactivity properties of diindolylmethane-phenylboronic acid hybrids. Molecules 22(10): 1744–1769. https://doi.org/10.3390/molecules22101744.

Fragoso-Medina, A.J., F. López-Saucedo, G.G. Flores-Rojas and E. Bucio. 2021. Sonochemical synthesis of inorganic nanomaterials. pp. 263–279 In: Inamuddin, R. Boddula, M.I. Ahamed and A.M. Asiri [eds.]. Green Sustainable Process for Chemical and Environmental Engineering and Science. Elsevier. https://doi.org/10.1016/B978-0-12-821887-7.00008-2.

Franco, E.E., J.M. Meza and F. Buiochi. 2011. Measurement of elastic properties of materials by the ultrasonic through-transmission technique. DYNA 78(168): 58–64. http://www.scielo.org.co/scielo.php?script=sci_arttext&pid=S0012-73532011000400007&lng=en&nrm=iso>. ISSN 0012-7353.

Friščić, T., C. Mottillo and H.M. Titi. 2020. Mechanochemistry for synthesis. In Angewandte Chemie-International Edition 59(3): 1018–1029. https://doi.org/10.1002/anie.201906755.

Fuentes, M., A. Sotomayor, F. García, E. Moreno and P. Acevedo. 2006. Design and construction of a blood flow detector probe for a continuous wave bidireccional doppler ultrasound system. Ingeniería Investigación y Tecnología 7(2): 97–103. https://doi.org/10.22201/fi.25940732e.2006.07n2.008.

Fuentes, M.A. and C.F. Amábile-Cuevas. 2013. El agua en bioquímica y fisiología. Acta Pediátrica de México 34(2): 86–95. http://www.redalyc.org/articulo.a?id=423640341010.

[GPE] Greenpeace España, Un Mediterráneo lleno de plástico Estudio sobre la contaminación por plásticos, impactos y soluciones, Article 2017. http://archivo-es.greenpeace.org/espana/Global/espana/2017/documentos/oceanos/Mediterranean%20plastic%20report-LR.pdf.

García, C.-F.F. 2009. Parámetros para el análisis de las reacciones en química sostenible. Anales de La Real Sociedad Española de Química 1(105): 42–49. https://www.researchgate.net/publication/47548301_Parametros_para_el_analisis_de_las_reacciones_en_quimica_sostenible.

García, F. 2007. Simulación de una cadena se suministro en el área farmacéutica. Visión Gerencial 1: 47–60. http://www.redalyc.org/articulo.oa?id=465545875003.

Garrido, R.F.Y. 2016. reutilización de residuos sólidos como alternativa de formación en la conservación del ambiente elaborando nuevos materiales para el docente de educación inicial. Revista Scientific 1(1): 169–189. https://doi.org/10.29394/scientific.issn.2542-2987.2016.1.1.10.169-189.

Girón, A. 2016. Objetivos del desarrollo sostenible y la agenda 2030: frente a las políticas públicas y los cambios de gobierno en américa latina. Problemas del Desarrollo 47(186): 3–8. https://doi.org/10.1016/j.rpd.2016.08.001.

González-Labrada, K., I. Quesada-Peñate, C. Julcour-Lebigue, H. Delmas, G. Cruz González and U.J. Jáuregui-Haza. 2010. El empleo del ultrasonido en el tratamiento de aguas residuales. Revista CENIC. Ciencias Químicas 41:1–11. http://www.redalyc.org/articulo.oa?id=181620500034.

Grande, I.E. and L.C. Ramírez Ramírez. 2015. Uso del ultrasonido terapéutico pulsado en el tratamiento de personas con osteoartritis de rodilla. Rev. Univ. Ind. Santander. Salud 47(473): 337–348. http://www.scielo.org.co/scielo.php?script=sci_arttext&pid=S0121-08072015000300010&lng=en&nrm=iso.

Gutiérrez-Mosquera, L.F., S. Arias-Giraldo and D.F. Cardona-Naranjo. 2019. Cavitación hidrodinámica: un enfoque desde la ingeniería y la agroindustria. Scientia et Technica. 24(2): 283–304. https://doi.org/10.22517/23447214.19921.

Herrera, S.Y., I.C.D. Zampini, R. D'Almeida, H. Boguetti and M.I. Isla. 2007. Extracción con fluido supercrítico de compuestos con capacidad antioxidante de *baccharis incarum*: comparación con métodos convencionales. Boletín Latinoamericano y Del Caribe de Plantas Medicinales y Aromáticas 6(5): 250–251. http://www.redalyc.org/articulo.oa?id=85617508059.

Hing, L.N.M. and R.C. Hing. 2008. La historia de la química y el desarrollo de la sociedad. Tecnología Química XXVIII(3): 15–27. http://www.redalyc.org/articulo.oa?id=445543757002.

Jiménez, G.J. and H. Loaiza C. 2010. Detección y caracterización de defectos en tuberías metálicas en pruebas ultrasónicas por inmersión. El Hombre y La Máquina. 34: 56–67. http://www.redalyc.org/articulo.oa?id=47817108006.

Kamble, M.P., S.A. Chaudhari, R.S. Singhal and G.D. Yadav. 2017. Synergism of microwave irradiation and enzyme catalysis in kinetic resolution of (R, S)-1-phenylethanol by cutinase from novel isolate Fusarium ICT SAC1. Biochemical Engineering Journal 117: 121–128. https://doi.org/10.1016/j.bej.2016.09.007.

Kappe, C.O. 2004. controlled microwave heating in modern organic synthesis. Angewandte Chemie International Edition 43(46): 6250–6284. https://doi.org/10.1002/anie.200400655.

Kar, S., H. Sanderson, K. Roy, E. Benfenati and J. Leszczynski. 2022. Química verde en la síntesis de productos farmacéuticos. Chem. Rev. 122(3): 3637–3710. https://doi.org/10.1021/acs.chemrev.1c00631.

Kashiwagi, T., F. Amemiya, T. Fuchigami and M. Atobe. 2012. In situ electrogeneration of o-benzoquinone and high yield reaction with benzenethiols in a microflow system. Chemical Communications 48(22): 2806. https://doi.org/10.1039/c2cc17979b.

Kowsari, E. and M. Mallakmohammadi. 2011. Ultrasound promoted synthesis of quinolines using basic ionic liquids in aqueous media as a green procedure. Ultrasonics Sonochemistry 18(1): 447–454. https://doi.org/10.1016/j.ultsonch.2010.07.020.

[LJL] La jornada en línea, El Instituto de Física desarrolla ecomateriales para ahorrar energía, Journalistic article, 2013. UNAM, México. http://www.jornada.unam.mx/2013/01/11/ciencias/a03n2cie.

Larhed, M., C. Moberg and A. Hallberg. 2002. Microwave-accelerated homogeneous catalysis in organic chemistry. Accounts of Chemical Research 35(9): 717–727. https://doi.org/10.1021/ar010074v.

Lebreton, L., B. Slat, F. Ferrari, B. Sainte-Rose, J. Aitken, R. Marthouse et al. 2018. Evidence that the great pacific garbage patch is rapidly accumulating plastic. Scientific Reports 8(1): 4666. https://doi. org/10.1038/s41598-018-22939-w.

Leiva, L.A., G. Fernández, S., R. Villaroel V. and E. Quezada V. 2005. Radar de onda superficial de HF (HFSWR). Revista Facultad de Ingeniería - Universidad de Tarapacá 13(3): 11–23. https://doi. org/10.4067/s0718-13372005000300003.

Leon, J.C. and C.J.E. Romero. 2015. The methodology non-destructive test inspection by ultrasound for predict the beginning of the failure of a material subject to rolling contact fatigue. Revista INGENIERÍA UC 22(1): 16–25. http://www.redalyc.org/articulo.oa?id=70735858003.

Llamosa, L.E. 2011. Fundamentos para la implementación de un laboratorio que certifique niveles de intensidad de cem-ni en colombia. Scientia et Technica 1(47): 163–168. https://doi. org/10.22517/23447214.539.

López-Saucedo, F., E. Bucio, G.G. Flores-Rojas, C. Flores-Morales, D. Martínez-Otero and N. Zúñiga-Villarreal. 2020. Gamma rays: an alternative energy source for the preparation of manganese carbonyls-based new materials. Applied Radiation and Isotopes 156: 108983. https://doi. org/10.1016/j.apradiso.2019.108983.

[MTERD] Ministerio para la transición ecológica y el reto demográfico, resultados de la COP21, 2021. https://www.miteco.gob.es/es/cambio-climatico/temas/cumbre-cambio-climatico-cop21/resultados-cop-21-paris/default.aspx.

Mallén, R.C. 2018a. Rachel carson, 50 años de romper el silencio. Revista Mexicana de Ciencias Forestales 3(14): 3–9. https://doi.org/10.29298/rmcf.v3i14.470.

Mallén, R.C. 2018b. Tres siglos de la invención de la sostenibilidad. Revista Mexicana de Ciencias Forestales, 4(20): 4–7. https://doi.org/10.29298/rmcf.v4i20.365.

Mani, G., R. Danasekaran and K. Annadurai. 2017. Endocrine disrupting chemicals: a challenge to child health. The Journal of Pediatric Research 4(1): 39–41. https://doi.org/10.4274/jpr.97830.

Mansur, H.S. 2010. Quantum dots and nanocomposites. wiley interdisciplinary reviews: Nanomedicine and Nanobiotechnology 2(2): 113–129. https://doi.org/10.1002/wnan.78.

Marín, S.L.T., C.P. Marín O. and J.S. Ospina Á. 2017. Laboratorio virtual de química: una experiencia de diseño interdisciplinar. Revista Virtual Universidad Católica Del Norte 51: 98–110. https:// revistavirtual.ucn.edu.co/index.php/RevistaUCN/article/view/845.

Martínez, A.J., J. Sierra A. and P. Martínez Y. 2011. Metabolitos secundarios en el guayacán amarillo y en el guayacán rosado. Scientia et Technica XVII (47): 297–301. https://doi.org/10.22517/23447214.1169.

Martínez, R., J.A., J. Vitola O. and S. del P. Sandoval C. 2007. Fundamentos teórico-prácticos del ultrasonido. Revista Tecnura 10(20): 4–18. https://doi.org/10.14483/22487638.6256.

Mason, T.J. 1986. Use of ultrasound in chemical synthesis. Ultrasonics 24(5): 245–253. https://doi. org/10.1016/0041-624X(86)90101-0.

Mason, T.J. 1997. Ultrasound in synthetic organic chemistry. Chemical Society Reviews 26(6): 443. https://doi.org/10.1039/cs9972600443.

Mason, T. and D. Peters. 2002. Practical sonochemistry power ultrasound uses and applications. In Practical Sonochemistry. Ellis Horwood. ISBN: 978 1 898563 83 9.

Mavares, F.P.J. 2020. Evaluación de un radar fmcw como herramienta didáctica en las carreras de Ingeniería Automotriz y Telecomunicaciones. Ingenius 25: 70–80. https://doi.org/10.17163/ings. n25.2021.07.

Mazo, P., L. Rios and G. Restrepo. 2007. Métodos alternativos para la obtención de biodiesel, microondas y ultrasonido. Revista ION 20(1): 51–57. http://www.redalyc.org/articulo.oa?id=342030278009.

Meléndrez, M.F. and C. Vargas-Hernández. 2013. Ultrasound assisted synthesis of ZnO nanorods on flexible substrates. Superficies y Vacio 26(3): 100–106. http://www.redalyc.org/articulo. oa?id=94229715007.

Miranda, R., O. Noguez, B. Velasco, G. Arroyo, G. Penieres, J.O. Martínez et al. 2009. Irradiación infrarroja: una alternativa para la activación de reacciones y su contribución a la química verde. Educación Química 20(4): 421–425. https://doi.org/10.1016/S0187-893X(18)30045-4.

Morales, R., A.C., M. Pérez F., J.R. Pérez G. and S. De León Almaraz. 2017. Energías renovables y el hidrógeno: un par prometedor en la transición energética de méxico. Investigación y Ciencia de La Universidad Autónoma de Aguascalientes 25(70): 92–101. https://doi.org/10.33064/ iycuaa2017701856.

Mukherjee, R. and R. Zanin. 2022. Advances in very high energy astrophysics. The Science Program of the Third Generation IACTs for Exploring Cosmic Gamma Rays. https://doi.org/10.1142/11141. ISBN: 978-981-3275-71-3.

Muñoz-Batista, M.J., D. Rodriguez-Padron, A.R. Puente-Santiago and R. Luque. 2018. Mechanochemistry: toward sustainable design of advanced nanomaterials for electrochemical energy storage and catalytic applications. ACS Sustainable Chemistry and Engineering 6(8): 9530–9544. https://doi.org/10.1021/acssuschemeng.8b01716.

Mutobola, C., Kabongo, Kanga M. Albano and Pedro G. João. 2020. Síntese do biodiesel a partir do óleo de mufuko por transesterificação. Tecnología Química 40(1): 35–51, http://www.redalyc.org/articulo.oa?id=445562743003.

Naula, I. 2010. Modelo de contaminación del aire. Enfoque UTE 1(1): 62–73. https://doi.org/10.29019/enfoqueute.v1n1.17.

Navarro, D., L.H. Ríos and H. Parra. 2004. Sensores de ultrasonido usados en robótica móvil para la medición de distancias. Scientia et Technica X (25): 35–40. https://doi.org/10.22517/23447214.7183.

Ochoa-Pérez, L. and A. Cardozo-Ocampo. 2015. Ultrasound applications in the central nervous system for neuroanaesthesia and neurocritical care. Colombian Journal of Anesthesiology 43(4): 314–320. https://doi.org/10.1016/j.rcae.2015.04.006.

Ovalle, R.J.O., A. Sáenz G., C.M. Pérez B., L.I. López L. and L. Barajas B. 2010. Síntesis de la *N,N*-diisopropiletanoamida y *N,N*-dibutiletanoamida mediante métodos de activación no convencionales microondas y ultrasonido. Avances En Química 5(3): 177–183. http://www.redalyc.org/articulo.oa?id=93315850006

Páramo, A.L.A., H.D., Delgado S. and C. K. Ríos G. 2021. Potencial del laboratorio de biotecnología del PIESA-UNI para desarrollar bioprocesos ambientales, agrícolas e industriales. Nexo Revista Científica 34(02): 534–546. https://doi.org/10.5377/nexo.v34i02.11540.

Pawliszyn, J., D., Barceló F. Arduini, L. Mondello, Z. Ouyang, P. Mateusz Nowak et al. Green analytical chemistry-a new Elsevier's journal facing the realities of modern analytical chemistry and more sustainable future. 2022. Green Analytical Chemistry 1: 100001–100003. https://doi.org/10.1016/j.greeac.2022.100001.

Pellón C., R.F. and M.L. Docampo P. 2008. Método sencillo para la obtención de ácidos 2-fenoxibenzoicos asistido por ultrasonido en dimetilformamida como disolvente. Revista CENIC. Ciencias Químicas 39(1): 35–39. http://www.redalyc.org/articulo.oa?id=181615032003.

Peng, Y. and G. Song. 2002. Combined microwave and ultrasound assisted williamson ether synthesis in the absence of phase-transfer catalysts. Green Chemistry 4(4): 349–351. https://doi.org/10.1039/b201543a.

Quiroga, J.M. 2018. Primeros desarrollos de tecnología radar en los principales beligerantes de la II guerra mundial. Un análisis desde la perspectiva Ciencia, Tecnología y Sociedad. Ciencia, Docencia y Tecnología 29(57): 36–59. https://doi.org/10.33255/2957/424.

Rauh, P., H. Neye, K. Mönkemüller, P. Malfertheiner and S. Rickes. 2010. The impact of high-resolution ultrasound in the differential diagnosis of non-hemolytic jaundice. Acta Gastroenterológica Latinoamericana 40(4): 328–331. http://www.redalyc.org/articulo.oa?id=199317295007.

Reséndiz, R.G. and P.A. Pastrana. 2014. El surgimiento de la industria farmacéutica en México (1917–1940). Revista Mexicana de Ciencias Farmacéuticas 45(2): 55–68. http://www.scielo.org.mx/scielo.php?script=sci_arttext&pid=S1870-01952014000200007&lng=es&nrm=iso 1870-0195.

Revueltas A., M., I. Avila R., R. Baqués M. and R.C. Beltrán Requena. 2014. Los campos electromagnéticos de frecuencia extremadamente baja y su impacto sobre la salud de los seres humanos extremely low frequency electromagnetic fields and their impact on human health. Revista Cubana de Higiene y Epidemiología 52(2): 210–227. http://scielo.sld.cu/scielo.php?script=sci_arttext&pid=S1561-30032014000200007&lng=es&nrm=iso.

Reyes-Cruz, J., G. Ruiz-Chavarría, R. Lambert-Sánchez, A. Turro-Breff, E. Torres-Tamayo and S. Hernández-Zapata. 2016. Dinámica de las burbujas de cavitación en fluidos amoniacales trasegados con bombas centrífugas. Minería y Geología 32(3): 128–146. https://www.redalyc.org/journal/2235/223547677009/html/.

Robles, O.L.E. and L.A. Ochoa-Martínez. 2012. Ultrasonido y sus aplicaciones en el procesamiento de alimentos. Revista Iberoamericana de Tecnología Postcosecha. 13(2): 109–122. http://www.redalyc.org/articulo.oa?id=81325441002.

Rodríguez D., V., M. Diez R. and O. Rodríguez G. 2012. Simulador de radar meteorológico basado en modelo de reflectividades en el espacio; weather radar simulator based on space reflectivity distribution. Ingeniería Energética XXXIII(2): 103–112. http://www.redalyc.org/articulo. oa?id=329127750003.

Rolando, E., M. Suárez, E. Ochoa, A. Peñamaría and F. Salfran. 2005. Síntesis de 1,4-dihidropiridinas utilizando energía de microondas en ausencias de solventes. Revista Cubana de Química XVII(3): 188–189. http://www.redalyc.org/articulo.oa?id=443543687073.

Sánchez-Rodríguez, E.P., A.J. Fragoso-Medina, E. Ramírez-Meneses, M. Gouygou, M.C. Ortega-Alfaro and J.G. López-Cortés. 2018. [N,P]-pyrrole-phosphine ligand: An efficient and robust ligand for Ru-catalyzed transfer hydrogenation microwave-assisted reactions. Catalysis Communications 115: 49–4. https://doi.org/10.1016/j.catcom.2018.07.009.

Serna-Galvis, E.A., J. Silva-Agredo, A.L. Giraldo-Aguirre, O.A. Flórez-Acosta and R.A. Torres-Palma. 2016. High frequency ultrasound as a selective advanced oxidation process to remove penicillinic antibiotics and eliminate its antimicrobial activity from water. Ultrasonics Sonochemistry 31: 276–283. https://doi.org/10.1016/j.ultsonch.2016.01.007.

Shen, X.-F. 2009. Combining microwave and ultrasound irradiation for rapid synthesis of nanowires: a case study on Pb(OH)Br. Journal of Chemical Technology and Biotechnology 84(12): 1811–1817. https://doi.org/10.1002/jctb.2250.

Sierra, A., L. Meléndez, A. Ramírez-Monroy and M. Arroyo. 2014. La química verde y el desarrollo sustentable. Revista Iberoamericana para la Investigación y el desarrollo sostenible 5(9): 1–16. https://www.redalyc.org/pdf/4981/498150317001.pdf.

Sillanpää, M., T-D. Pham and R.A. Shrestha. 2011. Ultrasound technology in green chemistry. pp. 1–21. *In*: Sillanpää, M., T.-D. Pham and R.A. Shrestha. [eds.]. Ultrasound Technology in Green Chemistry. Press. Springer Dordrecht. https://doi.org/10.1007/978-94-007-2409-9_1.

Silva, G. and L.C. di Serio. 2016. The sixth wave of innovation: are we ready?. RAI Revista de Administração e Inovação 13(2): 113–128. https://doi.org/10.1016/j.rai.2016.03.005.

Soto de Prado y Otero, C. 2010. Hacia una cultura de la sostenibilidad a través de los textos de goethe. revista de gilología alemana. 18: 57–77. http://www.redalyc.org/articulo.oa?id=321827635003.

[TWAQIPT] The World Air Quality Index Project Team, Air Pollution in World, Real-time Air Quality Index Visual Map, 2022. https://aqicn.org/map/world.

Tanaka, K. and F. Toda. 2000. Solvent-free organic synthesis. Chemical Reviews 100(3): 1025–1074. https://doi.org/10.1021/cr940089p.

[UN CNU] United Nations, Carta de las Naciones Unidas, 2021. https://www.un.org/es/about-us/un-charte.

[UN WCED] United Nations; World Commission on Environment and Development, Report, 1987. https://undocs.org/es/A/42/427.

[UNDP] United Nations Development Program, ¿Que son los Objetivos de Desarrollo Sostenible? Informative article, 2022. https://www1.undp.org/content/undp/es/home/sustainable-development-goals.html.

[UNDP LAC] United Nations Development Program in Latin America and the Carribbean, Next generation UNDP, Informative text 2022. https://www.latinamerica.undp.org/content/rblac/en/home/our-focus.html.

[UNESCO] United Nations Educational, Scientific and Cultural Organization; Blue Shield Australia Symposium: Cultural Heritage, Climate Change and Natural Disasters, Symposium, 2018. https://en.unesco.org/events/2018-blue-shield-australia-symposium-cultural-heritage-climate-change-and-natural-disasters.

[UNESCO BP] United Nations Educational, Scientific and Cultural Organization, UNESCO Building peace in the minds of men and women, Informative document, 2021. https://en.unesco.org/70years/building_peace.

[UNESCO CMEDS] United Nations Educational, Scientific and Cultural Organization; La Conferencia Mundial sobre la Educación para el Desarrollo Sostenible pide un compromiso renovado de todos los países, Conference, 2017. http://www.unesco.org/new/es/media-services/single-view-tv-release/news/world_conference_on_education_for_sustainable_development_ca.

[UNESCO DNU] United Nations Educational, Scientific and Cultural Organization; El Decenio de las Naciones Unidas para la EDS, Informative text, 2021. https://es.unesco.org/themes/educacion-desarrollo-sostenible/comprender-EDS/decenio-onu.

[UNESCO EDS] United Nations Educational, Scientific and Cultural Organization; ¿Qué es la Educación para el Desarrollo Sostenible? Informative document, 2021. https://es.unesco.org/themes/educacion-desarrollo-sostenible/comprender-EDS.

[UNESCO CFEAP] United Nations Educational, Scientific and Cultural Organization; La ciencia, factor esencial para el acuerdo de París sobre el clima y el desarrollo sostenible, afirma la Junta de Asesoramiento Científico de las Naciones Unidas, Informative text, 2021. https://www.unesco.org/es/articles/la-ciencia-factor-esencial-para-el-acuerdo-de-paris-sobre-el-clima-y-el-desarrollo-sostenible-afirma.

[UNESCO ODM] United Nations Educational, Scientific and Cultural Organization; ¿Qué son los Objetivos de Desarrollo del Milenio?, Informative text, 2017. http://www.unesco.org/new/es/culture/achieving-the-millennium-development-goals/mdgs/.

[UNESCO ODS] United Nations Educational, Scientific and Cultural Organization; La UNESCO y los Objetivos de Desarrollo Sostenible, Informative text, 2021. https://es.unesco.org/sdgs.

[UTAH DEQ WM&RC] UTAH Department of Environmental Quality, Waste Management and Radiation Control, Report of the review, 2021. Salt Lake City, EEUU. https://deq.utah.gov/waste-management-and-radiation-control/radiation-basics.

Vaitsis, C., M. Mechili, N. Argirusis, E. Kanellou, P.K. Pandis, G. Sourkouni et al. 2020. Ultrasound-assisted preparation methods of nanoparticles for energy-related applications. pp. 1–27. *In:* Mousumi Sen [ed.]. Nanotechnology and the Environment. EBOOK ISBN: 978-1-78985-672-9. https://doi.org/10.5772/intechopen.92802.

Vargas A., E.O. and L.P. Ruiz P. 2007. Química verde en el siglo XXI; química verde, una química limpia. Revista Cubana de Química XIX(1): 29–32. http://www.redalyc.org/articulo.oa?id=443543706009.

Vigueras-Carmona, S.E., G. Zafra-Jiménez, M. García-Rivero, M. Martínez-Trujillo and J. Pérez-Vargas. 2013. Efecto del pretratamiento sobre la biodegradabilidad anaerobia y calidad microbiológica de lodos residuales secundarios. Revista Mexicana de Ingeniería Química 12(2): 293–301. http://www.redalyc.org/articulo.oa?id=62030721009.

[WHO, AP] World Health Organization, Air pollution, Informative text, 2022. https://www.who.int/health-topics/air-pollution#tab=tab_1.

Wang, S.-Y., S.L. Ji and X.-M. Su. 2008. A meldrum's acid catalyzed synthesis of bis(indolyl)methanes in water under ultrasonic condition. Chinese Journal of Chemistry 26(1): 22–24. https://doi.org/10.1002/cjoc.200890029.

Zarta, Á.P. 2018. La sustentabilidad o sostenibilidad: un concepto poderoso para la humanidad. Tabula Rasa 28: 409–423. https://doi.org/10.25058/20112742.n28.18.

CHAPTER 2

Sonochemistry Applied to Different Areas of Knowledge

Diego F. Rodríguez,[1] *Sebastián Morales-Guerrero*[1]
and *Flavia C. Zacconi*[1,2,3,]*

1. Introduction

Ultrasound (US) energy allows for the generation of highly localized extreme conditions through cavitation processes in a solvent, providing a valuable energy source that enables the optimization of a great variety of processes (Pollet and Ashokkumar 2019). From organic and organometallic reactions to the synthesis of bioactive compounds, from implementation in the food industry to the remediation of environmental pollutants: sonochemistry is applied and utilized for the improvement of a vast number of processes and products (Luche and Damiano 1980, Casadonte et al. 2007, Collings et al. 2010, Alarcon-Rojo et al. 2015).

The greatest progress in sonochemistry has been the innovation and implementation of more efficient approaches toward valuable industrial commodities, transforming it from a basic research method into a standard technology. The use of US methods is operationally simple and involves low process temperatures. Sonochemistry methods can be divided into two categories: low intensity and high intensity processes. It enables the generation of sound waves, which can be modulated depending on the field of application. Thus, high intensity, low frequency (20 to 40 kHz) US is used for energy processes that range from chemical synthesis to biomass treatment. While low intensity, high frequency (one to 20 MHz) US can be used for therapeutic and medical purposes (Mason and Peters 2002, Xin et al. 2016).

---

[1] Departmento de Química Orgánica, Facultad de Química y de Farmacia, Pontificia Universidad Católica de Chile, Santiago 7820436, Chile.

[2] Institute for Biological and Medical Engineering, Schools of Engineering, Medicine and Biological Sciences, Pontificia Universidad Católica de Chile, Santiago 7820436, Chile.

[3] Centro de Investigación en Nanotecnología y Materiales Avanzados, CIEN-UC, Pontificia Universidad Católica de Chile, Santiago 7820436, Chile.

* Corresponding author: fzacconi@uc.cl

The use of US energy allows to decrease reaction times, reduce the necessity of toxic chemicals, increase product yields and overall increase process sustainability (Puri et al. 2013).

The implementation of US in bioprocesses has been a topic of outstanding interest recently, allowing for the production of bioproducts of high value in environmental friendly conditions (Rokhina et al. 2009).

The application of US energy in the food industry increases the value of foods and can help to enhance attributes like freshness, safety and impact through a greener processes alternative improves efficiencies as compared to conventional production methods (Chavan et al. 2022).

The development of efficient US assisted methods for the remediation of contaminated environments is another area of great potential for sonochemistry (Naidu et al. 2021). Currently, the remediation of pollution generated by civilization is one of the most critical tasks to guarantee its future integrity in the years to come. The generation, application, and disposal of hazardous chemicals and contaminants have led to the pollution of entire ecosystems, and the consequences for our civilization are still unknown (Serna-Galvis et al. 2022).

This chapter aims to provide a review of the main applications of US in various areas of knowledge with an emphasis on i) organic and organometallic synthesis, ii) pharmaceutical industry, iii) bioprocesses, iv) food industry, and iv) remediation (Figure1). Finally, the use of US as emerging green technology is emphasized, which allows the generation of naturally efficient processes and its potential to contribute to the development of sustainable methods.

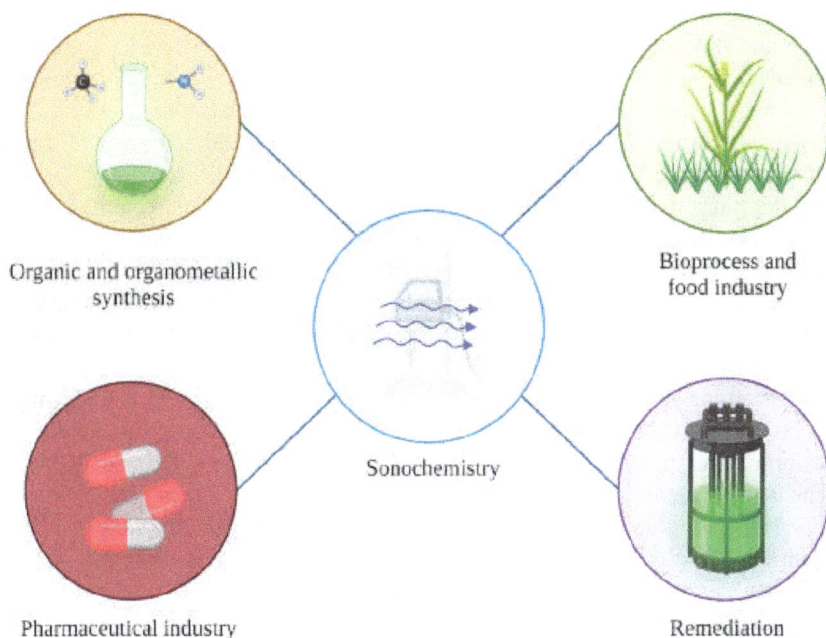

Organic and organometallic synthesis

Bioprocess and food industry

Sonochemistry

Pharmaceutical industry

Remediation

Figure 1. Sonochemistry applied to different areas of research addressed in this chapter.

2. Sonochemistry Applied to Different Areas of Knowledge

2.1 Organic and Organometallic

The generation of extreme local reaction conditions through solvent cavitation provides an invaluable energy source for the optimization of a variety of organic and organometallic reactions. Among the main effects of US assisted syntheses are a decrease in reaction time, an increase in yields, the reduction of metal activation for particular mechanistic pathways, and the overall implementation of more benign reaction conditions (Kimura 2015, Chatel 2016).

Furthermore, sonochemistry fulfills many goals of the Green Chemistry agenda. Improved yields, selectivities, and more efficient activation modes lead to a decrease in waste generation, fewer reagents necessary, decreased complexity of workup processes (e.g., extraction, recrystallization, chromatography) and an overall reduction of the E factor. In addition, the reduction of reaction times and the fact that US activation is very benign, allows for the reduction of energy consumption (Chatel 2016).

Due to these characteristics, sonochemistry has become a highly attractive tool for scientists working on organic and organometallic synthesis, and some examples of the application of US will be shown in the following sections. For more specific examples other chapters of this book can be consulted.

One of the first and most important examples of sonochemical activation in organic synthesis is a study carried out by the group of Luche in 1980 (Luche and Damiano 1980). The authors pointed out, that US was still rarely used by chemists at the time, with isolated examples in ester hydrolysis, carbon halogen bond fission and the mercury induced reduction of α,α'-dibromo ketones. Luche et al., reported the application of sonochemistry for the formation of organolithium and organo-magnesium compounds and their respective use in the Barbier reaction. Their study improve the methodology enabling lithium as the metal activation agent and increased the practicability and use of technical grade solvents.

After these seminal studies, sonochemistry became more common in the repertoire of techniques used by synthetic organic chemists. For example, the sonochemical activation of metal surfaces did not remain exclusive to the preparation of Grignard or organolithium reagents, but also found utility in the synthesis of organosilicon compounds, in Simmons-Smith reaction towards cyclopropanes and in Reformatsky reaction (Mason 1997).

Concerning organometallic chemistry, recently the use of sonochemistry was extended towards cross coupling reactions. These are among the most useful transformations for the construction of complex molecules and are widely applied in medicinal chemistry and the pharmaceutical industry. Generally, they require prolonged heating times and often form a series of undesired byproducts (Cella and Stefani 2018). In this context, US has brought all its advantages into the realm of cross coupling reactions and has contributed to the generation of more sustainable procedures.

A great example of US mediated cross couplings is the Heck reaction, which has been adapted to heterogeneous benign reaction conditions, resulting in good regioselectivities and yields (Cella and Stefani 2018). The Suzuki-Miyaura coupling was also optimized with the application of sonochemistry, generating good yields and leading to reduced reaction times. Furthermore, it could be carried out in a heterogeneous fashion in water and even ionic liquids, representing some of the major goals of Green chemistry (Casadonte et al. 2007, Puri et al. 2013, Cella and Stefani 2018). Other C-C cross couplings successfully improved by US assisted methods are the Stille, Hiyama, Sonogashira, and Ullman-Goldberg reactions respectively, although other metal mediated reactions like olefin metathesis were also reported along these lines (Casadonte et al. 2007, Puri et al. 2013, Chatel 2016, Said and Salem 2016, Cella and Stefani 2018, Lévêque et al. 2018, Andrade and Martins 2020).

Another area that has benefited from the application of US is the synthesis of heterocycles. These compounds are specially useful in the pharmaceutical industry due to their biological activity and their extraction is achieved only from natural sources. Hence the generation of new synthetic procedures addressing heterocycles is of utmost importance (Puri et al. 2013). Sonochemistry has become very useful for the generation of new synthetic methodologies and the optimization of existing ones, with the main advantages corresponding to the reduction of reaction times from days down to hours or minutes, the increase of yields, lower overall costs, the simplicity of work up procedures and higher product purities (Chatel 2016). Heterocycles successfully synthesized by methods featuring US activation are pyrroles (Zhang et al. 2008, Wang et al. 2015, Li et al. 2016, Peglow et al. 2019), pyrazoline (Li et al. 2007, Pacheco et al. 2013, Trilleras et al. 2013) imidazoline (Mirkhani et al. 2006, Sant'Anna et al. 2009, Chen et al. 2014) pyrimidine (Mittersteiner et al. 2021), amongst many others (Patil et al. 2013, Puri et al. 2013, Dandia et al. 2020, Draye et al. 2020).

Ultrasound has also been commonly applied in polymer science, both for polymerization and depolymerization procedures. The extreme conditions generated due to cavitation are ideal for breaking chemical bonds in macromolecules, initiating radical polymerization processes, with the US acting as an initiator. It is important to consider the general reaction mechanism here: as the initiating radicals are formed, the polymer chains which are already present can also degrade; therefore short reaction times favor the formation of the polymer, whereas long periods of sonication favor depolymerization (Bhanvase and Sonawane 2014, Chatel 2016, McKenzie et al. 2019).

An especially green approach for the preparation of macromolecules corresponds to the US mediated emulsion polymerization. This approach enables the polymerization of methyl methacrylate, styrene, n-butyl methacrylate, n-butyl acrylate and other monomers smoothly (Bhanvase and Sonawane 2014). Sonochemically assisted radical polymerization not only involves hydrocarbons, but can also be used to generate poly(organosilicons) (Puri et al. 2013). Additionally, condensation polymerizations for the formation of polyamides, polyesters and polyurethanes have been achieved under US assisted conditions (Price 2003, Puri et al. 2013, Tomke et al. 2017, Huang et al. 2020).

The potential of sonochemistry is even more extensive, and it has found its application in oxidation and reduction reactions, protection and deprotection operations (e.g., alcohols, amines or carboxylic acids), sonophoto organic transformations, and in the synthesis of metal organic frameworks (MOFs), among many others (Mason 1997, Casadonte et al. 2007, Puri et al. 2013, Kimura 2015, Chatel 2016, Draye and Kardos 2016, Cella and Stefani 2018, Lévêque et al. 2018, Draye et al. 2019, Dandia et al. 2020).

2.2 Pharmaceutical Industry

The development of sustainable processes for the synthesis of new pharmaceutically active ingredients is one of the great challenges of medicinal chemistry. Although the activities of the pharmaceutical industry have generated many beneficial products for humanity and public health, so far the major concern was the product itself and not a sustainable production process (Pineiro and Calvete 2018). One of the most important goals for the pharmaceutical industry is the drug design and an efficient and cost attractive synthesis, focusing on the reduction of expenses and the maximization of return.

Commonly these processes require the use of a large amounts of chemicals and solvents, and hence generate huge amounts of waste material (Draye et al. 2020). In fact, the main residues produced by the pharmaceutical industries are solvents, with up to 80 percent of waste related to solvent materials. In addition, the excessive use of energy in many of the actual processes directly impacts the environment (Pineiro and Calvete 2018).

Quite recently the pharmaceutical industry has started to use practices that are in agreement with the green chemistry principles, for the discovery, development and manufacture of their drugs and drug candidates (Pineiro and Calvete 2018). In this context, sonochemistry has had an incredibly special impact, mainly on the synthesis of drugs, since it has allowed generating new methodologies considering green solvents, ionic liquids, or even solvent free methods. Additionally the energy expenditure in many of the processes has drastically decreased (Draye et al. 2020).

One of the most used compounds in pharmaceutical industry is heterocyclic compounds, where the use of US has brought great advantages as mentioned previously. A powerful example is the synthesis of 3,4-dihydropyrimidin-2($1H$)-ones, which can be obtained by the Biginelli reaction. These products are widely used in medicinal chemistry as calcium channel blockers or hypertensive. It has been shown that the classical Biginelli reaction can be accelerated by a factor of > 40 by applying US, resulting in milder reaction conditions, higher yields and easier work up procedures (Cella and Stefani 2018).

As mentioned previously, US can be used in a wide spectrum of reactions and is of course not limited to the generation of heterocycles. For example, in 2018 Mane et al., developed an US assisted synthesis of paracetamol (acetaminophen) from hydroquinone. The authors described the improved preparation in a single synthetic step in only 150 min reaction time at a significantly lower temperature of 60°C, without the formation of undesired salts and with water as a by-product (Mane et al. 2018).

Sonochemistry in drug synthesis also found application in catalytic transformations. US mediated catalytic synthesis allows to work under milder reaction conditions, while increasing efficiency at three distinct stages: (i) the preparation of the supported catalysts, (ii) the activation of the catalyst and (iii) the enhancement of the reaction rate. Activation by US can be used in metal catalysis, organocatalysis and biocatalysis alike, which all are applied in the pharmaceutical industry (Draye et al. 2020).

The widespread use of US in the syntheses of drugs and drug candidates has led to the generation of more environmentally friendly methodologies. The examples for the successful use of sonochemistry in procedures relevant for the pharmaceutical industry are very varied, with some examples already mentioned earlier (Draye et al. 2020, Dandia et al. 2020).

An additional application of US assisted chemistry related to the pharmaceutical industry is the degradation of pharmaceutical products after their use (Thangavadivel et al. 2011). Over the past decades, a wide range of pharmaceutical and personal care products (PPCPs) have been repeatedly detected in the environment. Among them are antibiotics, analgesics, steroids, antidepressants, antipyretics, antimicrobials, and cosmetics. Repeated and continuous exposure to low subtoxic concentrations of certain PPCPs can cause undesired effects to non target species and ecosystems (Sui et al. 2015). In this context, the application of US in sonolytic irradiation methods is an interesting alternative for remediation for these pollutants.

An interesting example is diclofenac, as one of the most popular antiphlogistic, produced in great quantities. Due to its resistance towards biodegradation its presence in aquatic environments is quite common. Sonolytic irradiation over a TiO_2 catalyst has enabled the elimination of diclofenac directly from water. Within the first 30 min of irradiation, the relative concentration of diclofenac decreased from 100 to 16 percent. Sonolytic irradiation has also been used for the remediation of other pharmaceutical contaminants, such as dyes and hormones (Thangavadivel et al. 2011).

Additionally, US has found applications in the pharmaceutical sector well beyond the facilitation of chemical reactions: improvement of extraction processes, photoacoustic quality control of tablets, enhancement of chemotherapy, cell therapy, drug delivery technologies, sonocristallization, sonophoresis, solution atomization, synthesis of polymer nanocomposites and microspheres, among others (Farooq 2011, Chohadia et al. 2018).

2.3 Bioprocesses

The implementation of US techniques into bioprocesses has been successfully realized recently, leading to the production of high value bioproducts in environmentally friendly conditions. Noteworthy examples of this approach can be found in diverse research fields, such as protein synthesis and extraction (Yu et al. 2009, Avhad and Rathod 2014), biofuels production (Malani et al. 2019), extraction of bioactive compounds (Ojha et al. 2020), bioremediation (Pellis et al. 2016) and others (Rokhina et al. 2009, Delgado-Povedano and Luque de Castro 2015).

At first glance, the application of ultrasonic waves to bioprocesses could be considered controversial, especially due to multiple reports of protein denaturation and cell lysis under these conditions (Islam et al. 2014). However, a careful experimental design considering the US parameters, such as frequency, power, sonication time, and intensity (Delgado-Povedano and Luque de Castro 2015), in combination with the optimization of key parameters in the bioprocesses, has resulted in the generation of interesting synergistic effects (Ladole et al. 2018). Despite the advances in the field of US assisted bioprocesses, some of these effects remain poorly understood; this is mainly due to the lack of detailed information, use of inadequate equipment or the omission of crucial details in the protocols, thus still hampering reproducibility to some extent.

2.3.1 General Ultrasound Effects in Microorganisms and Enzymes

The effect of US on the acceleration of microbial cell proliferation and the enhancement of catalytic activity of enzymes has been recently revised (Huang et al. 2017). In general, the stimulation of growth assisted by US depends on the intensity and the frequency during application. For low intensity US waves, the cell damage is reversible and cell metabolism can be stimulated. Additionally, sonoporation of the membrane increases the permeability towards nutrients and oxygen (Akdeniz and Akalın 2020). Regarding enzymatic activation, the ultrasonic waves generate changes in the three dimensional structure of proteins, promoting alterations of the enzyme-substrate interactions (Pallarés et al. 2021). These conformational changes can increase the desired exposure to the substrate, while maintaining the enzymatic activity and enhancing mass transfer in the enzymes, increasing their overall catalytic efficiency. Additionally, these structural modifications can alter the kinetic parameters of the enzyme's activity (Tran et al. 2018), and in some cases even increase the enantioselectivity to certain reactions (An et al. 2016).

2.3.2 Ultrasound Bioprocess like Eco friendly Processes

The application of US in different bioprocesses has emphasized its potential to achieve sustainability goals in energy intensive processes. For example, the pretreatment of lignocellulose and the production of biofuels are processes in which a significant decrease in energy consumption has a major impact (Veljković et al. 2012, Bussemaker and Zhang 2013). The cooperation of US technology with bioprocesses aligns perfectly with the principles of green chemistry (Badgujar et al. 2021), representing an important tool for the development of new bioproducts replacing conventional petrochemical raw materials and promoting circular economies (Ubando et al. 2020).

2.3.2.1 Bioconversion of Lignocellulosic Biomass

Lignocellulosic biomass is one of the most abundant renewable materials on earth and a main pillar of bioeconomy, representing a valuable feedstock for different bioproducts (Zhou et al. 2011). Enzymatic processing of lignocellulosic material has been developed into a viable alternative for conventional chemical treatments. Its advantages feature mild operating conditions (low pressure and temperature), high efficiency and a wide availability of suitable natural and non-natural biocatalysts.

These catalysts can be adapted to the different types of existing biomass, and the overall productivity of the processes can be improved (Lopes et al. 2018). However, the recalcitrant nature of lignocellulosic material (mainly composed of cellulose, hemicellulose and lignin), hampers its pretreatment and a subsequent enzymatic transformation into valuable products to a certain extent (Chettri et al. 2021). One of the most promising approaches to overcome this problem is the use of a combination of US waves and enzymes, which has already proven to be effective for the treatment of lignocellulosic biomass, leading to increased yields of sugars and decreased process times (de Carvalho Silvello et al. 2020).

In 2009, Yachmenev et al., reported that the combined use of enzymes and US in the processing of corn stover towards glucose, led to an increase in efficiency of up to 22 percent after only one hour. The authors proposed several reasons for this beneficial effect: (i) a decrease in mass transfer effects, (ii) a more significant exposure to lignocellulosic biomass caused by the cavitation effect facilitating its enzymatic processing, (iii) an improved effect of the cavitation process in the present heterogeneous materials up to several hundred times and (iv) concordance between the optimum cavitation temperature and the optimum temperature of the enzymes activity (Chatel et al. 2014).

In general, the results obtained by treating lignocellulosic biomass with low frequency US suggests that the main effect is of a mechanoacoustic type (Chatel et al. 2014). The cavitation generated by the US waves exerts mainly mechanical effects on the biomass. Alterations of the structure at a macro and microscopic level are generated, thus facilitating its accessibility towards the enzymatic action, while not generating extensive hydrolysis of the biomass.

2.3.2.2 Biodiesel Productions

Biodiesel is obtained mainly by the transesterification of vegetable oils and fats with low molecular weight alcohols, such as methanol and ethanol (Basu 2018). This biofuel has characteristics similar to conventional diesel and can be used in a blend or in its pure form in combustion engines (Vignesh and Barik 2019). Currently, the largest biodiesel producer is the European Union, with 41.2 million tons in 2018, mainly obtained from vegetable oils and catalyzed by different alkali salts (Pasha et al. 2021). This type of catalysis requires in many cases a pretreatment of the feedstocks or oils with low acid indexes, to avoid the generation of undesired by products. The presence of major amounts of by products and other pollutants makes an energy intensive biofuel purification process necessary before the product can be marketed. In general, these processes are polluting and water intensive, decreasing the overall sustainability of the biofuel.

Currently, there are several alternatives for biodiesel production. The enzymatic production of biodiesel, however, is the most promising one considering sustainability aspects. It exhibits considerable advantages over conventional processes, since it generates fewer by products, requires less intensive purification methods, occurs under mild operating conditions (low pressure and temperature), features biodegradable catalysts, enables the catalyst recycling through immobilization and can be carried out with lower quality raw materials without the need for complex pretreatments (Fjerbaek et al. 2009). The catalytic efficiency of these biocatalysts is often lower than that of conventional chemical catalysts, leading to the requirement of longer reaction times. To overcome this shortcoming, different approaches have been used, amongst which the application of US has proven to be excellently suited to complement this type of bioprocess, lowering reaction times significantly.

In 2016, Subhedar et al., reported the synthesis of biodiesel assisted by US starting from waste cooking oil and methyl acetate using the immobilized lipase from *Thermomyces lanuginosus* (Lipozyme TLIM). Under optimized US conditions (20 kHz, 80 W with 60 percent duty cycle), a significant reduction in reaction time was observed (from 24 to three hours), along with an increase in yield from 90 to 96 percent, and a decrease in catalyst loading (Subhedar and Gogate 2016).

In 2018, Poppe et al., reported the US assisted transesterification of used frying oil and soybean oil with combi lipases (a combination of enzymes with different substrate specificities), allowing for an increase in the enzymatic production of biodiesel from both raw materials. This improvement, using immobilized enzyme mixtures with US, was mainly attributed to a decrease in mass transfer effects (Poppe et al. 2018), especially important at the beginning of the reaction where the alcohol and oil phases are not miscible.

A promising example of US assisted biodiesel production featured a *Proteus vulgaris* lipase, specifically modified by protein engineering to improve its thermostability and tolerance towards methanol and subsequently immobilized on functionalized polysulfone beads (Gupta et al. 2020). US mediated transesterification processes using this new biocatalyst resulted in fast reaction times (30 min) and high yields (> 99 percent), comparing extremely well with conventional methods.

2.4 Food Industry

The most important achievements from chemical research are those finally translated into profitable technologies. The use of sonochemistry and US techniques is very well established, involves low temperatures and is an innovative technology due to its potential in many industrial applications. This methodology implies the use of sound waves divided into two categories: low intensity (five to 10 MHz) and high intensity waves (> 20 kHz) (Lee and Feng 2011, Ikeda et al. 2016).

For example, the interaction between the ultrasonic waves, a liquid medium and dissolved gas leads to an interesting phenomenon known as acoustic cavitation (AC). Acoustic cavitation facilitates chemical reactions and physical forces, including microjets, shear forces, shock waves, vibration, heating, and turbulence. Ultrasound offers an advantage that is fundamentally different as compared to conventional extraction, processing, or preservation techniques (Ashokkumar et al. 2012, Shestakov et al. 2011). Several studies report on the combined use of ultrasound methodologies with elevated temperature, elevated pressure, or both, which correspond to ultrasonication (US), thermosonication (TS), manosonication (MS), and monothermosonication (MTS) respectively (Mason et al. 1996, De Jong et al. 1999, Zheng and Sun 2006).

The utilization of ultrasound in the food industry is on the rise due to the necessity of freshness, high quality, and microbiological safety of foods. In the case of sensible products, the utilization of this method for the inactivation of enzymes or microorganisms allows for a treatment without producing any kind of product damage (Bates and Patist 2010).

Taking into account, that some ingredients are sensitive to temperature and susceptible to damage through chemicals and physical or microbiological variations, ultrasound represents a sustainable and interesting technology for its application in the food industry, with a focus on diminishing the use of water, solvents, non renewable energy and minimizing the production of harmful side products (Arvanitoyannis et al. 2017). Additionally, US techniques can be applied in very different areas of the industry, such as mixing, homogenization, filtration, extraction, dehydration, freezing, brining, thawing, crystallization, fermentation, antifoaming, reduction of particle sizes, viscosity modifications, dispersion, inactivation of enzymes and microorganisms, and many more (Gallo et al. 2018).

In the food industry, the application of US energy enhances the value of processed foods, resulting in sustained freshness, while also improving the sustainability of the applied processes. Some of the main applications of ultrasound in the food industry are listed below.

2.4.1 Crystallization and Freezing

High intensity ultrasound can be used to control the nucleation in crystallization processes. The combination of US with low temperatures diminishes the crystallization time and improves the integrity and homogeneity of the product (Zheng and Sun 2006).

In the ice cream industry, US is commonly applied to prevent the formation of significant ice crystals during the mixing process of the main ingredients (milk, sugars, emulsifiers, additives, flavors, etc.), assuring the quality of final product (Kiani et al. 2011). Additionally, application of high power US during the freezing procedure of the ice cream, allows for decreased freezing times, improving the production process, and enhancing the consistency and morphology of the final product (Cook and Hartel 2010). In this regard, crystallization and freezing are highly interconnected.

Considering the general importance related to the control of nucleation rates, crystal sizes, and crystal growth for the improvement of product quality, US technologies have been used in several other industries, such as the production of milk fat, vegetable oils, fine chemicals, and pharmaceutics. Noteworthy, the US procedures inhibit the production of larger crystals and enhance the crystallization with an efficient cooling process (Cheng et al. 2015).

2.4.2 Cooking and Meat Tenderization

In regular cooking processes, foods are often subjected to elevated temperatures, overcooking the external part, while leaving the interior parts uncooked, lowering overall product quality. US enhances heat transfer, thus avoiding these problems and resulting in faster cooking with beneficial moisture preservation and a more sustainable energy utilization (Jayasooriya et al. 2004). These characteristics suggest the application of US technologies in the convenience food industry, especially for prepared meat products, due to the improved myofibrillar tenderness governed by the water binding physical properties of this protein.

In the industry concerning convenience meat products, tenderness is one of the most valuable attributes, having direct impact on the consumer satisfaction. Currently, ageing, or mechanical pounding are used to improve tenderness of meats, but these processes are usually slow, decreasing the quality of the meat (Alarcon-Rojo et al. 2015).

High intensity US exposure acts through cavitation, causing a reduction of the highly connective tissue content and a myofibrillar disruption, consequently enhancing tenderness. In case the US methodology is conducted in solution, the proteolysis of the meat could increase these desired characteristics even further.

In summary, US enhances the desired properties of meat products, such as tenderness, water binding capacity, color and cohesiveness (Kiani et al. 2012).

2.4.3 Deaeration; Defoaming; Degassing

Deaeration, defoaming, and degassing treatments of liquids are another attractive field of US applications (Knorr et al. 2004). The US releases entrapped gas and decreases the amount of gas dissolved in the liquid. This characteristic makes US especially attractive for the beverage industry, fermentation processes, cosmetic industry, and other fields where foam formation deteriorates the quality of products. The addition of chemical anti foaming agents or the use of mechanical breakers are the current state of the art to control foaming in general (Dedhia et al. 2004, Bilek and Turantaş 2013). The use of US energy has been well established, but actual industrial applications are still lacking. One of the most relevant examples is the use of high intensity US waves to regulate the excess of foam occurring in high speed bottling and canning lines of carbonated beverages. On the other hand, low intensity US techniques reduce the fermentation time of alcoholic beverages (beer, wine, and sake), representing an attractive alternative too (Sun et al. 2015).

The necessity to remove the air from liquids to diminish the presence of bacteria and oxygen, as well as maintaining appropriate organoleptic characteristics made US assisted degassing procedures the most attractive option for aqueous systems. Quite remarkably, US methodologies reduce overflow and the quantity of broken bottles during the industrial beverage processing (Lukić et al. 2019).

2.4.4 Depolymerization

The US assisted degradation of polymers is one of the first applications. This process is successful mainly due to cavitation, implying two different mechanisms: chemical degradation or mechanical degradation (Szent-GyÖrgyi 1933). The depolymerization of starch is an important application of US energy techniques (Schmid and Rommel 1939, Zuo et al. 2009). Furthermore, it also represents an attractive approach for the depolymerization, solubility enhancement and foaming optimization of whey proteins (Jambrak et al. 2008, 2009, Geng et al. 2021).

The application of low intensity US results in a viscosity reduction of polymeric liquids; high intensity US, however, will depolymerize these liquids, resulting in the corresponding rheology changes (Jambrak et al. 2010, Prajapat et al. 2016). The application of US energy for the treatment of raw materials is a promising field of research, since it can be used to produce value added products, while focusing on a decrease of the environmental impact of the industrial processes (Vodeničarová et al. 2006).

2.4.5 Extraction

Several organic compounds, minerals, polyphenols, and scents have effectively been isolated from different natural sources using US (Caldeira et al. 2004). The utilization has several advantages: diminishing the extraction time, decreasing the amount of energy needed in the extraction process, reducing CO_2 emissions (compared to the conventional techniques), lowering the environmental impact, and increasing yields of products in less time and at lower temperatures (Chemat et al. 2011). Remarkably, this methodology has enabled the extraction of pesticides from food samples, once again facilitated through mechanical effects produced by the cavitation (Vinatoru 2001). In this specific example, the decomposition of the cell walls allows for a better access of the solvent into the sample, effectively detaching the undesired compounds from the natural matrix for the obtention of olive oil or isolation of polyphenols from citrus peels (Toma et al. 2001, Achat et al. 2012, Clodoveo et al. 2013, Nayak et al. 2015).

2.4.6 Fermentation

Low intensity US enhances the reagents and products mass and heat transfer through the interface of liquid and cell wall, depending on the matrix of interest. High intensity US, on the other hand, leads to denaturation and collapse of cell walls (Chisti 2003). The careful combination of low intensity and high intensity US technology has been used to optimize the fermentation process, analyze the alcohol content of beverages, (Morata and Suárez-Lepe 2015), detect the maturation progress of cheeses and milk products, control the concentration of several chemical compounds in feedstocks and to stimulate the production of probiotics (Wu et al. 2000, Shershenkov and Suchkova 2015).

2.4.7 Demoulding

Industrial demoulding methods for pre-cooked foods are difficult to achieve, mainly due to the adhesion caused by the cooking process. Considering that the moulds are fabricated with a surface coating or covered by a grease layer, recurrent replacement is necessary, mainly because of the intensive use over extended periods (Knorr et al. 2004, Mousavi et al. 2007).

US can be used to improve the detachment of the mould from the product, generating a major improvement of product quality and presentation (Scotto 1998).

2.4.8 Drying

Conventional methods for drying of food products use forced streams of hot air, with advantages considering costs, but severe disadvantages concerning long exposure time, high temperatures harmful to the product, potential changes of taste and color and changes in the final nutritional composition (Musielak et al. 2016). However, the application of US energy coupled to an air drying system enables lower temperatures and shorter treatment times, without affecting the desired qualities of the final product (Mulet et al. 2003, de la Fuente-Blanco et al. 2006). Successful applications have been reported from the cheese, meat, and vegetable industries respectively (Jambrak et al. 2007, Fernandes et al. 2008, García-Pérez et al. 2009, Tao and Sun 2015).

2.4.9 Emulsification–Homogenization

US techniques are frequently used to obtain or optimize emulsions, juices, sauces, and other food products when homogenization is necessary (Behrend and Schubert 2001). These methods are utilized in the milk industry, enhancing the homogeneity and production of more stable products (Povey and Mason 1998, Wu et al. 2000). The cavitation helps to form optimize emulsions through the disruption of fat globules, while at the same time decreasing the size and volume of milk proteins.

To obtain the best final food characteristics, a US energy ranging from 20 kHz to two MHz is used for a more efficient production (Jafari et al. 2007, Shanmugam and Ashokkumar 2014).

Other foods that could benefit from US techniques are fruit juices, ice cream, mayonnaise, and many more (Wu et al. 2000, Canselier et al. 2002, O'Sullivan et al. 2015).

2.4.10 Pasteurization-Sterilization

Currently, thermal sterilization and pasteurization methods are used by the food industries to safely inactivate enzymes and microorganisms, thus enhancing the shelf life of the products (Mason et al. 1996, De Jong et al. 1999). These processes involve high energy intensity, affect the quality and properties of the final product, and increase the cost of the production (Knorr et al. 2004).

The utilization of high intensity US technologies has been reported to inactivate enzymes and microorganisms in fruit juices and dairy products (Crudo et al. 2014, Saeeduddin et al. 2015). Additionally, thermosonication (TS) has been applied in the dairy, yeast, and juice industry to reduce the aerobic bacteria levels and maintain product qualities unaltered (Piyasena et al. 2003, Ercan and Soysal 2011, Lee et al. 2013).

2.5 Environmental Remediation

The generation and careless disposal of pollutants and hazardous compounds has led to the contamination of entire ecosystems, with drastic consequences on our planet (Naidu et al. 2021). Currently, the remediation of pollutants generated by human activity is one of the most critical challenges for the future. The development of efficient US assisted remediation methods is of high potential and can address the contamination problematic generated by inefficient methodologies utilized in the past and present. Environmental remediation assisted by US waves has proven to be highly efficient and versatile (De Andrade et al. 2021). The adaptability of the technology has allowed for its use in combination with other methods, enabling the remediation of contaminants from complex matrices such as soil, sediments, water, and even special materials like plastics.

2.5.1 Ultrasound Assisted Soil Remediation

Ultrasound assisted soil remediation represents a viable alternative for the treatment of soils contaminated with hydrocarbons (Flores et al. 2007), persistent organic pollutants (Collings et al. 2010), heavy metals (Park and Son 2017), and substances originating from electronic waste (Chen et al. 2016). In this field, ultrasonic waves improve mass transfer, increase desorption, and trigger chemical degradation of contaminants in the soil, with the effects being highly dependent on the applied frequency. However, the combination of US with other methods is very promising for highly efficient remediation processes, showing significant synergistic effects for decontamination (Effendi et al. 2019).

Recently, a protocol concatenating different methodologies for soil treatment was reported (Li et al. 2020), combining persulfate (PS), US and Fe^0 (Fe) for the removal of petroleum based hydrocarbons in soil. The resulting US/Fe/PS system performed superior as compared to its individual components and their respective combination in pairs, such as US/Fe and Fe/PS. Under optimized conditions (PS one M, 200 W, initial hydrogen potential = six, soil/liquid ratio of three:two), the degradation of petroleum based hydrocarbons reached up to 82.23 percent and the total organic carbon (TOC) mineralization degree was 54 percent, with the efficiency of the method being proportional to the US power applied. The synergistic effects of the US/Fe/PS system were attributed to the sono activation of PS, allowing for the efficient removal of organic pollutants and the sono regeneration of Fe^0 surfaces with a subsequent promotion of the bulk radical reactions. This synergy effects were observed previously for the degradation of the antibiotic sulfadiazine (Zou et al. 2014).

2.5.2 Ultrasound Assisted Plastic Remediation

Plastics are synthetic materials used in diverse products deemed essential for humanity. In 2018, the global plastics production reached 359 million tons (Prata et al. 2021). Hence, its widespread distribution, durability and low overall recyclability make plastics a ubiquitous contaminant. Currently, nano and micro plastics formed by the degradation of polymers, are commonly found in animals and humans (Schwabl et al. 2019), with the long term effects still unknown (Toussaint et al. 2019). It has been shown that plastic nanoparticles can promote platelet aggregation, potentially triggering different cardiovascular diseases (Lett et al. 2021).

The development of efficient methods for plastic remediation is of critical importance, and various approaches seek to make plastic remediation processes more efficient and potentially allow for a circular economy (Schwarz et al. 2021, Meys et al. 2020). However, the recalcitrant nature of plastic requires harsh processing conditions for its treatment. Recently, a new eco-friendly plastic degradation method combining *Thermobifida cellulosilytica* cutinase one and US has been developed (Pellis et al. 2016). In this approach, the applied ultrasonic waves increased the enzymes activity after 10 min of sonication of polyethylene terephthalate (PET) samples. Additionally, the use of protein engineering for the development of more efficient enzymes suited for plastic degradation is of high interest, potentially re generating monomers, which can then be reused for a circular economy of PET (Tournier et al. 2020). Quite obviously, the combination of these technologies with ultrasonic activation could further increase their efficiency.

2.6 Some Examples of Application of US in Different Research Areas

2.6.1 Synthesis of Orgmanylselanyl Pyrroles

Ultrasonic radiation promoted one pot synthesis of mono and bis substituted organylselanyl pyrroles in high yields; 20 kHz, 130 W, rt 0.5 to 3.5 hr, CuI, 10 to 15 mol percent; Mechanism: Improved mass transport (Peglow et al. 2019).

2.6.2 Paracetamol Synthesis

The US assisted synthesis of paracetamol from hydroquinone allowed reduced reaction times (150 min) and low temperatures (60°C), compared to standard conditions 15 hr and 220°C; 22 kHz, 125 W, 50 percent duty cycle, 300 ppm; Mechanism: Cavitation effects generate localized temperatures, which contribute to the nucleophilic substitution reaction. (Mane et al. 2018).

2.6.3 Enzymatic Biodiesel Synthesis

US assisted enzymatic catalyzed interesterification of waste cooking oil, reduced reaction time (30 min), enzyme loading (3 percent w/v), and allowed high conversions (96 percent); 20 kHz, 80 W, 60 percent duty cycle, 50°C; Mechanism: The application of US increases the interaction between enzyme and substrate (Subhedar and Gogate 2016).

2.6.4 Food Industry

The US utilization in food preservations against *Escherichia coli;* 20 kHz, 100 W, 40 to 61°C, 100 to 500 kPa, 0.25 to four min; Mechanism: Utilization of several techniques: sonication, MS, TS, MTS (Chemat et al. 2011).

2.6.5 Soil Remediation

The System (US/Fe/PS), a combination of activated persulfate (PS) with US and Fe, enhancement of hydrocarbon degradation efficiency achieved up 82.23 percent; 200 W, persulfate one M, initial hydrogen potential = six, soil/liquid (S/L) ratio of three: two; Mechanism: The synergistic effect might be because of the acceleration of the persulfate decomposition, radical sulfate production, and destroying the passivation layer of Fe^0 surface caused by the sonochemical effect (Li et al. 2020).

3. Conclusions

US technologies are versatile and complementary on raising efficiencies for various processes. Moreover, US is an emerging green technology that has been successfully incorporated into several critical research areas. Nowadays, its field of action includes synthetic chemistry, in which it has allowed a decrease in reaction times, higher yields, less use of harsh conditions, metal activation, and specificity or selectivity for a particular mechanism route; allowing the generation of new synthetic procedures that are more efficient and friendly to the environment. This approach goes very much in line with the principles of green chemistry, especially considering energy efficiencies.

Additionally, the pharmaceutical industry has benefited from more efficient and green processes in the synthesis of active pharmaceutical ingredients, also in bioprocesses through the reduction in operation times in biocatalyzed processes such as biofuel production and treatment of lignocellulosic materials. The food industry, where the successful coupling in processes such as crystallization, degassing, extraction, drying, or fermentation, has allowed the generation of more efficient processes. Also, US is a helpful tool in the remediation of soils and materials that are difficult to degrade and potentially dangerous and toxic for nature and human health, such as plastic.

The overall endeavors made through these innovative sonochemical technologies and their unique advantages for different areas of the knowledge are continuously explored to enhance their applications.

Finally, the sonochemistry field presents important challenges to explore novel and essential methodologies for the scientific and industrial community.

Acknowledgments

FZ is grateful to ANID/CONICYT/FONDECYT Regular N° 1210763. DR thanks CONICYT-PCHA/Doctorado Nacional/2018-21180422. S.M-G. thanks ANID/Doctorado Nacional/21210627. We also want to thank Lars Ratjen PhD for the English revision of this manuscript.

References

Achat, S., V. Tomao, K. Madani, M. Chibane, M. Elmaataoui O. Dangles et al. 2012. Direct enrichment of olive oil in oleuropein by ultrasound-assisted maceration at laboratory and pilot plant scale. Ultrason. Sonochem. 19(4): 777–786.

Akdeniz, V. and A.S. Akalın. 2020. Recent advances in dual effect of power ultrasound to microorganisms in dairy industry: activation or inactivation. Crit. Rev. Food Sci. Nutr. 62(4): 889–904.

Alarcon-Rojo, A.D., H. Janacua, J.C. Rodriguez, L. Paniwnyk and T.J. Mason. 2015. Power ultrasound in meat processing. Meat Sci. 107: 86–93.

An, B., H. Fan, Z. Wu, L. Zheng, L. Wang, Z. Wang et al. 2016. Ultrasound-assisted enantioselective esterification of ibuprofen catalyzed by a flower-like nanobioreactor. Molecules 21(5): 565.

Andrade, M.A. and L.M.D.R.S. Martins. 2020. New trends in C–C cross-coupling reactions: the use of unconventional conditions. Molecules 25(23): 5506–5541.

Arvanitoyannis, I.S., K. V Kotsanopoulos and A.G. Savva. 2017. Use of ultrasounds in the food industry– methods and effects on quality, safety, and organoleptic characteristics of foods: a review. Crit. Rev. Food Sci. Nutr. 57(1): 109–128.

Ashokkumar, M., O.N. Krasulya, S. Shestakov and R. Rink. 2012. A new look at cavitation and the applications of its liquid-phase effects in the processing of food and fuel. Appl. Phys. Res. 4(1): 19–29.

Avhad, D.N. and V.K. Rathod. 2014. Ultrasound stimulated production of a fibrinolytic enzyme. Ultrason. Sonochem. 21(1): 182–188.

Badgujar, V.C., K.C. Badgujar, P.M. Yeole and B.M. Bhanage. 2021. Investigation of effect of ultrasound on inmobilized C. rugosa lipase: synthesis of biomass based furfuryl derivative and green metrics evaluation study. Enzyme and Microbia Technology 144: 109738–109752.

Basu, P. 2018. Production of synthetic fuels and chemicals from biomass. pp. 415–443. *In*: P. Basu [ed.]. Biomass Gasification, Pyrolysis and Torrefaction. Academic Press., Cambridge, USA.

Bates, D. and A. Patist. 2010. Industrial applications of high-power ultrasonics in the food, beverage and wine industry. pp. 119–138. *In*: Doona, C.J., K. Kustin and F.E. Feeherry [eds.]. Case Studies in Novel Food Processing Technologies. Woodhead Publishing, Sawston, UK.

Behrend, O. and H. Schubert. 2001. Influence of hydrostatic pressure and gas content on continuous ultrasound emulsification. Ultrason. Sonochem. 8(3): 271–276.

Bhanvase, B.A. and S.H. Sonawane. 2014. Ultrasound assisted in situ emulsion polymerization for polymer nanocomposite: a review. Chem. Eng. Process. Process Intensif. 85: 86–107.

Bilek, S.E. and F. Turantaş. 2013. Decontamination efficiency of high-power ultrasound in the fruit and vegetable industry, a review. Int. J. Food Microbiol. 166(1): 155–162.

Bussemaker, M.J. and D. Zhang. 2013. Effect of ultrasound on lignocellulosic biomass as a pretreatment for biorefinery and biofuel applications. Ind. Eng. Chem. Res. 52(10): 3563–3580.

Caldeira, I., R. Pereira, M.C. Clímaco, A.P. Belchior and R. Bruno de Sousa. 2004. Improved method for extraction of aroma compounds in aged brandies and aqueous alcoholic wood extracts using ultrasound. Anal. Chim. Acta 513(1): 125–134.

Canselier, J.P., H. Delmas, A.M. Wilhelm and B. Abismaïl. 2002. Ultrasound emulsification—an overview. J. Dispers. Sci. Technol. 23: 333–349.

Casadonte, D.J., Z. Li and D.M.P. Mingos. 2007. Applications of sonochemistry and microwaves in organometallic chemistry. pp. 307–339. *In*: Mingos, D.M.P. and R.H. Crabtree [eds.]. Comprehensive Organometallic Chemistry III. Elsevier Science, Amsterdam, Netherlands.

Cella, R. and H.A. Stefani. 2018. Ultrasonic reactions. pp. 343–371. *In*: Zhang, W. and B. Cue [eds.]. Green Techniques for Organic Synthesis and Medicinal Chemistry. Wiley. Hoboken.

Chatel, G., K. De Oliveira Vigier and F. Jérôme. 2014. Sonochemistry: what potential for conversion of lignocellulosic biomass into platform chemicals? Chem. Sus. Chem. 7(10): 2774–2787.

Chatel, G. 2016. Sonochemistry: new opportunities for green chemistry. World Scientific, London.

Chavan, P., P. Sharma, S.R. Sharma, T.C. Mittal and A.K. Jaiswal. 2022. Application of high-intensity ultrasound to improve food processing efficiency: a review. Foods. 11: 122–139.

Chemat, F., Zill-e-Huma and M.K. Khan. 2011. Applications of ultrasound in food technology: processing, preservation and extraction. Ultrason. Sonochem. 18(4): 813–835.

Chen, G.-F., H.-Y. Li, N. Xiao, B.-H. Chen, Y.-L. Song, J.-T. Li et al. 2014. Efficient synthesis of 2-imidazolines in the presence of molecular iodine under ultrasound irradiation. Aust. J. Chem. 67(10): 1516–1521.

Chen, F., B. Yang, J. Ma, J. Qu and G. Liu. 2016. Decontamination of electronic waste-polluted soil by ultrasound-assisted soil washing. Environ. Sci. Pollut. Res. 23(20): 20331–20340.

Cheng, X., M. Zhang, B. Xu, B. Adhikari and J. Sun. 2015. The principles of ultrasound and its application in freezing related processes of food materials: a review. Ultrason. Sonochem. 27: 576–585.

Chettri, D., A.K. Verma, L. Sarkar and A.K. Verma. 2021. Role of extremophiles and their extremozymes in biorefinery process of lignocellulose degradation. Extremophiles 25(3): 203–219.

Chisti, Y. 2003. Sonobioreactors: using ultrasound for enhanced microbial productivity. Trends Biotechnol. 21(2): 89–93.

Chohadia, A.K., Yasmin and N. Kunwar. 2018. Industrial applications. pp. 354–359. *In*: Ameta, S.C., A. Rakshit and G. Ameta [eds.]. Sonochemistry: An Emerging Green Technology. Apple Academic Press, Oakville.

Clodoveo, M.L., V. Durante and D. La Notte. 2013. Working towards the development of innovative ultrasound equipment for the extraction of virgin olive oil. Ultrason. Sonochem. 20(5): 1261–1270.

Collings, A.F., P.B. Gwan and A.P. Sosa-Pintos. 2010. Large scale environmental applications of high-power ultrasound. Ultrason. Sonochem. 17(6): 1049–1053.

Cook, K.L.K. and R.W. Hartel. 2010. Mechanisms of ice crystallization in ice cream production. Compr. Rev. Food Sci. Food Saf. 9(2): 213–222.

Crudo, D., V. Bosco, G. Cavaglià, S. Mantegna, L. Battaglia and G. Cravotto. 2014. Process intensification in the food industry: hydrodynamic and acoustic cavitation in fresh milk treatment. Agro Food Ind. Hi. Tech. 25: 55–59.

Dandia, A., S.L. Gupta, R. Sharma, D. Rathore and V. Parewa. 2020. Sonochemical protocol for stereoselective organic synthesis. pp. 71–93. *In*: Inamuddin, R. Boddula and A.M. Asiri [eds.]. Green Sustainable Process for Chemical and Environmental Engineering and Science. Elsevier, Amsterdam.

De Andrade, F.V., R. Augusti and G.M. de Lima. 2021. Ultrasound for the remediation of contaminated waters with persistent organic pollutants: a short review. Ultrason. Sonochem. 78: 105719.

De Carvalho Silvello, M.A., J. Martínez and R. Goldbeck. 2020. Low-frequency ultrasound with short application time improves cellulase activity and reducing sugars release. Appl. Biochem. Biotechnol. 191(3): 1042–1055.

De la Fuente-Blanco, S., E. Riera-Franco de Sarabia, V.M. Acosta-Aparicio, A. Blanco-Blanco and J.A. Gallego-Juárez. 2006. Food drying process by power ultrasound. Ultrasonics 44: e523–e527.

De Jong, P., E.H. Hamersveld and M. Villamiel. 1999. Review: effect of ultrasound processing on the quality of dairy products. Milchwissenschaft 54: 69–73.

Dedhia, A.C., P.V Ambulgekar and A.B. Pandit. 2004. Static foam destruction: role of ultrasound. ultrason. Sonochem. 11(2): 67–75.

Delgado-Povedano, M.M. and M.D. Luque de Castro. 2015. A review on enzyme and ultrasound: a controversial but fruitful relationship. Anal. Chim. Acta 889: 1–21.

Draye, M. and N. Kardos. 2016. Advances in green organic sonochemistry. top. Curr. Chem. 374(5): 74–102.

Draye, M., J. Estager and N. Kardos. 2019. Organic sonochemistry: ultrasound in green organic synthesis. pp. 1–93. *In*: Jean-Philippe, G., M. Malacria and C. Ollivier [eds.]. Activation Methods: Sonochemistry and High Pressure. Wiley Online Books, Hoboken.

Draye, M., G. Chatel and R. Duwald. 2020. Ultrasound for drug synthesis: a green approach. Pharmaceuticals. 13(2): 23–57.

Effendi, A.J., M. Wulandari and T. Setiadi. 2019. Ultrasonic application in contaminated soil remediation. Curr. Opin. Environ. Sci. Heal. 12: 66–71.

Ercan, S.Ş. and Ç. Soysal. 2011. Effect of ultrasound and temperature on tomato peroxidase. Ultrason. Sonochem. 18(2): 689–695.

Farooq, R. 2011. Applications of sonochemistry in pharmaceutical sciences. pp. 97–104. *In*: Chen, D., S.K. Sharma and A. Mudhoo [eds.]. Handbook on Applications of Ultrasound. CRC Press, Boca Raton.

Fernandes, F.A.N., F.E. Linhares and S. Rodrigues. 2008. Ultrasound as pre-treatment for drying of pineapple. Ultrason. Sonochem. 15(6): 1049–1054.

Fjerbaek, L., K.V Christensen and B. Norddahl. 2009. A review of the current state of biodiesel production using enzymatic transesterification. Biotechnol. Bioeng. 102(5): 1298–1315.

Flores, R., G. Blass and V. Domínguez. 2007. Soil remediation by an advanced oxidative method assisted with ultrasonic energy. J. Hazard. Mater. 140(1): 399–402.

Gallo, M., L. Ferrara and D. Naviglio. 2018. Application of ultrasound in food science and technology: a perspective. Foods 7:164–181.

García-Pérez, J.V., J.A. Cárcel, E. Riera and A. Mulet. 2009. Influence of the applied acoustic energy on the drying of carrots and lemon peel. Dry. Technol. 27(2): 281–287.

Geng, F., Y. Xie, Y. Wang and J. Wang. 2021. Depolymerization of chicken egg yolk granules induced by high-intensity ultrasound. Food Chem. 354: 129580–129579.

Gupta, S., P.B. Mazumder, D. Scott and M. Ashokkumar. 2020. Ultrasound-assisted production of biodiesel using engineered methanol tolerant proteus vulgaris lipase immobilized on functionalized polysulfone beads. ultrason. Sonochem. 68: 105211–105218.

Huang, G., S. Chen, C. Dai, L. Sun, W. Sun, Y. Tang et al. 2017. Effects of ultrasound on microbial growth and enzyme activity. ultrason. Sonochem. 37: 144–149.

Huang, W., J. Zhai, C. Zhang, X. Hu, N. Zhu, K. Chen et al. 2020. 100% Bio-based polyamide with temperature/ultrasound dually triggered reversible cross-linking. Ind. Eng. Chem. Res. 59(30): 13588–13594.

Ikeda, T., S. Yoshizawa, N. Koizumi, M. Mitsuishi and Y. Matsumoto. 2016. Focused ultrasound and lithotripsy. pp. 113–129. *In*: Escoffre, J.-M. and A. Bouakaz [eds.]. Therapeutic Ultrasound. Springer International Publishing, Cham.

Islam, M.N., M. Zhang and B. Adhikari. 2014. The inactivation of enzymes by ultrasound—a review of potential mechanisms. Food Rev. Int. 30(1): 1–21.

Jafari, S.M., Y. He and B. Bhandari. 2007. Production of sub-micron emulsions by ultrasound and microfluidization techniques. J. Food Eng. 82(4): 478–488.

Jambrak, A.R., T.J. Mason, L. Paniwnyk and V. Lelas. 2007. Accelerated drying of button mushrooms, brussels sprouts and cauliflower by applying power ultrasound and its rehydration properties. J. Food Eng. 81(1): 88–97.

Jambrak, A.R., T.J. Mason, V. Lelas, Z. Herceg and I.L. Herceg. 2008. Effect of ultrasound treatment on solubility and foaming properties of whey protein suspensions. J. Food Eng. 86(2): 281–287.

Jambrak, A.R., V. Lelas, T.J. Mason, G. Krešić and M. Badanjak. 2009. Physical properties of ultrasound treated soy proteins. J. Food Eng. 93(4): 386–393.

Jambrak, A.R., Z. Herceg, D. Šubarić, J. Babić, M. Brnčić, S.R. Brnčić et al. 2010. Ultrasound effect on physical properties of corn starch. carbohydr. Polym. 79(1): 91–100.

Jayasooriya, S.D., B.R. Bhandari, P. Torley and B.R. D'Arcy. 2004. Effect of high-power ultrasound waves on properties of meat: a review. Int. J. Food Prop. 7(2): 301–319.

Kiani, H., Z. Zhang, A. Delgado and D.-W. Sun. 2011. Ultrasound assisted nucleation of some liquid and solid model foods during freezing. Food Res. Int. 44(9): 2915–2921.

Kiani, H., D. Sun and Z. Zhang. 2012. The effect of ultrasound irradiation on the convective heat transfer rate during immersion cooling of a stationary sphere. Ultrason. Sonochem. 19(6): 1238–1245.

Kimura, T. 2015. Application of ultrasound to organic synthesis. pp. 171–186. *In*: Grieser, F., P.-K. Choi, N. Enomoto, H. Harada, K. Okitsu and K. Yasui [eds.]. Sonochemistry and the Acoustic Bubble. Elsevier, Amsterdam.

Knorr, D., M. Zenker, V. Heinz and D.-U. Lee. 2004. Applications and potential of ultrasonics in food processing. Trends Food Sci. Technol. 15(5): 261–266.

Ladole, M.R., R.R. Nair, Y.D. Bhutada, V.D. Amritkar and A.B. Pandit. 2018. Synergistic effect of ultrasonication and co-immobilized enzymes on tomato peels for lycopene extraction. Ultrason. Sonochem. 48: 453–462.

Lee, H. and H. Feng. 2011. Effect of power ultrasound on food quality. pp. 559–582. *In*: Feng, H., G. Barbosa-Canovas and J. Weiss [eds.]. Ultrasound Technologies for Food and Bioprocessing. Springer New York, New York, USA.

Lee, H., H. Kim, K.R. Cadwallader, H. Feng and S.E. Martin. 2013. Sonication in combination with heat and low pressure as an alternative pasteurization treatment–effect on *escherichia coli* k12 inactivation and quality of apple cider. Ultrason. Sonochem. 20(4): 1131–1138.

Lett, Z., A. Hall, S. Skidmore and N.J. Alves. 2021. Environmental microplastic and nanoplastic: exposure routes and effects on coagulation and the cardiovascular system. Environ. Pollut. 291: 118190.

Lévêque, J.-M., G. Cravotto, F. Delattre and P. Cintas. 2018. Efficient organic synthesis: what ultrasound makes easier. pp. 17–39. *In*: Lévêque, J.-M., G. Cravotto, F. Delattre and P. Cintas [eds.]. Organic Sonochemistry Challenges and Perspectives for the 21st Century. Springer International Publishing, Cham.

Li, J.-T., X.-H. Zhang and Z.-P. Lin. 2007. An improved synthesis of 1,3,5-triaryl-2-pyrazolines in acetic acid aqueous solution under ultrasound irradiation. Beilstein J. Org. Chem. 3: 13–16.

Li, Y.-H., L. Wang, Z. Wang, S. Yuan, S. Wu and S.-F. Wang. 2016. Ultrasound-assisted synthesis of novel pyrrole dihydropyrimidinones in lactic acid. ChemistrySelect 1(21): 6855–6858.

Li, Y.-T., D. Li, L.-J. Lai and Y.-H. Li. 2020. Remediation of petroleum hydrocarbon contaminated soil by using activated persulfate with ultrasound and ultrasound/fe. Chemosphere. 238: 124657–124664.

Lopes, A.M., E.X. Ferreira Filho and L.R.S. Moreira. 2018. An update on enzymatic cocktails for lignocellulose breakdown. J. Appl. Microbiol. 125(3): 632–645.

Luche, J.L. and J.C. Damiano. 1980. Ultrasounds in organic syntheses. 1. effect on the formation of lithium organometallic reagents. J. Am. Chem. Soc. 102(27): 7926–7927.

Lukić, K., M. Brnčić, N. Ćurko, M. Tomašević, D. Valinger, G.I. Denoya et al. 2019. Effects of high power ultrasound treatments on the phenolic, chromatic and aroma composition of young and aged red wine. Ultrason. Sonochem. 59: 104725–104738.

Malani, R.S., S.B. Umriwad, K. Kumar, A. Goyal and V.S. Moholkar. 2019. Ultrasound–assisted enzymatic biodiesel production using blended feedstock of non–edible oils: kinetic analysis. Energy Convers. Manag. 188: 142–150.

Mane, S.N., S.M. Gadalkar and V.K. Rathod. 2018. Intensification of paracetamol (acetaminophen) synthesis from hydroquinone using ultrasound. Ultrason. Sonochem. 49: 106–110.

Mason, T. J., L. Paniwnyk and J.P. Lorimer. 1996. The uses of ultrasound in food technology. Ultrason. Sonochem. 3(3): S253–S260.

Mason, Timothy J. 1997. Ultrasound in synthetic organic chemistry. Chem. Soc. Rev. 26 6: 443–451.

Mason, T. and D. Peters. 2002. An introduction to the uses of power ultrasound in chemistry. pp. 1–48. *In*: Mason, T. and D. Peters [eds.]. Practical Sonochemistry Power Ultrasound Uses and Applications. Woodhead Publishing, Sawston.

McKenzie, T.G., F. Karimi, M. Ashokkumar and G.G. Qiao. 2019. Ultrasound and sonochemistry for radical polymerization: sound synthesis. Chem.–A Eur. J. 25(21): 5372–5388.

Meys, R., F. Frick, S. Westhues, A. Sternberg, J. Klankermayer and A. Bardow. 2020. Towards a circular economy for plastic packaging wastes – the environmental potential of chemical recycling. Resources Conservation and Recycling 162: 105010–105019.

Mirkhani, V., M. Moghadam, S. Tangestaninejad and H. Kargar. 2006. Rapid and efficient synthesis of 2-imidazolines and bis-imidazolines under ultrasonic irradiation. Tetrahedron Lett. 47(13): 2129–2132.

Mittersteiner, M., F.F.S. Farias, H.G. Bonacorso, M.A.P. Martins and N. Zanatta. 2021. Ultrasound-assisted synthesis of pyrimidines and their fused derivatives: a review. Ultrason. Sonochem. 79: 105683–105717.

Morata, A. and J.A. Suárez-Lepe. 2015. New biotechnologies for wine fermentation and ageing. pp. 287–302. *In*: Rai, R.V. [ed.]. Advances in Food Biotechnology. Wiley Online Books, Hoboken.

Mousavi, S.A.A.A., H. Feizi and R. Madoliat. 2007. Investigations on the effects of ultrasonic vibrations in the extrusion process. J. Mater. Process. Technol. 187–188: 657–661.

Mulet, A., J.A. Cárcel, N. Sanjuán and J. Bon. 2003. New food drying technologies - use of ultrasound. Food Sci. Technol. Int. 9(3): 215–221.

Musielak, G., D. Mierzwa and J. Kroehnke. 2016. Food drying enhancement by ultrasound–a review. Trends Food Sci. Technol. 56: 126–141.

Naidu, R., B. Biswas, I.R. Willett, J. Cribb, B. Kumar Singh, C. Paul Nathanail et al. 2021. Chemical pollution: a growing peril and potential catastrophic risk to humanity. Environmental International 156: 106616–106627.

Nayak, B., F. Dahmoune, K. Moussi, H. Remini, S. Dairi, O. Aoun et al. 2015. Comparison of microwave, ultrasound and accelerated-assisted solvent extraction for recovery of polyphenols from citrus sinensis peels. Food Chem. 187: 507–516.

Ojha, K.S., R. Aznar, C. O'Donnell and B.K. Tiwari. 2020. Ultrasound technology for the extraction of biologically active molecules from plant, animal and marine sources. TrAC Trends Anal. Chem. 122: 115663–115672.

O'Sullivan, J., J. Beevers, M. Park, R. Greenwood and I. Norton. 2015. Comparative assessment of the effect of ultrasound treatment on protein functionality pre- and post-emulsification. Colloids Surfaces A Physicochem. Eng. Asp. 484: 89–98.

Pacheco, D.J., L. Prent, J. Trilleras and J. Quiroga. 2013. Facile sonochemical synthesis of novel pyrazolyne derivates at ambient conditions. Ultrason. Sonochem. 20(4): 1033–1036.

Pallarés, N., H. Berrada, E. Ferrer, J. Zhou, M. Wang, F.J. Barba et al. 2021. Ultrasound processing: a sustainable alternative. pp. 155–164. *In*: Lorenzo, J.M., P.E.S. Munekata and F.J. Barba [eds.]. Sustainable Production Technology in Food. Academic Press. Cambridge.

Park, B. and Y. Son. 2017. Ultrasonic and mechanical soil washing processes for the removal of heavy metals from soils. Ultrason. Sonochem. 35: 640–645.

Pasha, M.K., L. Dai, D. Liu, W. Du and M. Guo. 2021. Biodiesel production with enzymatic technology: progress and perspectives. Biofuels, Bioprod. Biorefining 15(5): 1526–1548.

Patil, R., P. Bhoir, P. Deshpande, T. Wattamwar, M. Shirude and P. Chaskar. 2013. Relevance of sonochemistry or ultrasound (US) as a proficient means for the synthesis of fused heterocycles. Ultrason. Sonochem. 20(6): 1327–1336.

Peglow, T.J., G.P. da Costa, L.F.B. Duarte, M.S. Silva, T. Barcellos, G. Perin et al. 2019. Ultrasound-promoted one-pot synthesis of mono- or bis-substituted organylselanyl pyrroles. J. Org. Chem. 84(9): 5471–5482.

Pellis, A., C. Gamerith, G. Ghazaryan, A. Ortner, E. Herrero Acero and G.M. Guebitz. 2016. Ultrasound-enhanced enzymatic hydrolysis of poly(ethylene terephthalate). bioresour. Technol. 218: 1298–1302.

Pineiro, M. and M.J.F. Calvete. 2018. Sustainable synthesis of pharmaceuticals using alternative techniques: microwave, sonochemistry and mechanochemistry. pp. 8–39. *In*: Pereira, M.M. and M.J.F. Calvete [eds.]. Sustainable Synthesis of Pharmaceuticals: Using Transition Metal Complexes as Catalysts. The Royal Society of Chemistry, London, UK.

Piyasena, P., E. Mohareb and R.C. McKellar. 2003. Inactivation of microbes using ultrasound: a review. Int. J. Food Microbiol. 87(3): 207–216.

Pollet, B.G. and M. Ashokkumar. 2019. Fundamental and applied aspects of ultrasonics and sonochemistry BT - introduction to ultrasound, sonochemistry and sonoelectrochemistry. pp. 1–19. *In*: B.G. Pollet and M. Ashokkumar [eds.]. Springer International Publishing, Cham.

Poppe, J.K., C.R. Matte, R. Fernandez-Lafuente, R.C. Rodrigues and M.A.Z. Ayub. 2018. Transesterification of waste frying oil and soybean oil by combi-lipases under ultrasound-assisted reactions. Appl. Biochem. Biotechnol. 186(3): 576–589.

Povey, M.J.W. and T.J. Mason. 1998. Ultrasound in food processing. springer, berlin.

Prajapat, A.L., P.B. Subhedar and P.R. Gogate. 2016. Ultrasound assisted enzymatic depolymerization of aqueous guar gum solution. Ultrason. Sonochem. 29: 84–92.

Prata, J.C., J.P. da Costa, I. Lopes, A.L. Andrady, A.C. Duarte and T. Rocha-Santos. 2021. A one health perspective of the impacts of microplastics on animal, human and environmental health. Sci. Total Environ. 777: 146094–146106.

Price, G.J. 2003. Recent developments in sonochemical polymerisation. Ultrason. Sonochem. 10 4: 277–283.

Puri, S., B. Kaur, A. Parmar and H. Kumar. 2013. Applications of ultrasound in organic synthesis - a green approach. Curr. Org. Chem. 17(16): 1790–1828.

Rokhina, E.V., P. Lens and J. Virkutyte. 2009. Low-frequency ultrasound in biotechnology: state of the art. Trends Biotechnol. 27(5): 298–306.

Saeeduddin, M., M. Abid, S. Jabbar, T. Wu, M.M. Hashim, F.N. Awad et al. 2015. Quality assessment of pear juice under ultrasound and commercial pasteurization processing conditions. LWT - Food Sci. Technol. 64(1): 452–458.

Said, K. and R. Ben Salem. 2016. Ultrasonic activation of suzuki and hiyama cross-coupling reactions catalyzed by palladium. Adv. Chem. Eng. Sci. 06: 111–123.

Sant'Anna, G. da S., P. Machado, P.D. Sauzem, F.A. Rosa, M.A. Rubin, J. Ferreira et al. 2009. Ultrasound promoted synthesis of 2-imidazolines in water: a greener approach toward monoamine oxidase inhibitors. Bioorg. Med. Chem. Lett. 19(2): 546–549.

Schmid, G. and O. Rommel. 1939. Rupture of macromolecules with ultrasound. Z. Phys. Chem. 185A(1): 97–139.

Schwabl, P., S. Köppel, P. Königshofer, T. Bucsics, M. Trauner, T. Reiberger et al. 2019. Detection of various microplastics in human stool. Ann. Intern. Med. 171(7): 453–457.

Schwarz, A.E., T.N. Ligthart, D. Godoi Bizarro, P. De Wild, B. Vreugdenhil and T. van Harmelen. 2021. Plastic recycling in a circular economy; determining environmental performance through an lca matrix model approach. Waste Manag. 121: 331–342.

Scotto, A. 1998. Device for demoulding industrial food products. FR. Patent # 2604063A1.

Serna-Galvis, E.A., J. Porras and R.A. Torres-Palma. 2022. A critical review on the sonochemical degradation of organic pollutants in urine, seawater, and mineral water. Ultrason. Sonochem. 82: 105861–105874.

Shanmugam, A. and M. Ashokkumar. 2014. Functional properties of ultrasonically generated flaxseed oil-dairy emulsions. Ultrason. Sonochem. 21(5): 1649–1657.

Shershenkov, B. and E. Suchkova. 2015. Upgrading the technology of functional dairy products by means of fermentation process ultrasonic intensification. Agron. Res. 13(4): 1074–1085.

Shestakov, S., O. Krasulya, J. Artemova and N. Tikhomirova. 2011. Ultrasonic water treatment. Dairy Indus. 5: 39–42.

Shirsath, S.R., S.H. Sonawane and P.R. Gogate. 2012. Intensification of extraction of natural products using ultrasonic irradiations—a review of current status. Chem. Eng. Process. Process Intensif. 53: 10–23.

Subhedar, P.B. and P.R. Gogate. 2016. Ultrasound assisted intensification of biodiesel production using enzymatic interesterification. Ultrason. Sonochem. 29: 67–75.

Sui, Q., X. Cao, S. Lu, W. Zhao, Z. Qiu and G. Yu. 2015. Occurrence, sources and fate of pharmaceuticals and personal care products in the groundwater: a review. Emerg. Contam. 1(1): 14–24.

Sun, Y., L. Zhong, L. Cao, W. Lin and X. Ye. 2015. Sonication inhibited browning but decreased polyphenols contents and antioxidant activity of fresh apple (malus pumila mill, cv. red fuji) juice. j. food sci. Technol. 52(12): 8336–8342.

Szent-GyÖrgyi, A. 1933. Chemical and biological effects of ultra-sonic radiation. Nature 131(3304): 278–278.

Tao, Y. and D.-W. Sun. 2015. Enhancement of food processes by ultrasound: a review. Crit. Rev. Food Sci. Nutr. 55(4): 570–594.

Thangavadivel, K., M. Megharaj and A. Mudhoo. 2011. Degradation of organic pollutants using ultrasound. pp. 447–474. In: Chen, D., S.K. Sharma and A. Mudhoo [eds.]. Handbook on Applications of Ultrasound: Sonochemistry for Sustainability. CRC Press, Boca Raton.

Toma, M., M. Vinatoru, L. Paniwnyk and T.J. Mason. 2001. Investigation of the effects of ultrasound on vegetal tissues during solvent extraction. Ultrason. Sonochem. 8(2): 137–142.

Tomke, P.D., X. Zhao, P.P. Chiplunkar, B. Xu, H. Wang, C. Silva et al. 2017. Lipase-ultrasound assisted synthesis of polyesters. Ultrason. Sonochem. 38: 496–502.

Tournier, V., C.M. Topham, A. Gilles, B. David, C. Folgoas, E. Moya-Leclair et al. 2020. An engineered PET depolymerase to break down and recycle plastic bottles. Nature 580(7802): 216–219.

Toussaint, B., B. Raffael, A. Angers-Loustau, D. Gilliland, V. Kestens, M. Petrillo et al. 2019. Review of micro- and nanoplastic contamination in the food chain. food addit. Contam. Part A 36(5): 639–673.

Tran, T.T.T., K.T. Nguyen and V.V.M. Le. 2018. Effects of ultrasonication variables on the activity and properties of alpha amylase preparation. Biotechnol. Prog. 34(3): 702–710.

Trilleras, J., E. Polo, J. Quiroga, J. Cobo and M. Nogueras. 2013. Ultrasonics promoted synthesis of 5-(Pyrazol-4-Yl)-4,5-dihydropyrazoles derivatives. Appl. Sci. 3: 457–468.

Ubando, A.T., C.B. Felix and W.-H. Chen. 2020. Biorefineries in circular bioeconomy: a comprehensive review. Bioresour. Technol. 299: 122585–122602.

Veljković, V.B., J.M. Avramović and O.S. Stamenković. 2012. Biodiesel production by ultrasound-assisted transesterification: state of the art and the perspectives. Renew. Sustain. Energy Rev. 16(2): 1193–1209.

Vignesh, G. and D. Barik. 2019. Toxic waste from biodiesel production industries and its utilization. pp. 69–82. *In*: Barik, D. [ed.]. Energy from Toxic Organic Waste for Heat and Power Generation. Woodhead Publishing, Sawston.

Vinatoru, M. 2001. An overview of the ultrasonically assisted extraction of bioactive principles from herbs. Ultrason. Sonochem. 8(3): 303–313.

Vodeničarová, M., G. Dřímalová, Z. Hromádková, A. Malovíková and A. Ebringerová. 2006. Xyloglucan degradation using different radiation sources: a comparative study. Ultrason. Sonochem. 13(2): 157–164.

Wang, S.-F., C.-L. Guo, K. Cui, Y.-T. Zhu, J.-X. Ding, X.-Y. Zou et al. 2015. Lactic acid as an invaluable green solvent for ultrasound-assisted scalable synthesis of pyrrole derivatives. Ultrason. Sonochem. 26: 81–86.

Wu, H., G.J. Hulbert and J.R. Mount. 2000. Effects of ultrasound on milk homogenization and fermentation with yogurt starter. Innov. Food Sci. Emerg. Technol. 1(3): 211–218.

Xin, Z., G. Lin, H. Lei, T.F. Lue and Y. Guo. 2016. Clinical applications of low-intensity pulsed ultrasound and its potential role in urology. Transl. Androl. Urol. 5(2): 255–266.

Yachmenev, V., B. Condon, K. Klasson and A. Lambert. 2009. Acceleration of the enzymatic hydrolysis of corn stover and sugar cane bagasse celluloses by low intensity uniform ultrasound. J. Biobased Mater. Bioenergy. 3: 1–7.

Yu, G., P. He, L. Shao and Y. Zhu. 2009. Enzyme extraction by ultrasound from sludge flocs. J. Environ. Sci. 21(2): 204–210.

Zhang, Z.-H., J.-J. Li and T.-S. Li. 2008. Ultrasound-assisted synthesis of pyrroles catalyzed by zirconium chloride under solvent-free conditions. Ultrason. Sonochem. 15(5): 673–676.

Zheng, L. and D.-W. Sun. 2006. Innovative applications of power ultrasound during food freezing processes—a review. Trends Food Sci. Technol. 17(1): 16–23.

Zhou, C.-H., X. Xia, C.-X. Lin, D.-S. Tong and J. Beltramini. 2011. Catalytic conversion of lignocellulosic biomass to fine chemicals and fuels. Chem. Soc. Rev. 40(11): 5588–5617.

Zou, X., T. Zhou, J. Mao and X. Wu. 2014. Synergistic degradation of antibiotic sulfadiazine in a heterogeneous ultrasound-enhanced fe0/persulfate fenton-like system. Chem. Eng. J. 257: 36–44.

Zuo, J.Y., K. Knoerzer, R. Mawson, S. Kentish and M. Ashokkumar. 2009. The pasting properties of sonicated waxy rice starch suspensions. Ultrason. Sonochem. 16(4): 462–468.

CHAPTER 3

Ultrasound-assisted Chemical Synthesis

G.G. Flores-Rojas,[1,2,3] *F. López-Saucedo,*[1,4] *L. Buendía-González,*[4]
R. Vera-Graziano,[3] *E. Mendizabal*[2,]* *and E. Bucio*[1,]*

1. Introduction

Currently, there are various reports where ultrasound equipment has been applied in chemical synthesis covering different areas of chemistry such as organic, organometallic, inorganic, polymers, material engineering, and catalysis (Cravotto and Cintas 2006, Li et al. 2021, Pokhrel et al. 2016, Zhang et al. 2009). In this regard, the catalog of sonochemistry in synthesis has grown in recent years more than other traditional energy sources. This increment in the interest for sonochemistry, has also benefited from and for the technological advances in ultrasound equipment. In general terms the idea of synthesis using this approach also carries with additional benefits such as efficiency of processes, and in case of materials the possibility to get a specific size or morphology, which is of interest for technological development. In the annals of sonochemistry it is found the reaction reported by Fry, who carried out the electrochemical reduction of α,α'-dibromoketones to a mixture of acetoxyketones, employing a dispersion of Hg(l) in acetic acid (Fry and Bujanauskas 1978, Montaña and Grima 2003). Derivative from this research, a considerable number of publications on similar chemical synthesis of molecules using ultrasound as energy source become more relevant (Mason 1986). Inside laboratory, a power range of ultrasound applied in the chemical synthesis is

[1] Departamento de Química de Radiaciones y Radioquímica, Instituto de Ciencias Nucleares, Universidad Nacional Autónoma de México, Circuito Exterior, Ciudad Universitaria, CDMX, C.P. 04510, México.
[2] Departamentos de Química e Ingeniería Química, Universidad de Guadalajara. Centro Universitario de Ciencias Exactas e Ingenierías, Blvd. M. García Barragán # 1451, Guadalajara, Jalisco, C.P. 44430, México.
[3] Instituto de Investigaciones en Materiales, Universidad Nacional Autónoma de México, Universidad Nacional Autónoma de México, Circuito Exterior, Ciudad Universitaria, CDMX, C.P. 04510, México.
[4] Facultad de Ciencias. Universidad Autónoma del Estado de México, Campus El Cerrillo Piedras Blancas Carretera Toluca-Ixtlahuaca Km 15.5, Toluca, Estado de México, C.P. 50200, México.
* Corresponding authors: lalomendizabal@hotmail.com; ebucio@nucleares.unam.mx

usually above 20 kHz, energy which is enough to activate chemical species, that with the help of other reaction conditions such as light, heat, and pressure, can allow reactions that would be impossible without the use of ultrasound (Saleh et al. 2017). The principle of ultrasound is to generate the cavitation process, formed in the rarefaction cycles of ultrasonic waves when the structure of the liquid breaks into microbubbles that subsequently collapse in compression cycles, producing this way high pressures in the order of 10,000 atm and high temperatures in the order of 5,000 K (Fragoso-Medina et al. 2021, Penteado et al. 2018).

This unconventional chemistry can offer not only a new synthesis methodology but also important advances in the design of more environmentally friendly synthetic methods (Cintas and Luche 1999, Draye and Kardos 2017, Rashmi Pradhan et al. 2019), especially in the area of "Green Chemistry" (Anastas and Kirchhoff 2002). The main advantages of this alternative energy source includes the possibility to offer the reduction of waste by optimization of reagents, shorter processes, as well as, in general, a shorter reaction time compared to conventional methods of synthesis (Banerjee 2017), promoting, in most cases, reactions that would otherwise require drastic conditions of temperature and pressure (Penteado et al. 2018).

2. Applications of Ultrasound in Chemical Synthesis

2.1 Heterogeneous Metallic Reactions

This reaction can be divided into two types: (i) where the metal is a reactant and is consumed in the reaction process, and (ii) where the metal is a catalyst. In both reaction types, sonication removes impurities from the metal surface by exposing a clean and reactive metal zone as intermediate reactants, or surface products, are removed, increasing the effective surface area of the metal for reaction. The increase in the active surface area of the metal is produced mainly by erosion through the cavitation implosion of microbubbles and the microflow of the solvent on the metal surface. This increase in the effective area is not possible using normal mechanical agitation (Cintas et al. 2010, Liu et al. 2021).

Reactions with metal powders or particles are also classified as heterogeneous because they involve solids dispersed in liquids. The reactivity of the solid phase depends mainly on the reactive surface area, increasing this by reducing the size of the particles, which is carried out by the continuous fragmentation of the particles and simultaneous activation by sonication, avoiding the use of phase transfer agents which help to promote the heterogeneous reaction. One of the most outstanding characteristics is that the particle size is reduced until particles are no longer affected by ultrasound. For example, ultrasound causes on Ni and Cu powders a radical change in the morphology of metal powders, and more important, the activity of Ni powder as a catalyst in hydrogenation reactions (with 1-nonene, 1-decane, and 1,5-decene) is enhanced under conditions of 273 K and one atm; also, the reactivity of Cu after pretreatment of ultrasound increased its reactivity as a stoichiometric reagent in the Ullman coupling of aryl iodides (Suslick et al. 1989).

2.2 Emulsion Reactions

The use of ultrasound in emulsion systems has proven to be extremely effective due to its ability to generate particularly fine emulsions. In such emulsions, there is a drastic increase in the interfacial contact area between the liquids and therefore an increase in the effective area in which any type of chemical reaction can take place. Similarly, to the reactions with powders, the use of transfer agents is not necessary. However, when using phase transfer catalysts combined with ultrasound, there is a significant improvement in the yield of the products compared to those obtained in any of the techniques separately (Cucheval and Chow 2008). Effects of ultrasound in the stabilization of emulsions are the most important aspect to prefer the use of this tool, just to mention some recent examples: water in oil emulsion (Zhang et al. 2022), soy protein isolate-pectin emulsion (Wang et al. 2022), β-carotene bulk emulsion gels (Geng et al. 2022), nanocellulose hydrogels stabilized in oil in water emulsions (Ni et al. 2022). So, ultrasound in emulsion reactions is a more topic.

2.3 Homogeneous Reactions

In the case of homogeneous reactions, the mechanical action of the collapse of the microbubbles formed in the sonication process does not significantly affect the reactivity of the chemical species. In this type of reaction, the phenomenon that directly affects the reactivity of the chemical species is the process of cavitational collapse, because the microbubbles contain solvent vapor and volatile reagents, which are subjected to high temperatures and pressures when collapsing, forming reactive species of the radical or carbene type, in addition to the fact that the collapse of the microbubble can alter the structure of the solvent, influencing the solvation of the reactive species and altering their reactivity (Tuulmets et al. 2010).

Therefore, a variety of effects can be expected from the cavitation process with the capacity of promoting and enhancing chemical reactions:

 (i) Acceleration of the chemical reaction or allow the use of non extreme reaction conditions.
(ii) Reduction in the number of steps in chemical synthesis in addition to allowing the use of lower purity reagents.
(iii) Removal of additives or initiators of the reaction.
(iv) Reduction in the induction time required to start the reaction. In other words a shorter reaction time.
 (v) Promote the chemical reaction via an alternative mechanism.

2.4 Ultrasound in Chemical Synthesis

2.4.1 Organometallic Compounds

The synthesis of organometallic compounds has been mainly focused on the chemistry of soft metals, being Li and Mg organometallics, in modified Barbier reaction there are several examples demonstrating that ultrasound improve coupling of alkyl halogenated compounds (Luche and Damiano 1980). Another important contribution of ultrasound for science is the acceleration of Grignard reagents with 1,3-dioxolanes of α,β-unsaturated aldehydes (Lu et al. 1998). The Grignard reagents obtained rapid reaction without the use of activators such as iodine with acceptable yields. Currently sonochemical protocols for Grignard reactions has been extensively proved in the activation of Mg surface to facilitate the RMgX formation under specific parameters of ultrasound irradiation (Smolnikov et al. 2020).

The use of sonication facilitates the synthesis of radical anionic salts of arene and alkali metals (Cao et al. 2020, Kinzyabaeva and Sabirov 2020), which have various applications in chemical synthesis. For example, in the ultrasound promoted synthesis of sodium isobenzoquinoline, which requires a long reaction time of 45 mins to complete, a considerably shorter time compared to 48 hrs under normal reaction conditions. Similarly, improvements were reported in obtaining sodium naphthalene in THF under non anhydrous conditions using an intensity of 36 kHz (Azuma et al. 1982). On the other hand, the use of ultrasound in chemical reactions has also allowed the synthesis of another organometallic compounds such as alkylaluminium halides where the use of dioxane as a solvent improved the performance compared to ethyl ether with a frequency range of 32 to 36 kHz, this result is attributed to the higher boiling point of dioxane which allows for a better cavitation effect. Other alkyl aluminum halide compounds were obtained by reacting bromoethane with Al in THF to give ethylaluminum sesquibromide ((Figure 1 (1)), R = Et, X = Br) with a short reaction time and at room temperature (Liou et al. 1985).

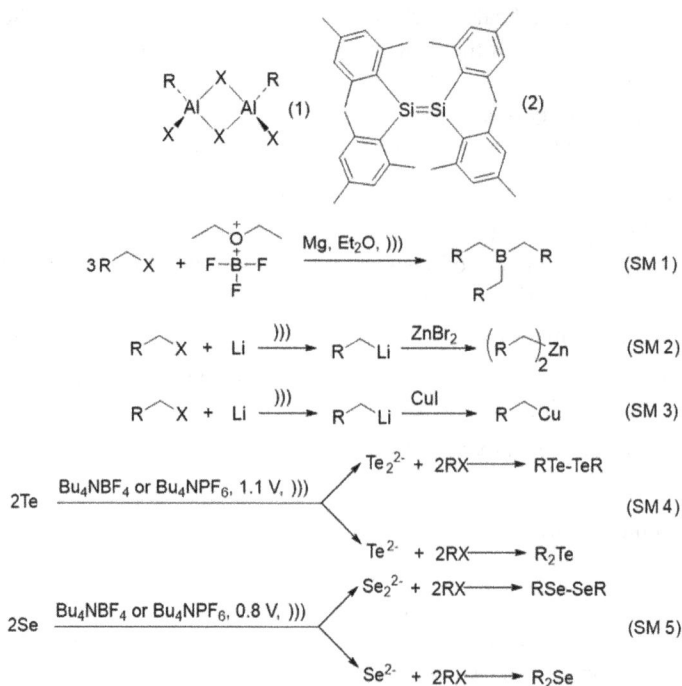

Figure 1. Compounds obtained by organometallic reactions.

Sonochemistry has also allowed the synthesis of organoborane compounds more efficiently from Grignard reagents obtained *in situ* (Figure 1 (SM 1)) (Tuulmets et al. 1995). Some of the organoborane compounds obtained were symmetrical trialkylboranes, Pr^n_3B with a purity of 99 percent, this purity being higher compared to those obtained by hydroboration of only 93 percent, similarly hindered compounds such as $(1\text{-naphthyl})_3B$ were also obtained with a yield of 93 percent and a reaction time of 15 mins.

Low intensity ultrasound (50 kHz) also facilitates the synthesis of compounds such as distananes and disilanes using lithium as a reducing agent, obtaining, for example, tetramesitylsilene compounds (Figure 1 (2)) with an approximate yield of 90 percent. However, it is very important to be meticulous in the reaction conditions because the slightest changes in the sonication parameters can give rise to different reaction products such as the formation of cyclic trisilanes (Masamune et al. 1983). Despite these drawbacks, the experiments carried out have generally shown a reduction in reaction time (Eaborn et al. 1984).

Currently, there are numerous publications on the synthesis of organozinc and organic copper reagents (Cintas et al. 2010), which were obtained by transmetallation reactions of metal halides with organolithium reagents assisted by low-intensity ultrasound (Figure 1 (SM 2, 3)). The experiments indicated that the use of solvents with high boiling points such as toluene improved the results up to 75 percent with short reaction times of between 20 to 30 min in obtaining organozinc compounds (Luche et al. 1983). Similarly, ultrasound promotes the formation of organic copper compounds through a subsequent transmetalation reaction of alkyl- or aryl-lithium and copper bromide (Luche et al. 1982). On the other hand, sonochemistry has provided novel synthesis routes for obtaining organoselenium and organotellurium compounds, which consist of ultrasound assisted electro reduction of Se or Te to their corresponding anions and later used in reactions with alkyl halides (Figure 1 (SM 4, 5)) (Gautheron et al. 1985).

2.4.1.1 Ligand Displacement Reactions

Substitution reactions in metal complexes proceed mainly by dissociation processes, in systems that are coordinately saturated and kinetically inert. The high punctual temperatures and pressures occurred during cavitation effectively promote these reactions, it means allowing the ligand exchange or displacement (Zhang et al. 2016). The effect of sonolysis was studied in $Fe(CO)_5$ within non-polar solvents. The findings revealed that the production of the trimetallic $Fe_3(CO)_{12}$ and $Fe(0)$ precipitates occurs in a ratio contingent upon the reaction conditions, specifically the vapor pressure of the solvent. The research indicated that the formation of $Fe_3(CO)_{12}$ is favored in solvents with high vapor pressure, such as heptane, resulting in high yields. Conversely, low yields were obtained in solvents with low vapor pressure, such as decalin (Suslick et al. 1996). The results show a dependence on the collapse energy of the cavitation bubbles, this being inversely proportional to the vapor pressure of the solvent because clustering is a process with lower activation energy and is favored in solvents with a lower boiling point that produces weaker cavitation,

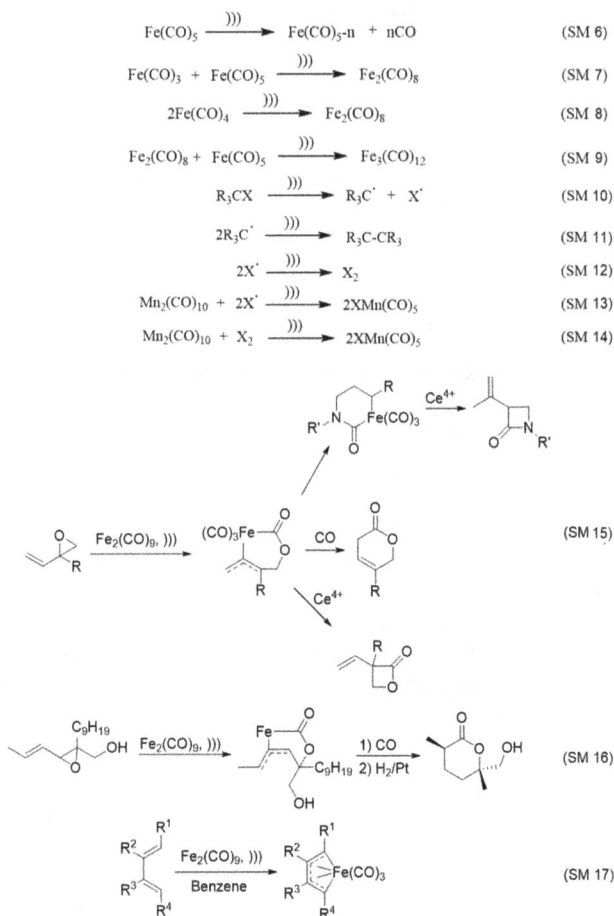

Figure 2. Substitution reactions of Fe and Mn carbonyl ligands.

while the formation of the compound $Fe_3(CO)_{12}$ arises mostly on coordinative unsaturated species such as $Fe(CO)_3$ (Figure 2 (SM 6–9)).

However, the presence of ligands such as phosphines and phosphites leads to the substitution of the CO ligand giving compounds of the type $LFe(CO)_4$, $L_2Fe(CO)_3$, in addition to the presence of $L_3Fe(CO)_2$ and Fe(0) precipitate using low vapor pressure solvents as reaction medium. The ratio of $LFe(CO)_4$ and $L_2Fe(CO)_3$ obtained remain unchanged even with a prolonged time of sonolysis, suggesting that the $LFe(CO)_4$ compound is very stable and does not undergo further substitutions. In other metal carbonyl compounds sonolysis substitution there are some effects on the kinetics, for example with decacarbonyl $Mn_2(CO)_{10}$ and $Re_2(CO)_{10}$ (Suslick and Schubert 1983), where the substitution of the dimers of $Mn_2(CO)_{10}$ and $Re_2(CO)_{10}$ proceeds without affecting the metal-metal bond, being more similar to the thermal method than to the photochemical one. On the other hand, the use of halogenated solvents in the sonolysis of $Mn_2(CO)_{10}$ and $Re_2(CO)_{10}$ compounds promotes the formation of halogen pentacarbonyl metal complexes through secondary reactions promoted by solvent-derived radicals (Figure 2 (SM 10–14)).

The facility to obtain some metal reagents by sonication can be combined with a variety of carbonyl, epoxide, or aryl sulfone reagents leading to the formation of four-, five-, and six-membered lactones and lactams via a π-allyl tricarbonyl iron intermediate lactones (ferry lactones) that can be isolated in the form of crystals. For example, sonolysis at room temperature of $Fe_2(CO)_9$ compounds in benzene and the presence of alkenyl oxides gives η^3-allylironcarbonyl lactone complexes as the main product in good yields, which can be easily oxidized to lactams and lactones (Figure 2 (SM 15, 16)) (Horton et al. 1984, Low 1995). Similarly, $Fe_2(CO)_9$ compounds in the presence of dienes result in the formation of $(n^4$-diene)$Fe(CO)_3$ complexes (Figure 3 (SM 18)) (Low 1995), because sonication of $Fe_2(CO)_9$ compounds liberates $Fe(CO)_5$ or $Fe(CO)_4$, which can react with alkenes and other double bonds, where these coordinatively unsaturated metal species act as radical equivalents or carbenes.

2.4.1.2 Powders and Metal Catalysts

The reduction of metallic salts employing ultrasound has been an important area of study because it produces a uniform dispersion, and the cavitation generates a clean surface with a greater number of dislocations considered active sites of catalysis. Therefore, obtaining metal powders from metal halides with lithium and THF as the reaction medium using low intensity ultrasound (50 kHz) produce metal catalysts with reactivities comparable to those synthesized by the Rieke method with a reaction time of about 40 mins compared to eight hrs (Rieke et al. 2013).

In broad terms, the ultrasonic method enables the production of transition metal salt powders, including Cu, Ni, and Pd, from the respective halide salts. The process involves a successful reaction in the Ullman reaction with Cu (Nelson and Crouch 2004), or Zn(0) (Ishihara and Hatakeyama 2012, Ross and Bartsch 2003), and In(0) (Lee et al. 2001) in Reformatsky rearrangements, using an electron transfer agent in the case of insoluble halides. The metallic powders obtained by this method have shown improved reactivity in these organic coupling reactions, obtaining high yields.

Mg(0) powders produced by ultrasound in the presence of anthracene and THF as reaction medium (Petrier et al. 1985), form a transfer complex (Figure 3 (SM 18)) which is an excellent reducing agent for metal salts as well as a useful route in the presence of ligands of Lewis bases for various types of organic transition metal complexes (n^5-cyclopentadienyl complexes Cp, M), alkene complexes and phosphine complexes. On the other hand, the lithium complex is also an alternative route to Grignard reagents, eliminating the coupling of allyl and MnX_2 with the initial alkyl halide, which is considered a common side reaction with conventional methods (Figure 3 (SM 19)) (Oppolzer and Schneider 1984). Reduction of transition metal halides with sodium in the presence of CO under sonication in THF is capable of providing reasonable yields of metal carbonyl anions such as $[W_2(CO)_{10}]^{2-}$ 47 percent, $[Mo_2(CO)_{10}]^{2-}$ 54 percent, $[Nb(CO)_6]^-$ 51 percent, $[V(CO)_6]^-$ 35 percent (Suslick and Johnson 1984), these results being notable because low yields are obtained through conventional heating and high pressures in autoclaves. The high yields by sonication are attributed to two mechanisms (i) to the interactions of CO

Mg(THF)$_3$

Mg $\longrightarrow$ Mg* (SM 18)

$$R^1X \ + \ R^2R^3CO \xrightarrow[\text{2) H}_2\text{O}]{\text{1) Li, THF,)))}} R^1R^2R^3COH \quad \text{(SM 19)}$$

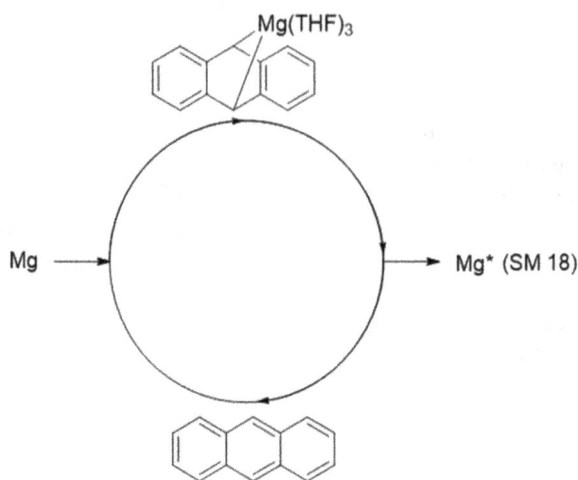

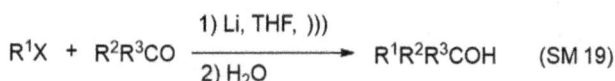

Figure 3. Alternative organometallic catalysts as Grignard reagent.

with partially reduced metal species on the sodium surface and (ii) to the excited CO species produced in the cavitation process.

The preparation of alkali metal suspensions becomes a simple method by low intensity sonication (35 kHz) where high boiling point solvents demonstrate their effectiveness due to higher cavitation power compared to lower boiling point solvents; for example, because xylene has a higher boiling point than toluene resulting in more powerful cavitation capable of dispersing sodium, it is possible to produce colloidal Na in xylene, whereas, in toluene, it is quite difficult. These systems are very useful in Dieckmann type reactions, Wittig type reactions, and desulfonation reactions (Yang and Xu 2021). Also, other emulsions obtained with Hg in acetic acid are effective in promoting reductive substitutions of α,α'-dibromoketones (Suslick and Johnson 1984), as well as Cu emulsions capable of improving the Ullmann coupling reactions (Lindley et al. 1986).

The preparation of heterogeneous Pd catalysts in the form of black powders has been widely carried out by sonication showed up to increase their catalytic activity in hydrogenation of 1-hexene, trans-1,3-hexadiene, trans-1,4-hexadiene, and 1,5-hexadiene (Tamai et al. 1999). This technique has not only allowed the synthesis of Pd catalysts but also allowed the synthesis of Ni catalysts by activating their corresponding powders, which are also widely applied in catalytic hydrogenation reactions of alkenes (Suslick et al. 1987).

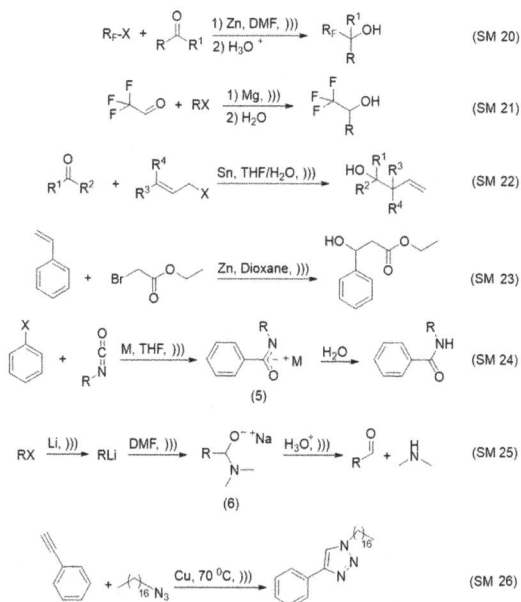

Figure 4. Metals as catalysts in organic reactions.

2.4.1.3 Ultrasonic Assisted Organic Synthesis

Some of the advantages of using ultrasound are that it allows the use of non anhydrous solvents such as commercial THF in reactions that include carbonyl functional groups, which are often carried out free of secondary reactions such as reductions and enolizations, which are commonly present in conventional methodologies such as in the Barbier reaction (Cenci et al. 2014).

In the case of reactions of carbonyl compounds with perfluoro halide units in their chemical structure, the use of Zn organometallics is preferred over those of Li and Mg, which are employed in Barbier reactions, because Li and Mg derivatives preferentially lead to perfluoroalkylations (Figure 4 (SM 20)) (Kitazume and Ishikawa 1985). However, to produce perfluoro Zn reagents is necessary the use of solvents with a high boiling point, such as DMF, to generate the necessary energy through cavitation to allow Zn to be activated where aldehydes react more quickly than ketones even in molecules that contain both functional groups. An alternative route for the synthesis of perfluoroalkyl alcohols is using perfluorinated aldehydes and Grignard reagents obtained *in situ* by ultrasound (Figure 4 (SM 21)) (Ishikawa et al. 1984), as well as allyl halides, which react rapidly in sonochemical conditions with ketones and aldehydes in the presence of Sn and aqueous THF, giving high yields of homoallylic alcohols (Figure 4 (SM 22)) (Petrier et al. 1985).

Ultrasound also significantly facilitates the Reformatsky reaction (Figure 4 (SM 23)) (Han and Boudjouk 1982) as is the example of the synthesis of ethyl

2-hydroxy phenylacetate which is obtained with a sonication time of five mins with a yield of 98 percent. This yield otherwise is obtained by other reaction methods in a minimum time of one hr. It is also possible to obtain good yields with perfluoroalkyl aldehydes, all promoted only using ultrasound (Ishikawa et al. 1984).

Other chemical groups that can be reacted by ultrasound under Barbier reaction conditions are aryl halides and alkyl isocyanates, giving secondary arylamides as products (Figure 4 (SM 24)) (Ishikawa et al. 1984), where is possible to obtain good yields of intermediate (5) using Mg compared to Li and Na. However, an advantage when using Na in the reaction system is the easy *ortho*-lithiation because it does not present ion exchange with Li as is the case with Mg, these intermediates being very useful in obtaining *ortho*-substituted arylamides by allowing the use of a wide range of electrophilic compounds (Figure 4 (SM 24)). Another application of ultrasound is in the synthesis of aldehydes using organolithium compounds in Bouveault type reactions (Figure 4 (SM 25)), where they have shown good results than those obtained using conventional methods since non ultrasound assisted reactions present numerous side reactions. The aldehyde yield increases with an increase in ultrasonic irradiation (Petrier et al. 1985).

In general, one of the main applications of ultrasonic activated zero valent metals is related to Barbier type reactions, which generally provide more sustainable results using aqueous media (Blomberg and Hartog 1977). Single electron transfer mechanisms are possible in electropositive metals as well as those with low ionization potential energy. However, one of the least dangerous and cheap metals useful in azide-alkyne cycloaddition reactions are Cu(I) salts, where sonochemistry has allowed the generation of ions *in situ* allowing the removal of byproducts of Cu(I) salt and avoiding the use of stabilizing agents. This type of reaction can be used with different reaction systems, giving high yields in shorter reaction times using a high-boiling polar solvent, allowing higher cavitation energy (Figure 4 (SM 26)) (Cintas et al. 2010). It has been reported that the cavitation process can produce a variety of complex organic compounds such as amino acids from carbon and nitrogen sources, as reported by the Grieser group, where a mixture of nitrogen, methane, water, and acetic acid promoted the formation of several amino acids such as glycine, ethyl glycine, and alanine, compounds that are formed through reactions via free radicals (Figure 5 (SM 27)) (Dharmarathne and Grieser 2016).

The use of surfactants in ultrasound reaction systems can increase the solubility of the reagents, having a positive effect on the inertial cavitation process, since it increases the growth rate of the bubbles. The use of cetyltrimethylammonium hydroxide as a cationic surfactant has been reported that increases the solubility of isatin and aryl ketones or heteroaryl ketones, whose condensation reaction forms quinolines with a yield greater than 75 percent at room temperature under sonication, increasing the reaction rate and avoiding the use of strong bases such as NaOH or KOH (More and Shankarling 2017). Other compounds used to solubilize hydrophobic reagents are hydrotropic compounds but unlike surfactants, these do not form micelles, because these compounds do not have a critical micellar concentration and only increase the solubility of hydrophobic compounds. Aqueous hydrotropic solutions combine low vapor pressure with increased viscosity relative to water, leading to a more intense cavitational collapse, thereby accelerating the reaction rate.

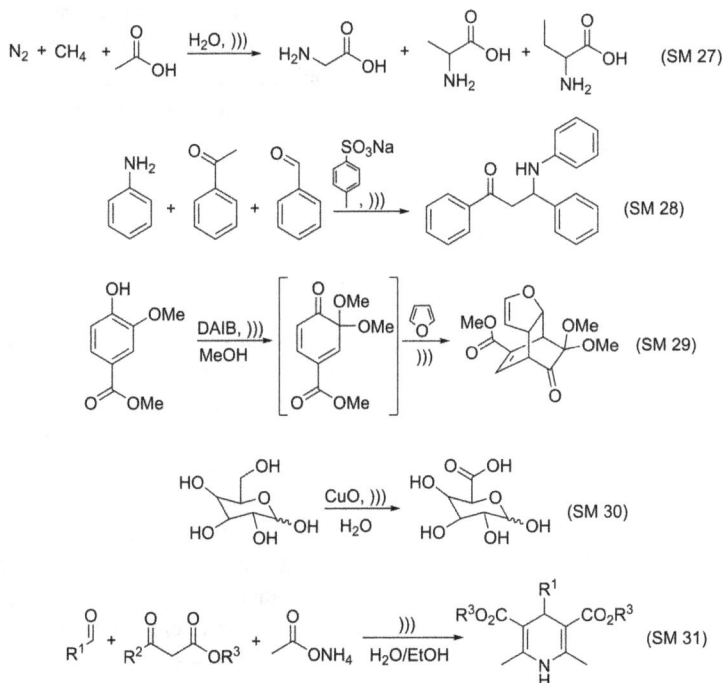

Figure 5. Organic reactions assisted by ultrasound.

For instance, a hydrotropic solution of sodium-toluenesulfonate enables maximum solubilization of acetophenones, aniline, and aryl aldehydes, which are facilitated when sonication is applied (refer to Figure 5 (SM 28)) (Kamble et al. 2012).

The addition of furan (as 2π component) to the masked *o*-benzoquinones, generated *in situ* by oxidation of the corresponding phenol with (diacetoxyiodo)benzene (DAIB) accelerated by sonication without affecting regiochemistry and diastereoselection (Figure 5 (SM 29)), where the yield of the cycloadducts showed a dependence on acoustic energy, temperature, and solvent composition, without being affected by free radical initiators, pressure, and non inert atmosphere. The results of studies carried out on this type of reaction indicate a Michael addition and not a concerted process (Avalos et al. 2003). The selective oxidation of glucose to glucuronic acid catalyzed by cupric oxide (Amaniampong et al. 2019) proceeds by high frequency sonolysis of water under an inert atmosphere leading to the formation of H• and OH• radicals, where the H• radical reacts with the oxygen from CuO, leaving a greater amount of OH• radicals on the CuO surface that selectively oxidizes glucose (Figure 5 (SM 30)). Ultrasound assisted chemistry can provide a more environmentally friendly synthesis, as is the Hantzch reaction in the synthesis of 1,4-dihydropyridines. These compounds are of great importance in the pharmacological area, due to their anticancer, anticoagulant, antithrombin, and antibiotic activity (Ahamed et al. 2018). The synthesis of these compounds assisted by ultrasound is based on the condensation of aldehydes, ammonium acetate, and 1,3-dicarbonyl compounds (Figure 5 (SM 31)) using nanocatalysts and green solvents such as ethanol and-or water, providing a new and green alternative (Moradi and Zare 2018).

In recent years, the value of metal free catalysis has increased because there are significant progresses that provide a real and sustainable alternative route of synthesis, and more important, using sonochemistry as energy source, as final note, metal free reactions will represent a benign compensation for environment (Borah and Chowhan 2022).

3. Conclusion

In conclusion, sonochemical conditions are capable of providing several benefits to chemical synthesis since, in most cases, they improve yields and reduce reaction times, which are determinant factors to prefer a source of energy. Reducing reaction times results in energy savings compared to conventional methods, while heating and stirring times are decreased. Similar kinetics and yields are obtained in most of the chemical reactions promoted by heating or ultrasound; however, the reactions promoted by ultrasound are carried out at lower temperatures, which is also an advantage in the case of thermally unstable molecules. Other advantages of using ultrasound are that it produces more pure products and the inhibition of side reactions leads to higher yields, reducing chemical waste and environmental impact and improving synthesis costs. These and other advantages can only be achieved when the sonication medium meets the requirements for effective cavitation processes and acoustic energy absorption, favoring sonochemical reactions.

The sonochemical conditions in the field of organic synthesis offer other convenient features, such as the excellent homogenization or mixing effect that it offers. In the case of heterogeneous systems or reactions that involve supported catalysts. Solid surfaces are cleaned by ultrasonic cavitation, enhancing their catalytic activity, in addition to facilitating the penetration of reagents into the catalyst structure due to better mixing, as well as better mass transfer and solid dispersion. The increasing number of reports show the rapid growth in the development of green chemistry using ultrasound as an alternative and efficient energy source with an environmentally friendly approach.

Acknowledgments

This work was supported by the Dirección General de Asuntos del Personal Académico (DGAPA), Universidad Nacional Autónoma de México under Grant IN204223. University of Guadalajara under PROSNI 2021. Support Program for Technological Research and Innovation Projects under PAPIIT IG100220 and National Council of Science and Technology under CONACyT CF-19 No 140617. Call for basic scientific research CONAHCyT 2017–2018 del "Fondo Sectorial de Investigación para la educación CB2017-2018" (A1-S-29789). GGFR (CVU 407270) and FLS (CVU 409872) thanks to CONACyT for the grant awarded during postdoctoral stay.

References

Ahamed, A., I.A. Arif, M. Mateen, R. Surendra Kumar and A. Idhayadhulla. 2018. Antimicrobial, anticoagulant, and cytotoxic evaluation of multidrug resistance of new 1,4-dihydropyridine derivatives. Saudi J. Biol. Sci. 25: 1227–1235. doi:10.1016/j.sjbs.2018.03.001.

Amaniampong, P.N., Q.T. Trinh, K. De Oliveira Vigier, D.Q. Dao, N.H. Tran, Y. Wang et al. 2019. Synergistic effect of high-frequency ultrasound with cupric oxide catalyst resulting in a selectivity switch in glucose oxidation under argon. J. Am. Chem. Soc. 141: 14772–14779. doi:10.1021/jacs.9b06824.

Anastas, P.T. and M.M. Kirchhoff. 2002. Origins, current status, and future challenges of green chemistry. Acc. Chem. Res. 35: 686–694. doi:10.1021/ar010065m.

Avalos, M., R. Babiano, N. Cabello, P. Cintas, M.B Hursthouse, J.L. Jiménez et al. 2003. Thermal and sonochemical studies on the Diels–Alder cycloadditions of masked o-benzoquinones with furans: new insights into the reaction mechanism. J. Org. Chem. 68: 7193–7203. doi:10.1021/jo0348322.

Azuma, T., S. Yanagida, H. Sakurai, S. Sasa and K. Yoshino. 1982. A facile preparation of aromatic anion radicals by ultrasound irradiation. Synth. Commun. 12: 137–140. doi:10.1080/00397918208063667.

Banerjee, B. 2017. Recent developments on ultrasound assisted catalyst-free organic synthesis. Ultrason. Sonochem. 35: 1–14. doi:10.1016/j.ultsonch.2016.09.023.

Blomberg, C. and F.A. Hartog. 1977. The barbier reaction - a one-step alternative for syntheses via organomagnesium compounds. Synthesis (Stuttg). 1977: 18–30. doi:10.1055/s-1977-24261.

Borah, B. and L.R. Chowhan. 2022. Ultrasound-assisted transition-metal-free catalysis: a sustainable route towards the synthesis of bioactive heterocycles. RSC Adv. 12: 14022–14051. doi:10.1039/D2RA02063G.

Cao, H., W. Zhang, C. Wang and Y. Liang. 2020. Sonochemical degradation of poly- and perfluoroalkyl substances – a review. Ultrason. Sonochem. 69: 105245. doi:10.1016/j.ultsonch.2020.105245.

Cenci, S.M., L.R. Cox and G.A. Leeke. 2014. Ultrasound-induced CO_2/H_2O emulsions as a medium for clean product formation and separation: the barbier reaction as a synthetic example. ACS Sustain. Chem. Eng. 2: 1280–1288. doi:10.1021/sc500112q.

Cintas, P. and J.-L.Luche. 1999. Green chemistry. Green Chem. 1: 115–125. doi:10.1039/a900593e.

Cintas, P., A. Barge, S. Tagliapietra, L. Boffa and G.Cravotto. 2010. Alkyne–azide click reaction catalyzed by metallic copper under ultrasound. Nat. Protoc. 5: 607–616. doi:10.1038/nprot.2010.1.

Cravotto, G. and P. Cintas. 2006. Power ultrasound in organic synthesis: moving cavitational chemistry from academia to innovative and large-scale applications. Chem. Soc. Rev. 35: 180–196. doi:10.1039/B503848K.

Cucheval, A. and R.C.Y. Chow. 2008. A study on the emulsification of oil by power ultrasound. Ultrason. Sonochem. 15: 916–920. doi:10.1016/j.ultsonch.2008.02.004.

Dharmarathne, L. and F. Grieser. 2016. Formation of amino acids on the sonolysis of aqueous solutions containing acetic acid, methane, or carbon dioxide, in the presence of nitrogen gas. J. Phys. Chem. A 120: 191–199. doi:10.1021/acs.jpca.5b11858.

Draye, M. and N. Kardos. 2017. Advances in green organic sonochemistry. pp. 29–57 *In*: Colmenares, J. and G. Chatel, [eds.]. Sonochemistry. Topics in Current Chemistry Collections. Springer, Cham,. doi:10.1007/978-3-319-54271-3_2.

Eaborn, C., P.B. Hitchcock and P.D. Lickiss. 1984. The crystal structure of tris(phenyldimethylsilyl) methane. J. Organomet. Chem. 269: 235–238. doi:10.1016/0022-328X(84)80308-3.

Fragoso-Medina, A.J., F. López-Saucedo, G.G. Flores-Rojas and E. Bucio. 2021. Sonochemical synthesis of inorganic nanomaterials. pp. 263–279 *In*: R. Inamuddin, M.I. Boddula and A.M. Asiri Ahamed. [eds.]. Green Sustainable Process for Chemical and Environmental Engineering and Science Elsevier. https://doi.org/10.1016/B978-0-12-821887-7.00008-2.

Fry, A.J. and J.P. Bujanauskas. 1978. Electrochemical and mercury-promoted reduction of alpha, alpha'-dibromophenylacetones in acetic acid. J. Org. Chem. 43: 3157–3163. doi:10.1021/jo00410a013.

Gautheron, B., G. Tainturier and C. Degrand. 1985. Ultrasound-induced electrochemical synthesis of the anions selenide (Se22-, Se2-), and telluride (Te22-, and Te2-). J. Am. Chem. Soc. 107: 5579–5581. doi:10.1021/ja00305a070.

Geng, M., Z. Wang, L. Qin, A.Taha, L. Du, X. Xu et al. 2022. Effect of ultrasound and coagulant types on properties of β-carotene bulk emulsion gels stabilized by soy protein. Food Hydrocoll. 123: 107146. doi:10.1016/j.foodhyd.2021.107146.

Han, B.H. and P. Boudjouk. 1982. Organic sonochemistry. sonic acceleration of the reformatsky reaction. J. Org. Chem. 47: 5030–5032. doi:10.1021/jo00146a044.

Horton, A.M., D.M. Hollinshead and S.V. Ley. 1984. Fe$_2$(CO)g in tetrahydrofuran or under sonochemical conditions as convenient practical routes to π-allyltricarbonyliron lactone complexes. Tetrahedron 40: 1737–1742. doi:10.1016/S0040-4020(01)91124-X.

Ishihara, J. and S. Hatakeyama. 2012. Recent developments in the reformatsky-claisen rearrangement. Molecules 17: 14249–14259. doi:10.3390/molecules171214249.

Ishikawa, N., G.K. Moon, T. Kitazume and K.C. Sam. 1984. Preparation of trifluoromethylated allylic alcohols from trifluoroacetaldehyde and organometallic compounds. J. Fluor. Chem. 24: 419–430. doi:10.1016/S0022-1139(00)83162-0.

Kamble, S., A. Kumbhar, G. Rashinkar, M. Barge and R. Salunkhe. 2012. Ultrasound promoted efficient and green synthesis of β-amino carbonyl compounds in aqueous hydrotropic medium. Ultrason. Sonochem. 19: 812–815. doi:10.1016/j.ultsonch.2011.12.001.

Kinzyabaeva, Z.S. and D.S. Sabirov. 2020. Sonochemical synthesis of novel C60 fullerene 1,4-oxathiane derivative through the intermediate fullerene radical anion. Ultrason. Sonochem. 67: 105169. doi:10.1016/j.ultsonch.2020.105169.

Kitazume, T. and N. Ishikawa. 1985. Ultrasound-promoted selective perfluoroalkylation on the desired position of organic molecules. J. Am. Chem. Soc. 107: 5186–5191. doi:10.1021/ja00304a026.

Lee, P.H., K. Bang, K. Lee, S. Sung and S. Chang. 2001. Ultrasound promoted synthesis of β-hydroxyesters by reformatsky reaction using indium metal. Synth. Commun. 31: 3781–3789. doi:10.1081/SCC-100108228.

Li, Z., J. Dong, H. Zhang, Y. Zhang, H. Wang, X. Cui et al. 2021. Sonochemical catalysis as a unique strategy for the fabrication of nano-/micro-structured inorganics. Nanoscale Adv. 3: 41–72. doi:10.1039/D0NA00753F.

Lindley, J., J.P. Lorimer and T.J. Mason. 1986. Enhancement of an ullmann coupling reaction induced by ultrasound. Ultrasonics 24: 292–293. doi:10.1016/0041-624X(86)90108-3.

Liou, K.-F., P.-H. Yang and Y.-T. Lin. 1985. Ultrasonic irradiation in the synthesis of triethylborane from ethyl bromide via ethylaluminum sesquibromide. J. Organomet. Chem. 294: 145–149. doi:10.1016/0022-328X(85)87463-5.

Liu, K., R. Qin and N. Zheng. 2021. Insights into the interfacial effects in heterogeneous metal nanocatalysts toward selective hydrogenation. J. Am. Chem. Soc. 143: 4483–4499. doi:10.1021/jacs.0c13185.

Low, C.M. 1995. Ultrasound in synthesis: natural products and supersonic reactions? ultrason. Sonochem. 2: S153–S163. doi:10.1016/1350-4177(95)00017-Z.

Lu, T.-J., S.-M Cheng and L.-J. Sheu. 1998. Ultrasound accelerated coupling reaction of grignard reagents with 1,3-dioxolanes of α,β-unsaturated aldehydes. J. Org. Chem. 63: 2738–2741. doi:10.1021/jo972048+.

Luche, J.L. and J.C. Damiano. 1980. Ultrasounds in organic syntheses. 1. effect on the formation of lithium organometallic reagents. J. Am. Chem. Soc. 102: 7926–7927. doi:10.1021/ja00547a016.

Luche, J.L., C. Petrier, A.L. Gemal and N. Zikra. 1982. Ultrasound in organic synthesis. 2. formation and reaction of organocopper reagents. J. Org. Chem. 47: 3805–3806. doi:10.1021/jo00140a054.

Luche, J.-L., C. Petrier, J.P. Lansard and A.E. Greene. 1983. Ultrasound in organic synthesis. 4. a simplified preparation of diarylzinc reagents and their conjugate addition to .alpha.-enones. J. Org. Chem. 48: 3837–3839. doi:10.1021/jo00169a055.

Masamune, S., S. Murakami and H. Tobita. 1983. Disilene system (R2Si:SiR2). the tetra-tert-butyl derivative. Organometallics 2: 1464–1466. doi:10.1021/om50004a041.

Mason, T.J. 1986. Use of ultrasound in chemical synthesis. Ultrasonics 24: 245–253. doi:10.1016/0041-624X(86)90101-0.

Montaña, A.M. and P.M. Grima. 2003. Generation of oxyallyl cations by reduction of α,α′-diiodoketones under sonochemical or thermal conditions: Improved methodology for the [4C(4π;)+3C(2π;)] cycloaddition reactions. Synth. Commun. 33: 265–279. doi:10.1081/SCC-120015712.

Moradi, L. and M. Zare. 2018. Ultrasound-promoted green synthesis of 1,4-dihydropyridines using fuctionalized MWCNTs as a highly efficient heterogeneous catalyst. Green Chem. Lett. Rev. 11: 197–208. doi:10.1080/17518253.2018.1458160.

More, P.A. and G.S. Shankarling. 2017. Energy efficient pfitzinger reaction: a novel strategy using a surfactant catalyst. New J. Chem. 41: 12380–12383. doi:10.1039/C7NJ01937H.

Nelson, T.D. and R.D. Crouch. 2004. Cu, Ni, and Pd mediated homocoupling reactions in biaryl syntheses: the ullmann reaction. pp. 265–555 *In*: Organic Reactions. John Wiley and Sons, Inc., Hoboken, NJ, USA,. doi:10.1002/0471264180.or063.03

Ni, Y., J. Wu, Y. Jiang, J. Li, L. Fan and S. Huang. 2022. High-internal-phase pickering emulsions stabilized by ultrasound-induced nanocellulose hydrogels. Food Hydrocoll. 125: 107395. doi:10.1016/j.foodhyd.2021.107395.

Oppolzer, W. and P. Schneider. 1984. Practical preparation and metallo-ene reactions of (2-alkenylallyl)-magnesium chlorides: Comparative study of magnesium activation. Tetrahedron Lett. 25: 3305–3308. doi:10.1016/S0040-4039(01)81370-8.

Penteado, F., B. Monti, L. Sancineto, G. Perin, R.G. Jacob, C. Santi et al. 2018. Ultrasound-assisted multicomponent reactions, organometallic and organochalcogen chemistry. Asian J. Org. Chem. 7: 2368–2385. doi:10.1002/ajoc.201800477.

Petrier, C., J. Einhorn and J.L. Luche. 1985. Selective tin and zinc mediated allylations of carbonyl compounds in aqueous media. Tetrahedron Lett. 26: 1449–1452. doi:10.1016/S0040-4039(00)99068-3.

Petrier, C., J. De Souza Barbosa, C. Dupuy and J.L. Luche. 1985. Ultrasound in organic synthesis. 7. preparation of organozinc reagents and their nickel-catalyzed reactions with .alpha.,.beta.-unsaturated carbonyl compounds. J. Org. Chem. 50: 5761–5765. doi:10.1021/jo00350a065.

Pokhrel, N., P.K. Vabbina and N. Pala. 2016. Sonochemistry: science and engineering. Ultrason. Sonochem. 29: 104–128. doi:10.1016/j.ultsonch.2015.07.023.

Rashmi Pradhan, S., R.F. Colmenares-Quintero and J.C. Colmenares Quintero. 2019. Designing microflowreactors for photocatalysis using sonochemistry: a systematic review article. Molecules 24: 3315. doi:10.3390/molecules24183315.

Rieke, V., R. Instrella, J. Rosenberg, W. Grissom, B. Werner, E. Martin et al. 2013. Comparison of temperature processing methods for monitoring focused ultrasound ablation in the brain. J. Magn. Reson. Imaging 38: 1462–1471. doi:10.1002/jmri.24117.

Ross, N.A. and R.A. Bartsch. 2003. High-intensity ultrasound-promoted reformatsky reactions. J. Org. Chem. 68: 360–366. doi:10.1021/jo0261395.

Saleh, T.S., A.S. Al-Bogami, A.E.M Mekky and H.Z. Alkhathlan. 2017. Sonochemical synthesis of novel pyrano[3,4-e][1,3]oxazines: a green protocol. Ultrason. Sonochem. 36: 474–480. doi:10.1016/j.ultsonch.2016.12.015.

Smolnikov, S.A., M.S. Shahari and A.V. Dolzhenko. 2020. Sonochemical protocols for grignard reactions. pp. 243–255 *In*: Inamuddin, R.B. and A.M. Asiri [eds.]. Green Sustainable Process for Chemical and Environmental Engineering and Science. Elsevier, Amsterdam, doi:10.1016/B978-0-12-819540-6.00009-7.

Suslick, K.S. and P.F. Schubert. 1983. Sonochemistry of dimanganese decacarbonyl (Mn2(CO)10) and dirhenium decacarbonyl (Re2(CO)10). J. Am. Chem. Soc. 105: 6042–6044. doi:10.1021/ja00357a014.

Suslick, K.S. and R.E. Johnson. 1984. Sonochemical activation of transition metals. J. Am. Chem. Soc. 106: 6856–6858. doi:10.1021/ja00334a073.

Suslick, K.S. and D.J. Casadonte, Green, M.L.H., Thompson, M.E., 1987. Effects of high intensity ultrasound on inorganic solids. Ultrasonics 25: 56–59. doi:10.1016/0041-624X(87): 90013-8.

Suslick, K., D. Casadonde and S. Docktycz. 1989. The effects of ultrasound on nickel and copper powders. Solid State Ionics 32–33: 444–452. doi:10.1016/0167-2738(89)90254-3.

Suslick, K.S., M. Fang and T. Hyeon. 1996. Sonochemical synthesis of iron colloids. J. Am. Chem. Soc. 118: 11960–11961. doi:10.1021/ja961807n.

Tamai, H., T. Ikeya and H. Yasuda. 1999. Hydrogenation of 1-hexene and hexadienes by ultrafine Pd particles supported on the surface of PrPO4 hollow particles. J. Colloid Interface Sci. 218: 217–224. doi:10.1006/jcis.1999.6412.

Tuulmets, A., K. Kaubi and K. Heinoja. 1995. Influence of sonication on grignard reagent formation. Ultrason. Sonochem. 2: S75–S78. doi:10.1016/1350-4177(95)00028-5.

Tuulmets, A., G. Cravotto, S. Salmar and J. Jarv. 2010. Sonochemistry of homogeneous ionic reactions. Mini. Rev. Org. Chem. 7: 204–211. doi:10.2174/157019310791384155.

Wang, T., N. Wang, N. Li, X. Ji, H. Zhang, D. Yu et al. 2022. Effect of high-intensity ultrasound on the physicochemical properties, microstructure, and stability of soy protein isolate-pectin emulsion. Ultrason. Sonochem. 82: 105871. doi:10.1016/j.ultsonch.2021.105871.

Yang, R.-Y. and B. Xu. 2021. Chemo-, regio- and stereoselective synthesis of monofluoroalkenes via a tandem fluorination–desulfonation sequence. Chem. Commun. 57: 7802–7805. doi:10.1039/D1CC03207K.

Zhang, K., B.-J. Park, F.-F. Fang and H.J. Choi. 2009. Sonochemical preparation of polymer nanocomposites. Molecules 14: 2095–2110. doi:10.3390/molecules140602095.

Zhang, M., L. Fan, Y. Liu and J. Li. 2022. Relationship between protein native conformation and ultrasound efficiency: for improving the physicochemical stability of water–in–oil emulsions. Colloids Surfaces A Physicochem. Eng. Asp. 651: 129737. doi:10.1016/j.colsurfa.2022.129737.

Zhang, P., M. Behl, X. Peng M.Y. Razzaq and A. Lendlein. 2016. Ultrasonic cavitation induced shape-memory effect in porous polymer networks. Macromol. Rapid Commun. 37: 1897–1903. doi:10.1002/marc.201600439.

CHAPTER 4

Sonochemical Organic Synthesis under Catalyst-Free Conditions and Nanoparticles as Efficient Catalyst

A Journey towards a Sustainable Future

Sasadhar Majhi,[1,*] *Sivakumar Manickam*[2,*] and *Giancarlo Cravotto*[3]

1. Introduction

Catalysis is one of the fundamental pillars in modern synthetic organic chemistry; catalysts profoundly affect chemical transformations in synthesizing promising organic molecules, including bioactive natural products, for accomplishing a high degree of molecular complexity from relatively simple building blocks (Kitanoso et al. 2018, Odagi and Nagasawa 2019). However, the most benign route in green chemistry is to design a methodology without any catalyst(s) because most traditional catalysts are toxic and expensive; conventional catalysts particularly homogeneous catalysts are difficult to separate from the reaction mixture by simple techniques such as centrifugation, filtration, magnetic separation, etc. (Baruah and Deb 2021). Ultrasound (US) mediated organic reactions fulfil the goal of sustainable chemistry under catalyst free conditions.

Nanomaterials (NPs) have been extensively studied in organic synthesis as multitalented catalysts because of their high surface area, high absorbing capability, and reactive morphologies (Sharma et al. 2016). The important perspective of clean chemistry is the utility of ecologically sound catalysts with the trouble free recovery of the catalyst after completing the transformation (Banerjee 2019). In this context, nanocatalysts have various merits compared to traditional catalytic systems, including simple preparation and modification, smooth recovery, and reusability (Sharma et al. 2016, Banerjee 2019). In addition, magnetic NPs have gained attention due to

[1] Department of Chemistry (UG & PG Dept.), Triveni Devi.
Bhalotia College, Raniganj, Kazi Nazrul University, West Bengal 713347, India.
[2] Petroleum and Chemical Engineering, Faculty of Engineering, Universiti Teknologi Brunei, Bandar Seri Begawan, BE1410, Brunei Darussalam.
[3] Department of Drug Science and Technology, University of Turin, Via P. Giuria 9, 10125 Turin, Italy.
* Corresponding authors: sasadharmajhi@gmail.com; manickam.sivakumar@utb.edu.bn

their better catalytic activity than conventional catalysts, with the simple recovery of magnetically active NPs employing an external magnet only.

Organic synthesis is a subdiscipline of synthesis that has remarkable contributions in industry and academia to produce promising organic molecules in fine chemicals, agrochemicals, pharmaceutical agents, and much more (Kotha and Khedkar 2012, Armaly et al. 2015, Nicolaou 2014). With the industrial revolution, the environment has also been polluted due to the discharge of several harmful substances and hazardous waste materials. As a result, our environment and living organisms, including humans, are not safe from pollution. To protect Mother Nature, chemists and engineers try to develop greener products and methodologies that should be safer, cheaper, nontoxic, as well as nondetrimental to the environment. Green or clean chemistry is related to creativity and innovative research development, and it tries to conduct at the molecular level to attain sustainability. This sustainability can be achieved by practicing 12 principles of green chemistry, which are fundamental pillars of clean chemistry, and a mentor to design new chemical products and processes (Anastas and Warner 1998, Tang et al. 2008). The sixth principle of green chemistry is devoted to "design for energy efficiency", which discloses that the need for energy should be minimum to carry out a chemical transformation with optimum productivity. It should have minimum influence on the economy and environment (Majhi 2020). Nowadays, sonochemistry has emerged as a greener tool for reducing the energy of specific chemical reactions with excellent productivity (Mason 1997).

The current trends in the synthesis of organic molecules are the evolution of novel, green, and quick synthetic protocols for targeted products, including synthetic and bioactive natural compounds (Mason 1997, Majhi 2021a, 2021b, 2021c, 2022, Cravotto and Cintas 2006, Majhi and Das 2021). Since the ultrasound (US) mediated methodologies as a source of clean energy provide excellent yields of the desired compounds with shorter reaction time, eliminating or reducing the use of toxic catalysts, the present chapter concentrates on the sonochemical synthesis of important organic molecules under catalyst free conditions and nanoparticles as efficient catalysts to reach the sustainability of the society.

2. Sonochemical Organic Synthesis under Catalyst Free Conditions and Nanoparticles toward Sustainable Society

Catalysis is a potential synthetic tool for synthesizing valuable organic molecules, and which further illustrates the relationship between innovation and sustainability (Anastas et al. 2000). However, traditional catalysts are normally linked with toxicity, high costs, and no reusability, thereby frequently creating wastes responsible for environmental pollution. In this contest, sonochemical organic synthesis under catalyst free conditions contributes remarkably to attaining sustainability (Baruah and Deb 2021). Hence, this section deals with US assisted chemical transformations under catalyst free conditions.

2.1 Synthesis of Important Organic Compounds Based on Pyrazole Scaffold

Five membered aza heterocyclic pyrazoles molecules have been found to show a wide spectrum of pharmacological and biological activities, including antidiabetic, antitumor, antihyperglycemic, antidepressive, etc., Nowadays, pyrazole motif bearing drugs such as Celecoxib and nonsteroidal drugs are employed to remedy arthritis and acute pain (Shabalala et al. 2020).

2.1.1 Synthesis of Bis-3-Methyl-1-Phenyl-1 H-Pyrazol-5-ols

An efficient, catalyst free, onepot synthesis of bis-3-methyl-1-phenyl-1*H*-pyrazol-5-ols was demonstrated by Jonnalagadda et al., by applying US irradiation as a nonpolluting source of energy from the reactions of various benzaldehydes, β-keto ester, and phenylhydrazine in aqueous ethanol at room temperature in excellent yields (92 to 99 percent) for five to eight min (Equation 1) (Shabalala et al. 2020). This study was conducted without column chromatography under mild reaction conditions but with selective transformation.

The plausible mechanism (Figure 1) for the synthesis of bis(3-methyl-1-phenyl-1 *H*-pyrazol-5-ol)s is given below:

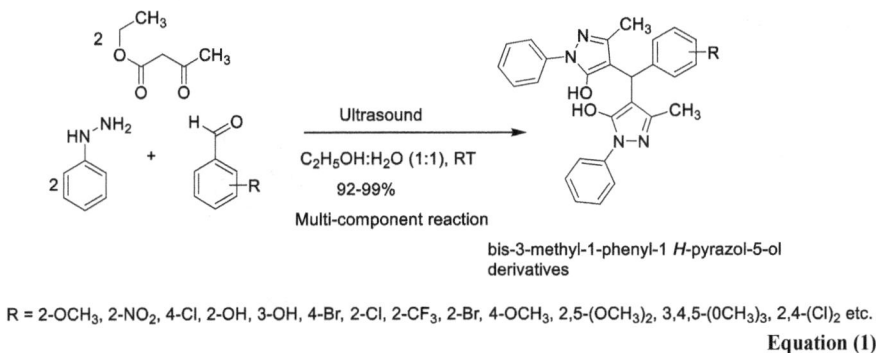

R = 2-OCH$_3$, 2-NO$_2$, 4-Cl, 2-OH, 3-OH, 4-Br, 2-Cl, 2-CF$_3$, 2-Br, 4-OCH$_3$, 2,5-(OCH$_3$)$_2$, 3,4,5-(OCH$_3$)$_3$, 2,4-(Cl)$_2$ etc.

Equation (1)

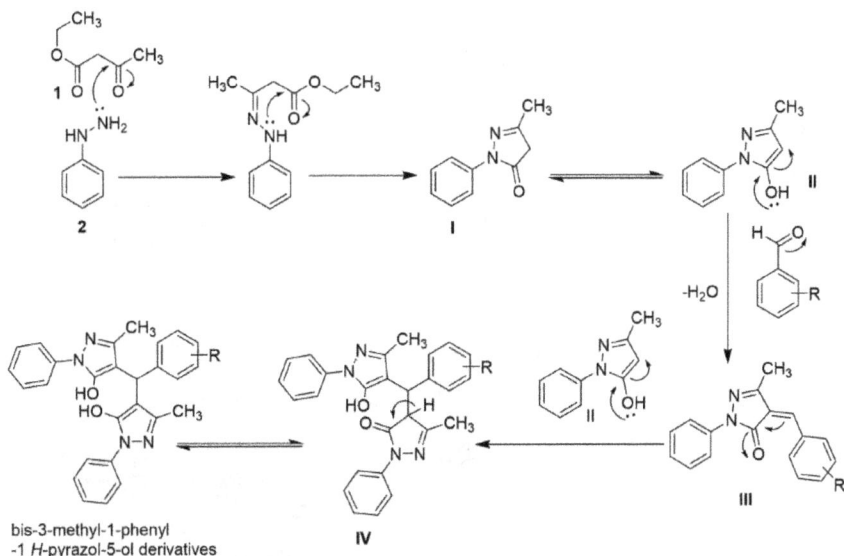

Figure 1. A plausible mechanism for the preparation of bis(3-methyl-1-phenyl-1 *H* -pyrazol-5-ol)s.

In another study, the same group described a remarkable, practical, single-pot technique for the rapid synthesis of pyrazoles by the multicomponent reaction of hydrazine monohydrate, various aryl aldehydes, ethyl acetoacetate, and malononitrile-ammonium acetate as easily obtainable starting materials under US irradiation as a green tool in water within 0.5 to 2.5 h (Shabalala et al. 2015).

2.1.2 Synthesis of 3,12b-Dihydropyrazolo[4',3':3,4]Pyrido[2,1-a]Isoquinolines

Tabarsaei et al., developed a new protocol for the preparation of pyrazole derivatives under ultrasonic irradiation in excellent yields in water as a green solvent through the catalyst-free one pot multicomponent reaction of activated acetylenic molecules, isoquinoline, triphenylphosphine, alkyl bromides, and hydrazine to follow in good harmony with some principles of green or clean chemistry (Equation 2) (Tabarsaei et al. 2020). The same investigators also examined the antioxidant activity of synthesized compounds by DPPH (2,2-diphenyl-1-picrylhydrazyl) radical scavenging and ferric reducing power analyses due to the presence of pyrazole scaffold together with isoquinoline scaffold in the synthesized molecules.

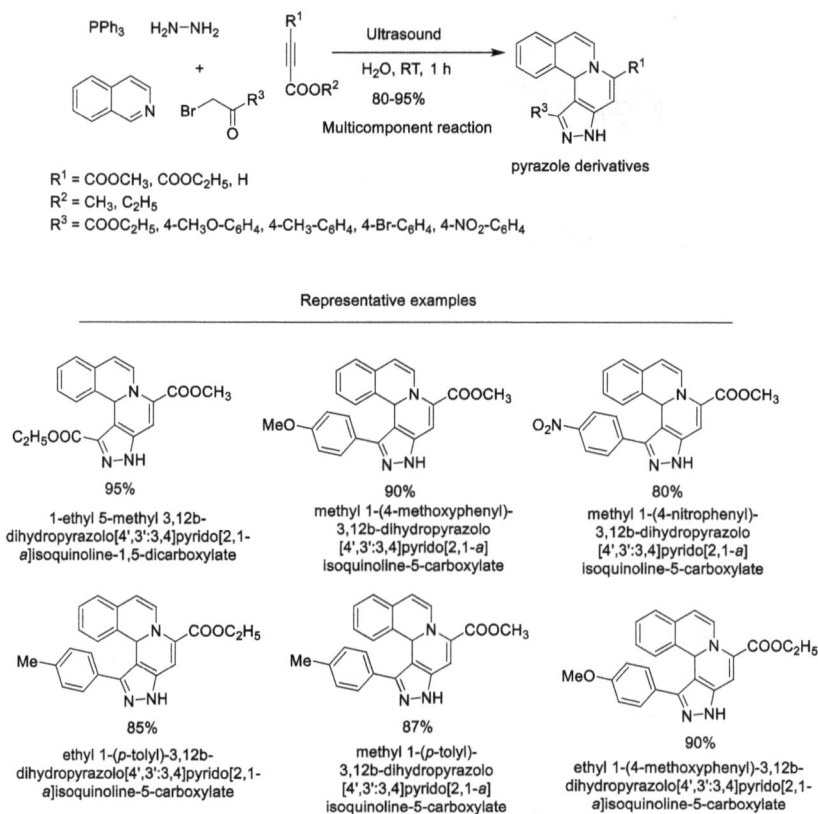

R^1 = COOCH$_3$, COOC$_2$H$_5$, H
R^2 = CH$_3$, C$_2$H$_5$
R^3 = COOC$_2$H$_5$, 4-CH$_3$O-C$_6$H$_4$, 4-CH$_3$-C$_6$H$_4$, 4-Br-C$_6$H$_4$, 4-NO$_2$-C$_6$H$_4$

Representative examples

95%

1-ethyl 5-methyl 3,12b-dihydropyrazolo[4',3':3,4]pyrido[2,1-a]isoquinoline-1,5-dicarboxylate

90%

methyl 1-(4-methoxyphenyl)-3,12b-dihydropyrazolo[4',3':3,4]pyrido[2,1-a]isoquinoline-5-carboxylate

80%

methyl 1-(4-nitrophenyl)-3,12b-dihydropyrazolo[4',3':3,4]pyrido[2,1-a]isoquinoline-5-carboxylate

85%

ethyl 1-(p-tolyl)-3,12b-dihydropyrazolo[4',3':3,4]pyrido[2,1-a]isoquinoline-5-carboxylate

87%

methyl 1-(p-tolyl)-3,12b-dihydropyrazolo[4',3':3,4]pyrido[2,1-a]isoquinoline-5-carboxylate

90%

ethyl 1-(4-methoxyphenyl)-3,12b-dihydropyrazolo[4',3':3,4]pyrido[2,1-a]isoquinoline-5-carboxylate

Equation (2)

2.1.3 Synthesis of 4,4-(Arylmethylene)Bis(3-Methyl-1-Phenyl-1H-Pyrazol-5-ols)

Pyrazoles and their analogues, as well as 4,4-(arylmethylene)bis(3-methyl-1-phenyl-1*H*-pyrazol-5-ols), display a wide spectrum of biological properties such as antipyretic, antiinflammatory, antibacterial, antidepressant, analgesic, etc. (McDonald et al. 2006, Faisal et al. 2019). Hasaninejed et al., accomplished a new protocol for the synthesis of 4,4-(arylmethylene)bis(3-methyl-1-phenyl-1*H*-pyrazol-5-ols through a single-pot pseudo five component reaction of β-ketoesters (two eq), aromatic aldehydes (one eq) and phenylhydrazine derivatives (two eq) under US irradiation as a green technique at room temperature in water: ethanol (one:one) without the need of any catalyst (Equation 3) (Hasaninejed et al. 2013). The main merits of the US assisted preparation of pyrazole derivatives include excellent yield, simple work up, ecologically friendly, and shorter reaction time.

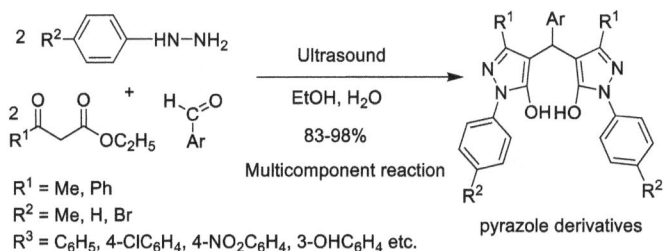

R^1 = Me, Ph
R^2 = Me, H, Br
R^3 = C_6H_5, 4-ClC_6H_4, 4-$NO_2C_6H_4$, 3-OHC_6H_4 etc.

Representative examples

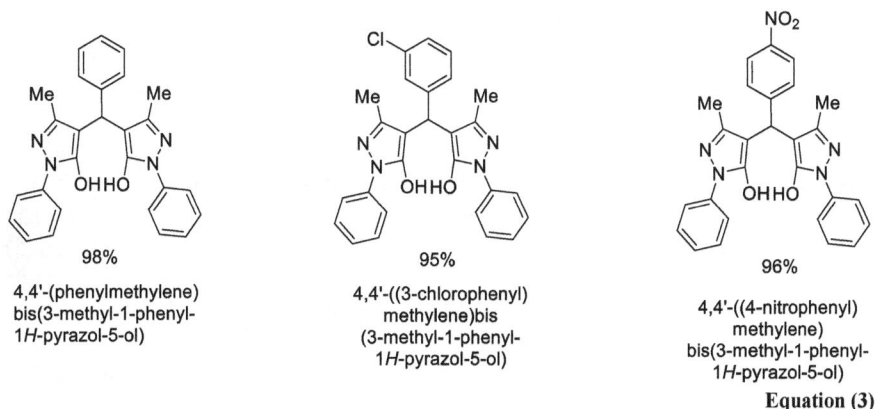

98%

4,4'-(phenylmethylene)
bis(3-methyl-1-phenyl-
1*H*-pyrazol-5-ol)

95%

4,4'-((3-chlorophenyl)
methylene)bis
(3-methyl-1-phenyl-
1*H*-pyrazol-5-ol)

96%

4,4'-((4-nitrophenyl)
methylene)
bis(3-methyl-1-phenyl-
1*H*-pyrazol-5-ol)

Equation (3)

2.1.4 Synthesis of Highly Substituted Pyrazoles

Nemati et al., disclosed a highly useful and ecologically sound methodology for synthesizing extremely substituted pyrazole derivatives through the multicomponent reaction of several aldehydes, phenylhydrazine or 4-phenylthiosemicarbazide, and malononitrile as economical starting materials under US irradiation in PEG-400 and water (Equation 4) (Nemati et al. 2015). Applying green solvents at room temperature under catalyst-free conditions makes this procedure attractive from environmental and economic viewpoints.

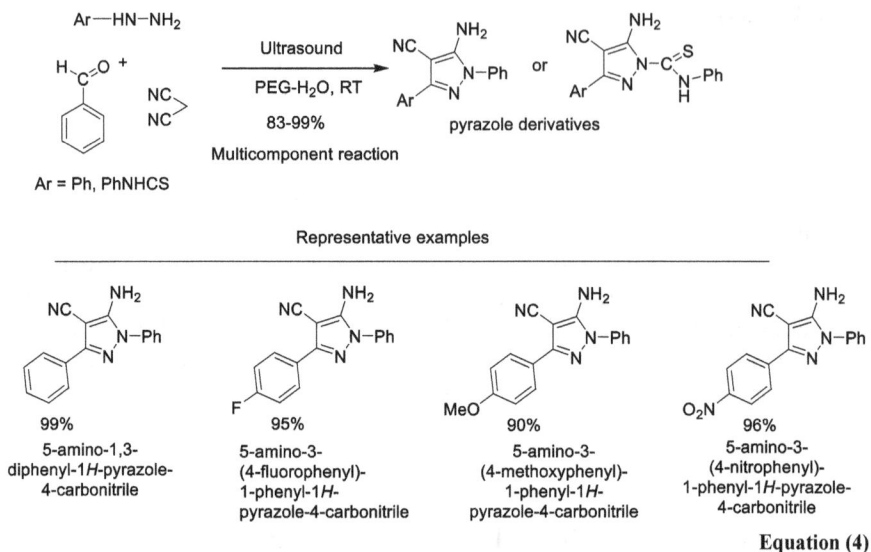

Equation (4)

2.2 Synthesis of Important Organic Compounds Based on Pyrimidine Scaffold

The common motif in various natural products and biologically active molecules is the pyridodiazine framework, particularly pyrido[2,3-d]pyrimidine; it contributes a privileged scaffold for constructing the desired molecule in drug discovery (Gupta et al. 1989). It has been investigated that pyrido[2,3-*d*]pyrimidines exhibited various biological and pharmacological activities (Ibrahim and El-Metwally 2010).

2.2.1 Synthesis of Pyrido-[2,3-d] Pyrimidines

Mamaghani et al., synthesized several new derivatives of pyrido-[2,3-*d*] pyrimidines through a catalyst-free one-pot multicomponent reaction under ultrasonic irradiation (80°C) as a green source of energy in good to high yields (79 to 92 percent) in ethylene glycol solvent as well as conventional heating (120°C) (Equation 5) (Barghi-Lish et al. 2016). The sonochemical method provided higher yields with shorter reaction times (25 to 40 min) than the conventional procedure (80 to 120 min). It has been evident from the work that the aldehydes bearing electron withdrawing groups furnished the targeted products with higher yields compared to electron donating groups.

The plausible mechanism (Figure 2) for the synthesis of pyrido-[2,3-*d*] pyrimidines is given below:

In another study, a one pot, three component and catalyst free protocol was described by the same group for the construction of indenopyrido[2,3-*d*]pyrimidine derivatives from ethylene glycol solution of aromatic aldehyde, 1,3-indanedione, and 6-amino-2-(alkylthio)-pyrimidin-4(3*H*)one under US irradiation at 65°C in good to high yields (82 to 97 percent) with short reaction times (10 to 33 min) (Mamaghani et al. 2014).

Ar = C_6H_5, 4-$NO_2C_6H_4$, 4-$CH_3OC_6H_4$, 3-HOC_6H_4, 4-ClC_6H_4, 2-thienyl etc.

pyrido–[2,3-*d*] pyrimidines

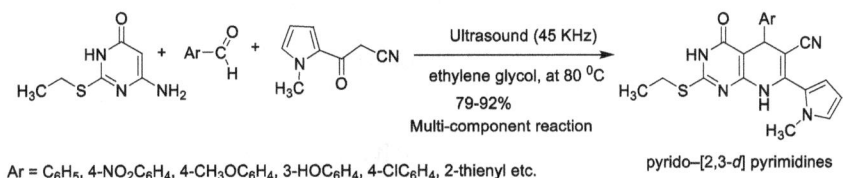

Representative examples

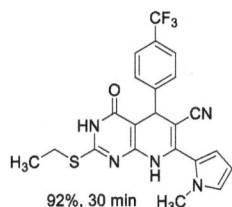

92%, 30 min

2-(ethylthio)-7-(1-methyl-
1*H*-pyrrol-2-yl)-4-oxo-5-(4-
(trifluoromethyl)phenyl)-
3,4,5,8-tetrahydropyrido
[2,3-*d*]pyrimidine-
6-carbonitrile

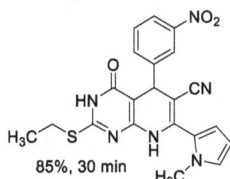

85%, 30 min

2-(ethylthio)-7-(1-methyl-1*H*-pyrrol-2-yl)-
5-(3-nitrophenyl)-4-oxo-3,4,5,8-
tetrahydropyrido[2,3-*d*]pyrimidine-6-
carbonitrile

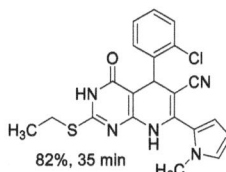

82%, 35 min

5-(2-chlorophenyl)-
2-(ethylthio)-7-
(1-methyl-1*H*-pyrrol-2-yl)-
4-oxo-3,4,5,8-
tetrahydropyrido
[2,3-*d*]pyrimidine-
6-carbonitrile

Equation (5)

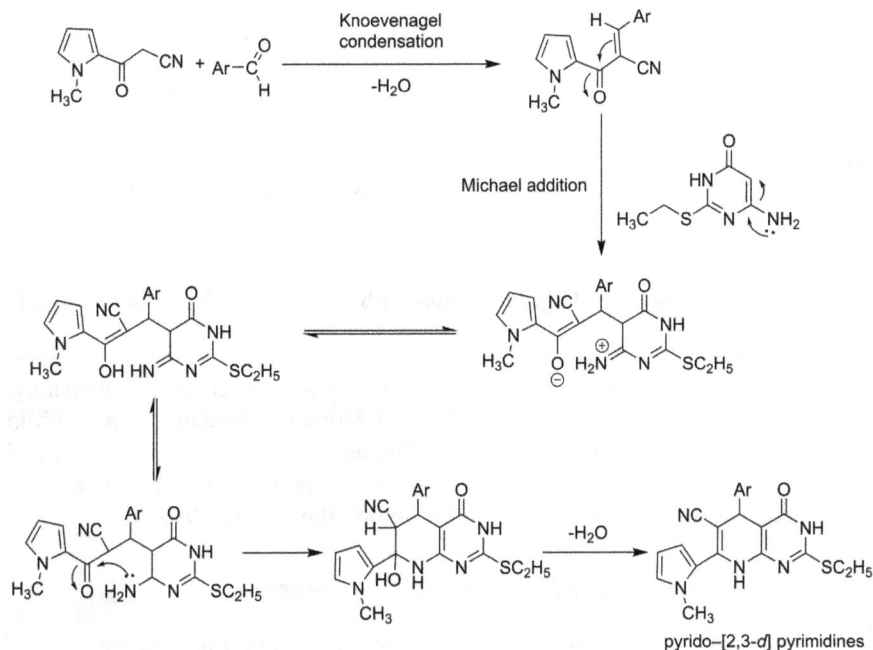

Figure 2. A plausible mechanism for the preparation of pyrido-[2,3-*d*] pyrimidines.

2.2.2 Synthesis of 7-Methyl-Substituted Pyrido[4,3-d]Pyrimidines

Prajapati et al., reported the synthesis of 7-methyl-substituted pyrido[4,3-*d*]pyrimidine derivatives using a catalyst free protocol by the multi component reaction of various aromatic aldehydes, 6-[2-(dimethylamino)prop-1-enyl]-1,3-dimethyluracil, and ammonium acetate in ethanol as a non toxic solvent for 30 min at room temperature (Equation 6) (Sarmah and Prajapati 2015).

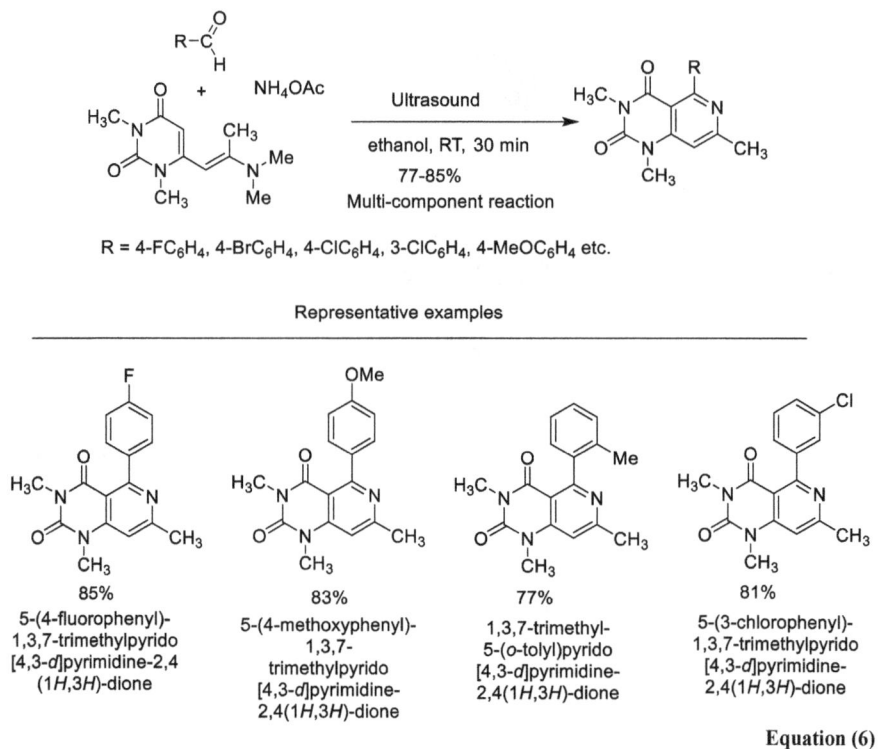

$R = 4\text{-}FC_6H_4, 4\text{-}BrC_6H_4, 4\text{-}ClC_6H_4, 3\text{-}ClC_6H_4, 4\text{-}MeOC_6H_4$ etc.

Representative examples

85%
5-(4-fluorophenyl)-
1,3,7-trimethylpyrido
[4,3-*d*]pyrimidine-2,4
(1*H*,3*H*)-dione

83%
5-(4-methoxyphenyl)-
1,3,7-
trimethylpyrido
[4,3-*d*]pyrimidine-
2,4(1*H*,3*H*)-dione

77%
1,3,7-trimethyl-
5-(*o*-tolyl)pyrido
[4,3-*d*]pyrimidine-
2,4(1*H*,3*H*)-dione

81%
5-(3-chlorophenyl)-
1,3,7-trimethylpyrido
[4,3-*d*]pyrimidine-
2,4(1*H*,3*H*)-dione

Equation (6)

2.3 Synthesis of Important Organic Compounds Based on Quinoxaline Scaffold

In the pharmaceutical industry, quinoxaline derivatives are very popular; they exhibit a wide spectrum of biological properties such as anticancer, antiinflammatory, antiviral, kinase inhibition, etc. (Kim et al. 2004). Moreover, the quinoxaline scaffold is an important part of many antibiotics, including echinomycin, actinomycin, and levomycin, which effectively inhibit the growth of Gram positive bacteria and are also efficient against several transplantable tumors (Bailly et al. 1999).

2.3.1 Synthesis of 11H-indeno[1,2-b]Quinoxalin-11-ones

Srivastava et al., demonstrated the synthesis of quinoxaline from the reactions of 1,2-diketones with 1,2-difunctionalized benzene-pyridine under US irradiation at room temperature using water as a green solvent without utilizing any catalyst (Equation 7) (Mishra et al. 2019).

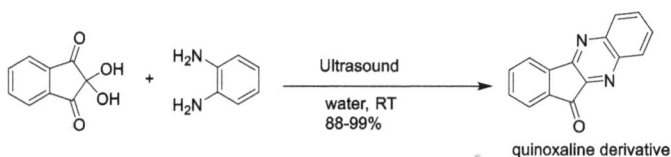

quinoxaline derivative

Representative examples

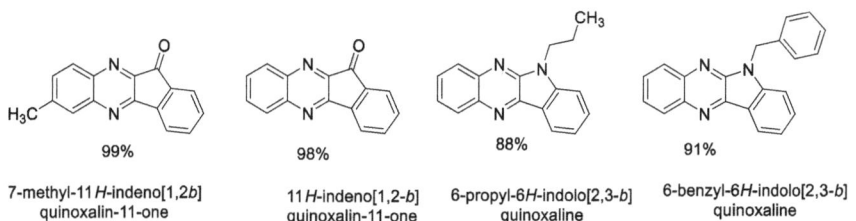

99%

7-methyl-11*H*-indeno[1,2*b*]
quinoxalin-11-one

98%

11*H*-indeno[1,2-*b*]
quinoxalin-11-one

88%

6-propyl-6*H*-indolo[2,3-*b*]
quinoxaline

91%

6-benzyl-6*H*-indolo[2,3-*b*]
quinoxaline

Equation (7)

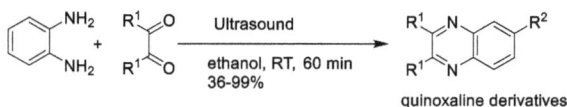

quinoxaline derivatives

Representative examples

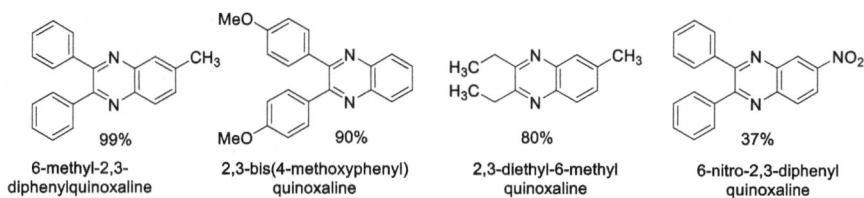

99%

6-methyl-2,3-
diphenylquinoxaline

90%

2,3-bis(4-methoxyphenyl)
quinoxaline

80%

2,3-diethyl-6-methyl
quinoxaline

37%

6-nitro-2,3-diphenyl
quinoxaline

Equation (8)

2.3.2 Synthesis of Dimethylquinoxaline Derivatives

Wu et al., established a catalyst free, highly effective, and facile protocol for synthesizing quinoxalines by reacting 1,2-diamines and several 1,2-diketones such as heterocyclic and aliphatic 1,2-diketones at room temperature under US irradiation (Equation 8) (Guo et al. 2009). The investigators also performed a systematic study to investigate the influence of reaction media and the electronic factors of the substrates on the outcome.

2.4 Synthesis of Important Organic Compounds Based on Quinazoline (Benzopyrazine) Scaffold

Quinazolinones are important heterocyclic molecules that include a broad range of anti cancer, anesthetic, muscle relaxant, antidepressant, and antimycobacterial activities (Tiwari et al. 2007).

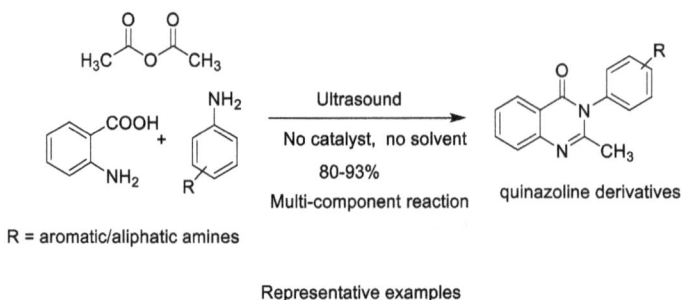

R = aromatic/aliphatic amines

Representative examples

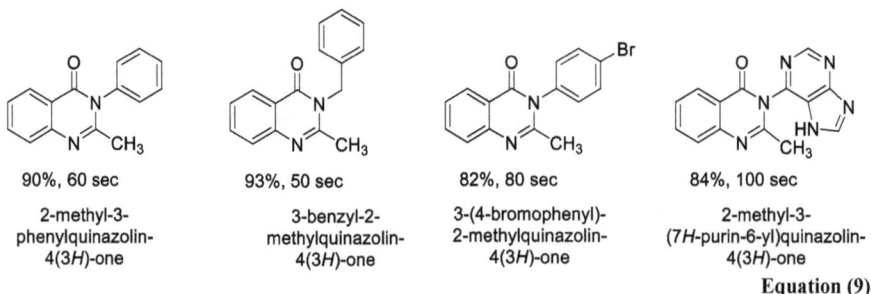

90%, 60 sec	93%, 50 sec	82%, 80 sec	84%, 100 sec
2-methyl-3-phenylquinazolin-4(3H)-one	3-benzyl-2-methylquinazolin-4(3H)-one	3-(4-bromophenyl)-2-methylquinazolin-4(3H)-one	2-methyl-3-(7H-purin-6-yl)quinazolin-4(3H)-one

Equation (9)

2.4.1 Synthesis of 2-Methyl-3-Phenylquinazolin-4(3H)-Ones

Khalafi-Nezhad et al., reported an effective and facile procedure for preparing quinazoline derivatives under the influence of ultrasonic waves at ambient conditions (Equation 9) (Purkhosro et al. 2019). The one pot reaction was executed to obtain novel quinazolinone derivatives in higher yields than the traditional thermal method through the reactions of acetic anhydride, anthranilic acid and primary amines; the authors also examined the vasorelaxant activity of the newly synthesized molecules.

2.5 Synthesis of Organic Compounds Based on Quinoline Scaffold

Nitrogen contained in heterocyclic moieties has a profound role in natural, medicinal, and synthetic chemistry because of their pharmaceutical, computational, and biological activities. Organic molecules having quinoline scaffolds have been widely studied due to their applications in medicinal, agrochemistry, and pharmaceuticals (Maddila et al. 2020). Quinoline derivatives display many biological properties, such as antileishmanial, antineoplastic, antidiabetic, antitubercular, anticancer, etc. (Upadhyay et al. 2019, Czarnecka et al. 2019). A vast family of medicinally significant molecules includes dihydroquinoline derivatives. They have also attracted attention in medicinal and pharmaceuticals as they produce antidiabetic, antihypertensive, and various other drugs (Hong et al. 2010).

2.5.1 Synthesis of 2-Methyl-5-oxo-Hexahydroquinoline-3-Carboxylates

Maddila et al., developed a sustainable methodology for the synthesis of quinoline derivatives by the multicomponent reactions of several aryl aldehydes, benzyl acetoacetate, 1,3- cyclohexanedione, and ammonium acetate at room temperature under US irradiation as a green source of energy using ethanol as a nontoxic solvent (Equation 10) (Devi et al. 2020). Hence, this methodology includes ethanol as a is an eco-friendly solvent and it proceeds under catalyst-free conditions as a cost-efficient process. Besides, this protocol offers several benefits, such as excellent yield (92 to 98 percent), operational simplicity, fast reaction time ($\leq$ 10 min), easy work up, and mild reaction conditions.

A plausible mechanism (Figure 3) for the synthesis of 2-methyl-5-oxo-hexahydroquinoline-3-carboxylate derivatives is given below:

2.5.2 Synthesis of 6H-1-Benzopyrano[4,3-b]Quinolin-6-Ones

Pal et al., reported the synthesis of novel 6*H*-1-benzopyrano[4,3-*b*]quinolin-6-ones through the reaction of 4-chloro-2-oxo-2*H*-chromene-3-carbaldehyde with several aromatic amines at room temperature in methanol under the US without employing any catalyst (Equation 11) (Mulakayala et al. 2012). Various synthesized compounds were examined for their antiproliferative activities *in vitro* against four cancer cell lines (K562, MDA-MB 231, Colo-205, and IMR32); various resultant molecules were observed to be potent.

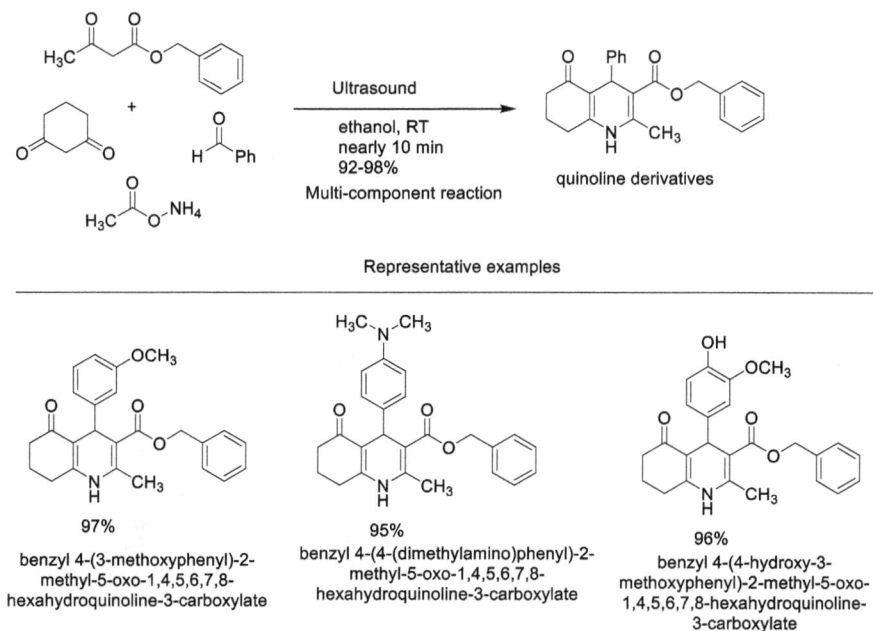

Representative examples

97%

benzyl 4-(3-methoxyphenyl)-2-methyl-5-oxo-1,4,5,6,7,8-hexahydroquinoline-3-carboxylate

95%

benzyl 4-(4-(dimethylamino)phenyl)-2-methyl-5-oxo-1,4,5,6,7,8-hexahydroquinoline-3-carboxylate

96%

benzyl 4-(4-hydroxy-3-methoxyphenyl)-2-methyl-5-oxo-1,4,5,6,7,8-hexahydroquinoline-3-carboxylate

Equation (10)

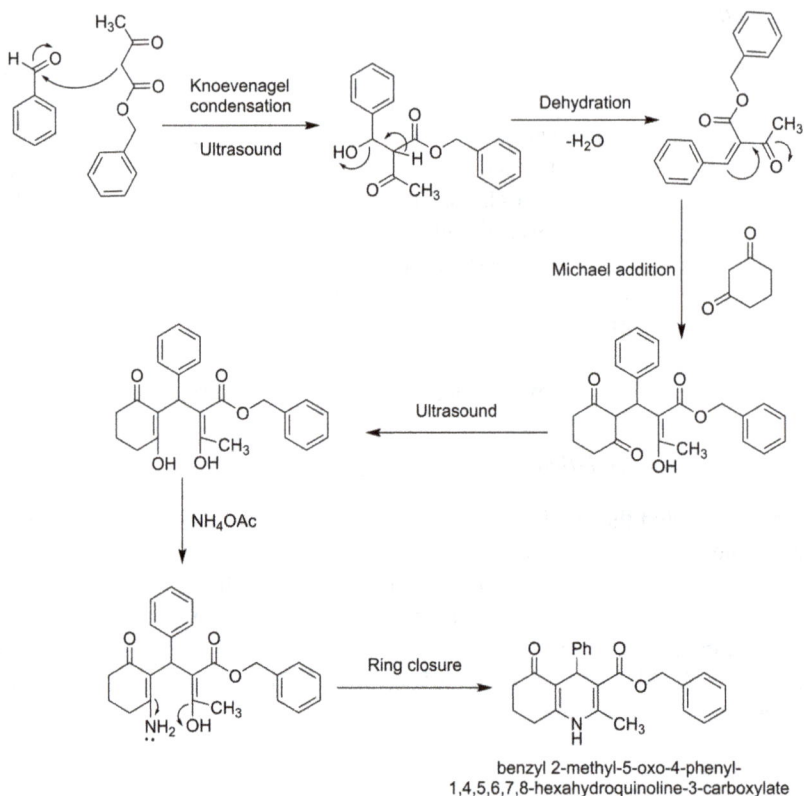

Figure 3. A plausible mechanism for the preparation of 2-methyl-5-oxo-hexahydroquinoline-3-carboxylate derivatives.

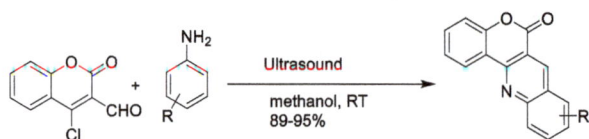

R = H, 2-OCH$_3$, 3-Br, 3-F, 2,3-(F)$_2$, 4-CH$_3$ etc.

Representative examples

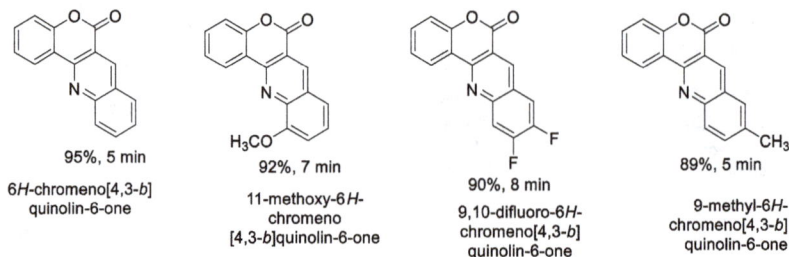

95%, 5 min

6*H*-chromeno[4,3-*b*]
quinolin-6-one

92%, 7 min

11-methoxy-6*H*-
chromeno
[4,3-*b*]quinolin-6-one

90%, 8 min

9,10-difluoro-6*H*-
chromeno[4,3-*b*]
quinolin-6-one

89%, 5 min

9-methyl-6*H*-
chromeno[4,3-*b*]
quinolin-6-one

Equation (11)

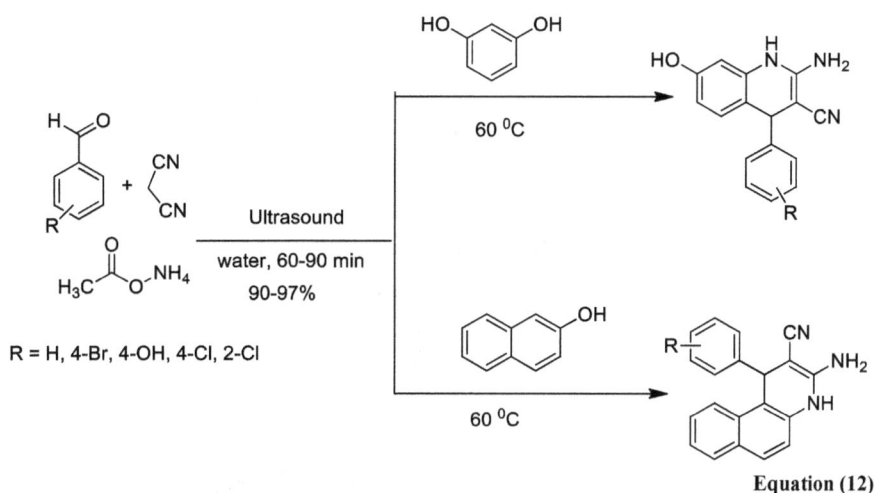

Equation (12)

2.5.3 Synthesis of Dihydroquinolines

Jonnalagadda et al., demonstrated a catalyst free multicomponent methodology for synthesizing valuable dihydroquinolines from the reactions of different aldehydes, ammonium acetate, malononitrile, and 2-naphthol or resorcinol in water as a green solvent under US irradiation as a green technique at 60°C. They obtained excellent yields (90 to 97 percent) with rapid reaction times (60 to 90 min) (Equation 12) (Pagadala et al. 2014).

2.6 Synthesis of Important Organic Compounds based on Xanthene Scaffold

Xanthene molecules contain an oxygen bearing central heterocyclic structure with two more cyclic structures; xanthene and its derivatives include diverse pharmaceutical activities such as Alzheimer, anti malarial, anti inflammatory, anti cancer, anti bacterial, etc. (Zelefack et al. 2009, Hafez et al. 2008, Ghahsare et al. 2019). Interestingly, xanthenedione is a structural unit in several secondary metabolites and exhibits various therapeutic and pharmacological activities (Wang et al. 2002).

2.6.1 Synthesis of 1,8-Dioxo-Octahydroxanthenes

In organic transformations, ionic liquids (ILs) have considerable significance as eco-friendly solvents because of their better thermal stability, negligible vapor pressure, and reusability. Raval et al., developed a methodology for the synthesis of octahydroxanthene derivatives using ionic liquid (1-carboxymethyl-3-methylimidazolium tetrafluoroborate [cmmim][BF$_4$]) as a green solvent from structurally diverse aldehydes and dimedone at room temperature in the absence of catalyst (Equation 13) (Dadhani et al. 2012).

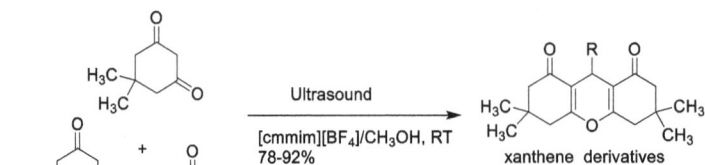

R = C_6H_5, 2-ClC_6H_4, 4-$MeOC_6H_4$, 4-MeC_6H_4, 2-$NO_2C_6H_4$, 2- Furyl, 2-Pyridyl, 2-Thienyl etc.

Representative examples

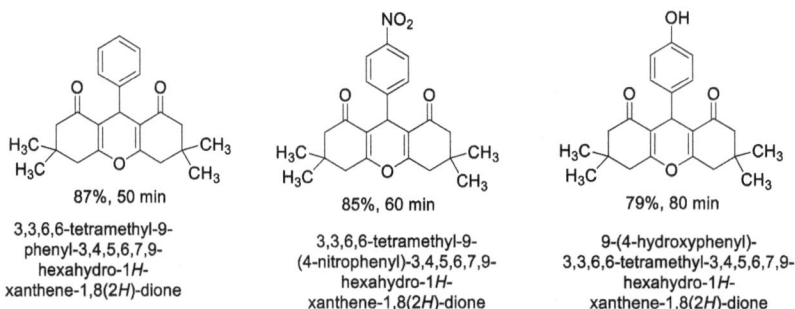

87%, 50 min

3,3,6,6-tetramethyl-9-
phenyl-3,4,5,6,7,9-
hexahydro-1*H*-
xanthene-1,8(2*H*)-dione

85%, 60 min

3,3,6,6-tetramethyl-9-
(4-nitrophenyl)-3,4,5,6,7,9-
hexahydro-1*H*-
xanthene-1,8(2*H*)-dione

79%, 80 min

9-(4-hydroxyphenyl)-
3,3,6,6-tetramethyl-3,4,5,6,7,9-
hexahydro-1*H*-
xanthene-1,8(2*H*)-dione

Equation (13)

2.7 Applications of Sonochemistry in Organic Synthesis using Nanoparticles (NPs)

Nanoparticles have been extensively employed in both the academic and pharmaceutical industries. Several organic transformations have been carried out using nanoparticles (NPs) as efficient catalysts. Nanocatalysts have replaced expensive traditional catalysts due to their better catalytic activity and selectivity, which permit greener, and waste minimized methodologies. Increasing attention is nowadays devoted to the US promoted chemical transformations using NPs (Banerjee 2019). Hence, this section deals with sonochemistry applications in organic synthesis employing several NPs with a greener approach in good to excellent yields.

2.8 Synthesis of Promising Organic Molecules Based on Pyridine Scaffold under the Influence of Ultrasonic Irradiation

Heterocyclic compounds exhibit a broad spectrum of biological as well as pharmaceutical properties. Among them, pyridine derivatives have attracted the attention of the synthetic and biological community because of their potential therapeutic properties, including anti tumor, anti Parkinsonian, anti inflammatory, inhibitors of HIV-1 integrase, anti microbial agents, etc. (Ziarani et al. 2021).

2.8.1 Synthesis of Dihydropyrimido[4,5-b]Quinolinetriones

Moradi et al., explored a green, facile, and environmentally sound protocol for the preparation of dihydropyrimido [4,5-*b*]quinolinetrione derivatives through the one pot four component reaction of dimedone, various aromatic aldehydes, barbituric acid, and different amines employing $CoFe_2O_4@SiO_2/PPA$ (PPA = polyphosphoric acid) as a reusable, efficient solid acid nanocatalyst in ethanol under ultrasonic irradiation (Equation 14) (Moradi and Mahdipour 2019). Some merits of this methodology consist of easy work up, excellent yields, environmentally benign conditions, and shorter reaction times.

2.8.2 Synthesis of Polysubstituted Tetrahydropyridine-3-Carboxylates

Recently, Fathima achieved an elegant, atom-efficient methodology for the preparation of a series of pharmacologically attractive polysubstituted tetrahydropyridine-3-carboxylates using NiO NPs as an efficient catalyst by the one-pot multi-component approach of aromatic aldehyde, Meldrum's acid, arylamine, and ethyl cyanoacetate in ethanol under ultrasonic irradiation for 20 min. The interesting features of this protocol comprise utilizing ecologically sound solvent, reusable catalysts, economic starting materials, good to excellent yield, and mild reaction conditions (Equation 15) (Fathima 2021).

In addition, the synthesis of pyridine derivatives employing several nanocatalysts, almost all of them, at room temperature (RT) was presented.

1. CNTs@meglumine as nanocatalyst; four reuse cycles; ethanol as solvent: at room temperature; 15 to 34 min, and 82 to 95 percent yield (Moradi and Zare 2018).

2. $Fe_3O_4@GA@IG$ as nanocatalyst; six reuse cycles: ethanol as solvent; at room temperature; 35 to 70 min, and 61 to 93 percent yield (Pourian et al. 2018).

3. ZnO NPs as nanocatalyst; ethanol as solvent: at 80°C; 27 to 38 min, and 88 to 93 percent of yield (Abaszadeh et al. 2015).

4. Fe_3O_4/SiO_2-PDA NPs as nanocatalyst: seven reuse cycles: ethanol as solvent; at room temperature; 10 min, and 68 to 91 percent yield (Taheri-Ledari et al. 2019).

5. $CuFe_2O_4$ NPs as nanocatalyst; four reuse cycles: water as solvent; at room temperature; four to 30 min, and 90 to 99 percent yield (Naeimi and Didar 2017).

2.9 Synthesis of Promising Organic Compounds Based on Pyrimidine Scaffold under Ultrasonic Irradiation

Literature reveals that 68 percent of marketed drugs include heterocycles and several bioactive drugs contain the pyrimidine scaffold (Mittersteiner et al. 2021). It has been examined that heterocycle containing pyrimidine scaffold display manifold biological as well as pharmacological activities such as anti tumor, anti malarial, anti bacterial, analgesic, anti hypertensive, etc. (Hilmy et al. 2010).

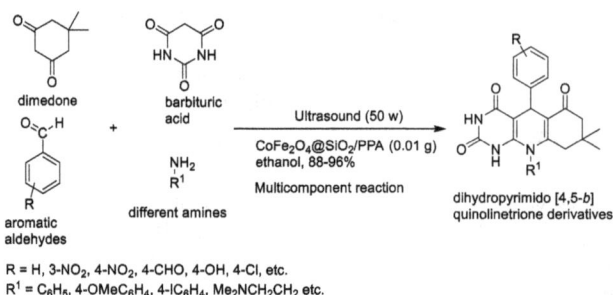

dimedone

barbituric acid

aromatic aldehydes

different amines

Ultrasound (50 w)

CoFe$_2$O$_4$@SiO$_2$/PPA (0.01 g)
ethanol, 88-96%

Multicomponent reaction

dihydropyrimido [4,5-*b*]
quinolinetrione derivatives

R = H, 3-NO$_2$, 4-NO$_2$, 4-CHO, 4-OH, 4-Cl, etc.
R^1 = C$_6$H$_5$, 4-OMeC$_6$H$_4$, 4-IC$_6$H$_4$, Me$_2$NCH$_2$CH$_2$ etc.

Representative examples

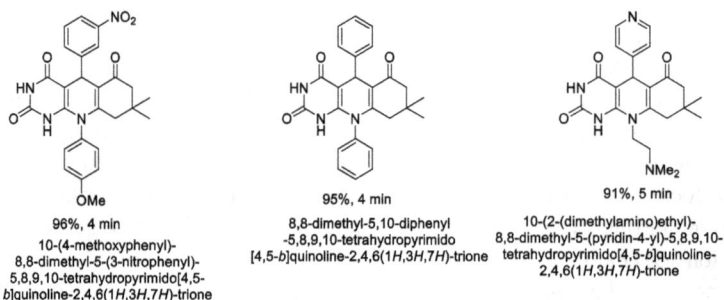

96%, 4 min

10-(4-methoxyphenyl)-
8,8-dimethyl-5-(3-nitrophenyl)-
5,8,9,10-tetrahydropyrimido[4,5-
b]quinoline-2,4,6(1*H*,3*H*,7*H*)-trione

95%, 4 min

8,8-dimethyl-5,10-diphenyl
-5,8,9,10-tetrahydropyrimido
[4,5-*b*]quinoline-2,4,6(1*H*,3*H*,7*H*)-trione

91%, 5 min

10-(2-(dimethylamino)ethyl)-
8,8-dimethyl-5-(pyridin-4-yl)-5,8,9,10-
tetrahydropyrimido[4,5-*b*]quinoline-
2,4,6(1*H*,3*H*,7*H*)-trione

Equation (14)

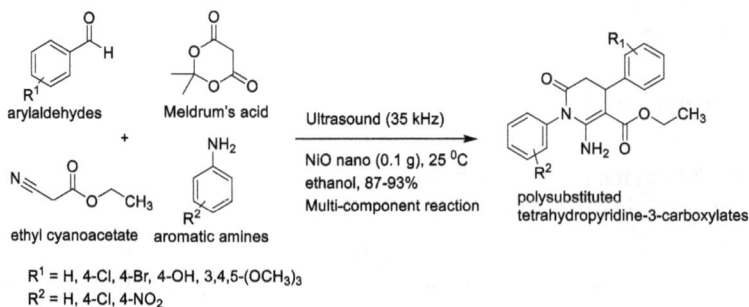

arylaldehydes

Meldrum's acid

ethyl cyanoacetate   aromatic amines

Ultrasound (35 kHz)

NiO nano (0.1 g), 25 ^{0}C
ethanol, 87-93%

Multi-component reaction

polysubstituted
tetrahydropyridine-3-carboxylates

R^1 = H, 4-Cl, 4-Br, 4-OH, 3,4,5-(OCH$_3$)$_3$
R^2 = H, 4-Cl, 4-NO$_2$

Representative examples

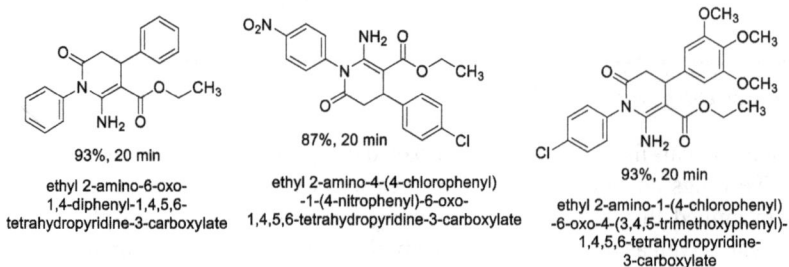

93%, 20 min

ethyl 2-amino-6-oxo-
1,4-diphenyl-1,4,5,6-
tetrahydropyridine-3-carboxylate

87%, 20 min

ethyl 2-amino-4-(4-chlorophenyl)
-1-(4-nitrophenyl)-6-oxo-
1,4,5,6-tetrahydropyridine-3-carboxylate

93%, 20 min

ethyl 2-amino-1-(4-chlorophenyl)
-6-oxo-4-(3,4,5-trimethoxyphenyl)-
1,4,5,6-tetrahydropyridine-
3-carboxylate

Equation (15)

2.9.1 Synthesis of 3,4-Dihydropyrimidin-2(1H)-Ones/thiones

Safaei-Ghomi et al., for the first time in 2018, disclosed an effective and improved synthetic strategy for the preparation of 3,4-dihydropyrimidin-2(1H)-ones/thiones employing Dendrimer-PWA[n] (dendrimer-encapsulated phosphotungstic acid NPs immobilised on nanosilica, Dendrimer@$H_3PW_{12}O_{40}$@SiO_2 NPs) as an efficient and economically viable nanocatalyst (0.004 g) under ultrasonic irradiation through a multi component Biginelli reaction between aromatic aldehydes, β-dicarbonyl compounds and urea or thiourea in ethanol at 50°C (Equation 16) (Safaei-Ghomi et al. 2018). Compared with traditional Biginelli reaction conditions, the merits of this new synthetic methodology consist of short reaction time, exceptionally good yields, recyclable green catalyst, simple experimental work up, and environmentally benign procedure.

The most important synthesis of ethyl 6-methyl-2-oxo-4-phenyl-1,2,3,4-tetrahydropyrimidine-5-carboxylate occurred through one pot Biginelli condensation between benzaldehyde, ethyl acetoacetate, and urea in the presence of different catalysts under different conditions, including the Dendrimer-PWA[n] nanocatalyst under sonication, (Safaei-Ghomi et al. 2018). To compare the advantages of Dendrimer-PWA[n] nanocatalyst with other catalysts, the Dendrimer-PWA[n] nanocatalyst was the best catalyst as most methods as several other catalysts in various solvents such as dioxane, acetonitrile, and acetic acid, etc., required a larger amount of catalyst with longer reaction times. The highest turnover number (TON; 1,011.53), as well as turnover frequencies (TOF; 5,950.17 h^{-1}), were obtained during the preparation of ethyl 6-methyl-2-oxo-4-phenyl-1,2,3,4-tetrahydropyrimidine-5-carboxylate as the model reaction using the Dendrimer-PWA[n] as the green catalyst compared to other catalysts, signifying the better catalytic efficiency of Dendrimer-PWA[n] nanocatalyst, this synthesis, with different catalyst is presented below:

1. 12-Tungstophosphoric acid as catalyst; under reflux conditions with acetic acid; for six hr; with reported values of turnover frequency (TOF) and turnover number (TON) of 91.75 h^{-1} and 550.50, respectively, and 75 percent yield (Heravi et al. 2005).

2. 40 percent of PW/SiO_2 as catalyst; at 80°C and acetonitrile; for two hr; with reported values TOF and TON of 160.01 h^{-1} and 320.02, respectively, and 95 percent yield (Rafiee and Shahbazi 2006).

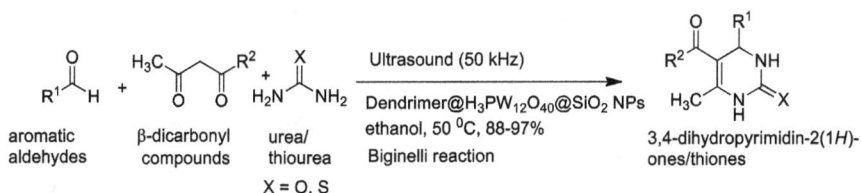

| aromatic aldehydes | β-dicarbonyl compounds | urea/ thiourea | Ultrasound (50 kHz) Dendrimer@$H_3PW_{12}O_{40}$@SiO_2 NPs ethanol, 50 °C, 88-97% Biginelli reaction | 3,4-dihydropyrimidin-2(1H)-ones/thiones |

X = O, S

R^1 = C_6H_5, 4-$CH_3OC_6H_4$, 2-$CH_3OC_6H_4$, 4-$CH_3C_6H_4$, 4-$NO_2C_6H_4$, 2-$NO_2C_6H_4$, 3-$NO_2C_6H_4$, 4-ClC_6H_4, 2-ClC_6H_4
PhCH=CH, 2-thiophene, 2-furyl
R^2 = CH_3, OC_2H_5, OCH_3

Equation (16)

3. PS-PEG-SO$_3$H as catalyst; at 80°C, dioxane and *iso*-propanol: for 10 hr; with reported values of TOF and TON of 40.04 h^{-1} and 400.45, respectively, and 80 percent yield (Quan et al. 2009).

4. Trichloroisocyanuric acid (TCCA) as catalyst; under reflux conditions with EtOH; for 12 hr; with reported values of TOF and TON of 23.36 h-1 and 280.35, respectively, and 94 percent yield (Bigdeli et al. 2007).

5. Bioglycerol based carbon as catalyst; at 75 to 80 0 C in CH3CN; for six hr; with reported values of TOF and TON of 95.04 h-1 and 570.29, respectively, and 91 percent yield (Konkala et al. 2012).

6. p-sulfonic acid calixarenes as catalyst; under reflux conditions with EtOH; for eight hr; with reported values of TOF and TON of 82.53 h-1 and 660.24, respectively, and 81 percent yield (da Silva et al. 2011).

7. (TiCl4-MgCl2/MgCl2·4CH3OH) as catalyst; at 100°C in solvent free; for three hr; with reported values of TOF and TON of 150.13 h-1 and 450.39, respectively, and 91 percent yield (Kumar and Maurya 2007).

8. Silica sulfuric acid as catalyst; under reflux conditions with EtOH; for six hr; with reported values of TOF and TON of 113.48 h-1 and 680.90, respectively, and 91 percent yield (Salehi et al. 2003).

9. Dendrimer-PWAn as catalyst; at 80°C and solvent free; for 40 min; with reported values of TOF and TON of 1,501.02 h-1 and 990.68, respectively, and 95 percent yield (Safaei-Ghomi et al. 2018).

10. Dendrimer-PWA[n] as catalyst: in conditions of US in ethanol at room temperature; for 10 min; with reported values of TOF and TON of 5,950.17 h^{-1} and 1,011.53, respectively, and 97 percent yield (Safaei-Ghomi et al. 2018).

Aromatic aldehydes carrying electron donating and electron withdrawing groups in the Biginelli reaction produced corresponding Biginelli compounds under ultrasonic irradiation (with ethanol at 50°C) and for traditional synthesis (in solvent free conditions at 80°C) in comparable yields but with shorter reaction time compared to traditional heating in solvent free conditions, as shown below: (Safaei-Ghomi et al. 2018).

1. The product obtained was ethyl 6-methyl-2-oxo-4-phenyl-1,2,3,4-tetrahydropyrimidine-5-carboxylate; in 10 min with 97 percent yield (for US irradiation), and in 40 min with 95 percent yield (for traditional synthesis).

2. The product obtained was ethyl 6-methyl-2-oxo-4-(*p*-tolyl)-1,2,3,4-tetrahydropyrimidine-5-carboxylate; in 10 min with 95 percent yield (for US irradiation), and in 50 min with 93 percent yield (for traditional synthesis).

3. The product obtained was ethyl 6-methyl -4-(*p*-nitrophenyl)-2-oxo-1,2,3,4-tetrahydropyrimidine-5-carboxylate; in five min with 96 percent yield (for US irradiation), and in 30 min with 95 percent yield (for traditional synthesis).

4. The product obtained was ethyl 4-(*p*-methoxiphenyl)-6-methyl-2-oxo-)-1,2,3,4-tetrahydropyrimidine-5-carboxylate; in 15 min with 95 percent yield (for US irradiation), and in 50 min with 94 percent yield (for traditional synthesis).

5. The product obtained was ethyl 6-methyl-2-oxo-4-(2*H*-1λ^3)-thiophen-2-yl)-1,2,3,4-tetrahydropyrimidine-5-carboxylate; in seven min with 95 percent yield (for US irradiation), and in 35 min with 94 percent yield (for traditional synthesis).

6. The product obtained was ethyl 4-(2*H*-1λ^3-furan-2-yl)-6-methyl-2-oxo-1,2,3,4-tetrahydropyrimidine-5-carboxylate; in 10 min with 91 percent yield (for US irradiation), and in 35 min with 92 percent yield (for traditional synthesis).

7. The product obtained was ethyl 4-(4-methoxyphenyl)-6-methyl-2-thioxo-1,2,3,4-tetrahydropyrimidine-5-carboxylate; in 15 min with 88 percent yield (for US irradiation), and in 50 min with 94 percent yield (for traditional synthesis).

8. The product obtained was 5-acetyl-6-methyl-4-phenyl-3,4-dihydropyrimidin-2(1*H*)-one; in five min with 92 percent yield (for US irradiation), and in 30 min with 90 percent yield (for traditional synthesis).

The positive incorporation of sonication could enhance the number of active cavitation bubbles and the size of individual bubbles to increase the rate of transformation (Safaei-Ghomi et al. 2018).

Figure 4 deals with the plausible mechanism for constructing 3,4-dihydropyrimidin-2(1*H*)-ones/thiones using the Dendrimer-PWA[n] nanocatalyst under US irradiation (Safaei-Ghomi et al. 2018). Herein, the Dendrimer-PWA[n] nanocatalyst is crucial in donating protons for activating the aldehyde as a Brønsted acidic catalyst. The mechanism is believed to start with rate determining nucleophilic addition of urea or thiourea to the aromatic aldehyde (Kappe et al. 1997) to form the iminium intermediate, followed by dehydration. Then, the electrophile iminium attacks the enol of the 1,3-dicarbonyl compound to yield the corresponding dihydropyrimidinones, followed by cyclisation and dehydration.

In another study, nano-$CoFe_2O_4$@SiO_2/PrNH$_2$ was prepared as an effective catalyst and was employed to synthesize 2,4-diamino-6-arylpyrimidine-5-carbonitrile derivatives (Barkhordarion-Mohammadi and Safaei-Ghomi 2018). The US, microwave, and conventional heating methods were applied to prepare pyrimidine derivatives through the one pot multi component reaction of malononitrile, various aromatic aldehydes, and guanidine hydrochloride in ethanol-water. US irradiation (40 W) as a green tool in ethanol-water provided the best results using amino functionalized $CoFe_2O_4$@SiO_2 NPs (eight mg); the needed catalyst in the US method was smaller than microwave irradiation since the catalyst acts as an internal heat source under microwave irradiation (Barkhordarion-Mohammadi and Safaei-Ghomi 2018).

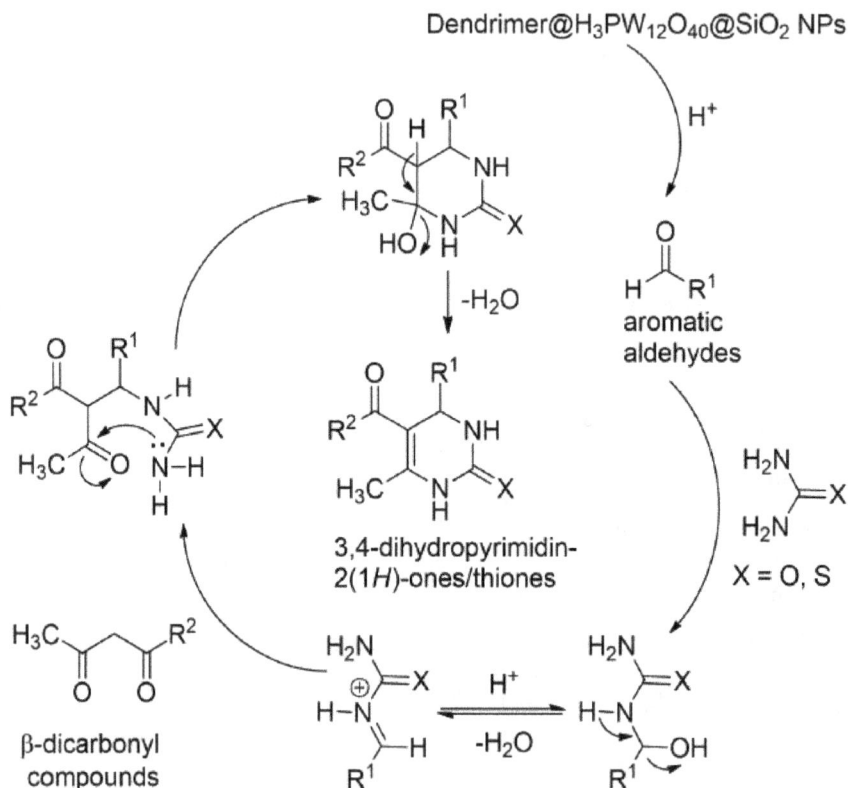

Figure 4. The plausible mechanism for the preparation of dihydropyrimidine derivatives.

2.10 Synthesis of Promising Organic Compounds Based on Quinolone Scaffold under Ultrasonic Irradiation

Quinolone and its derivatives carry great significance in pharmaceutical and medicinal chemistry. They have attracted further attention towards drug designing and biological evaluation in exploring novel drug molecules. Efforts have been made to invent quinolone based analogs together with quinolone hybrids because of their broad range of biological properties (Sharma et al. 2021).

2.10.1 Synthesis of Quinolone Hybrids

Singh et al., developed an eco compatible aqueous mediated sonochemical preparation of quinolone hybrids by the multi component reaction of Meldrum's acid, 3,4-methylenedioxy aniline, and several aromatic aldehydes using TiO_2 NPs as an efficient catalyst (Equation 17) (Bhardwaj et al. 2019). At first, TiO_2 NPs were prepared under sonication through a direct interaction of titanium (IV) isopropoxide and *Origanum majorana* (sweet marjoram) leaves extract to make the whole protocol more environment-friendly and convenient. Herein, *Origanum majorana*

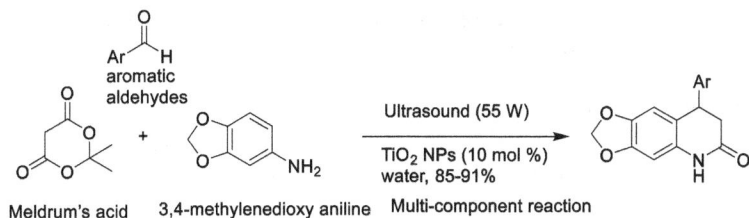

Meldrum's acid   3,4-methylenedioxy aniline   Multi-component reaction

Representative examples

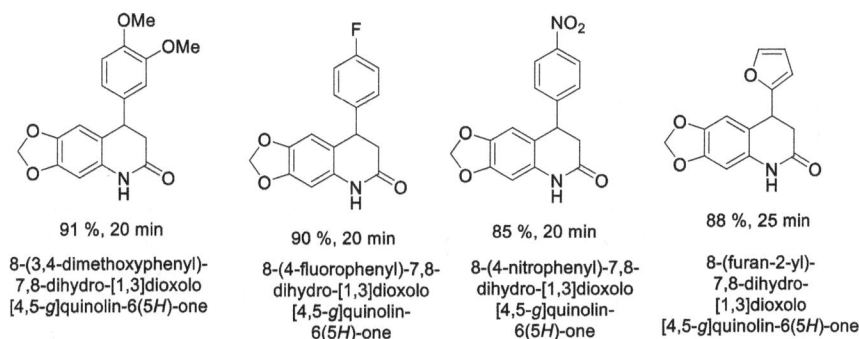

91 %, 20 min

8-(3,4-dimethoxyphenyl)-
7,8-dihydro-[1,3]dioxolo
[4,5-*g*]quinolin-6(5*H*)-one

90 %, 20 min

8-(4-fluorophenyl)-7,8-
dihydro-[1,3]dioxolo
[4,5-*g*]quinolin-
6(5*H*)-one

85 %, 20 min

8-(4-nitrophenyl)-7,8-
dihydro-[1,3]dioxolo
[4,5-*g*]quinolin-
6(5*H*)-one

88 %, 25 min

8-(furan-2-yl)-
7,8-dihydro-
[1,3]dioxolo
[4,5-*g*]quinolin-6(5*H*)-one

Equation (17)

leaves extract acts as a reducing and capping agent. A series of 8-aryl-7,8-dihydro-[1,3]-dioxolo[4,5-*g*]quinolin-6(5*H*)-ones were synthesized by simple procedure with high atom economy, high yields, mild reaction conditions in the presence of TiO$_2$ NPs under ultrasonic irradiation without any environmental issue. The synthesized TiO$_2$ NPs were recovered easily, and the catalytic activity of the nanocatalyst was unaltered after its application.

To verify the influence of US irradiation and its impact on power per amplitude, the synthesis of 8-(4-fluoro-phenyl)-7,8-dihydro-5*H*-[1,3]dioxolo[4,5-*g*]quinolin-6-one was carried out in the presence of optimized catalyst and solvent as a model reaction. It has been observed that the increasing US power from silent to 50 W increased the yield of the targeted product with a shorter reaction time, as presented below: (Bhardwaj et al. 2019).

1. In 160 min; at when the power or amplitude was silent, and 60 percent of yield.
2. In 40 min; at 30 W of power, and 60 percent of yield.
3. In 30 min; at 3 W of power, and 72 percent of yield.
4. In 20 min; at 30 W of power and 80 percent of yield.
5. In 15 min; at 45 W of power and 85 percent of yield.
6. In 15 min; at 50 W of power, and 90 percent of yield.
7. In 15 min; at 55 W of power, and 90 percent of yield.

2.11 Synthesis of Promising Organic Compounds Based on Triazole Scaffold under Ultrasonic Irradiation

1,2,3-Triazole and its derivatives have been found to comprise potential pharmacophores, which gained significant attention over the past few years. Triazole derivatives, including one or more triazole rings, display a broad range of biological activities such as anti microbial, anti fungal, anti tumor, anti HIV, anti allergic, etc. (Meldal and Tornøe 2008, Peyton et al. 2015).

2.11.1 Synthesis of Triazolo[1,2-a] Indazole-Triones

Verma et al., executed a US assisted synthesis of triazolo[1,2-*a*] indazole-triones (TAITs) involving one pot multi component condensation of various aromatic aldehydes, dimedone, 4-phenylurazole in the presence of SiO_2-coated ZnO NPs (ZnO@SiO_2 NPs) as an efficient heterogeneous, eco-friendly, and reusable catalyst in a green reaction medium water at nearly 60°C (Equation 18) (Verma et al. 2017). The advantages of this three-component reaction include good to excellent yields in short reaction duration, simple experimental work up, excellent catalytic activity, and environmentally sound methodology. ZnO@SiO_2 NPs displayed remarkable catalytic activities because of the oxygen deficiency induced negative charge; the nanocatalyst comprises a huge specific surface area of 97.5 m²/g, the smallest hydrodynamic diameter of 752 nm, and smart recyclability up to six runs without any significant loss of catalytic performance.

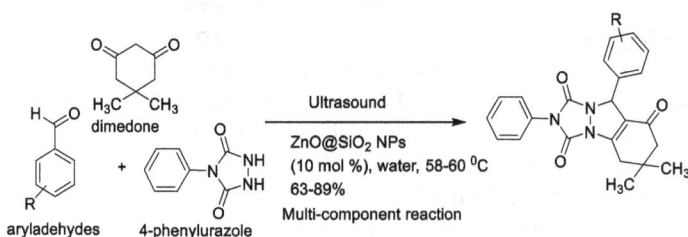

R = H, 3-NO_2, 4-NO_2, 4-CF_3, 4-Br, 4-Cl, 4-F, 4-CH_3, 4-CN, 4-OH, 2-Cl etc.

Representative examples

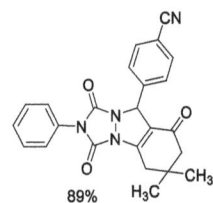

89%

4-(6,6-dimethyl-1,3,8-trioxo-2-phenyl-2,3,5,7,8,9-hexahydro-1*H*,6*H*-[1,2,4]triazolo[1,2-a]indazol-9-yl)benzonitrile

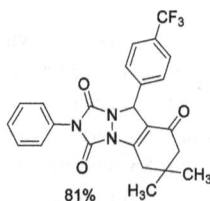

81%

6,6-dimethyl-2-phenyl-9-(4-(trifluoromethyl)phenyl)-5,6,7,9-tetrahydro-1*H*,8*H*-[1,2,4]triazolo[1,2-a]indazole-1,3,8(2*H*)-trione

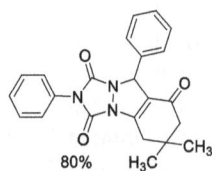

80%

6,6-dimethyl-2,9-diphenyl-5,6,7,9-tetrahydro-1*H*,8*H*-[1,2,4]triazolo[1,2-a]indazole-1,3,8(2*H*)-trione

Equation (18)

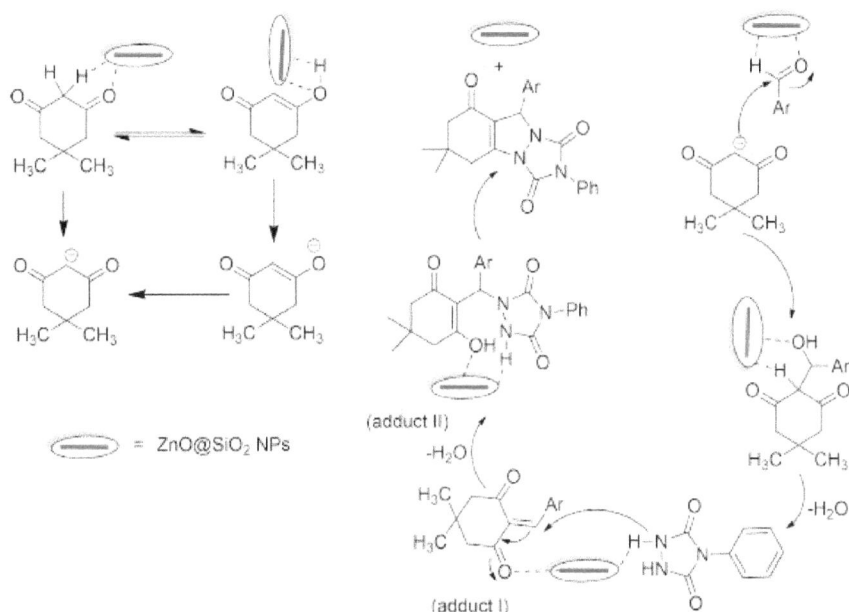

Figure 5. The plausible mechanism for the synthesis of triazolo[1,2-*a*]indazole-triones.

A reasonable mechanism for preparing triazolo[1,2-*a*]indazole-triones in the presence of ZnO@SiO$_2$ NPs is presented in Figure 5. ZnO NPs function as Lewis acids (Schweitzer et al. 2014) because of the electropositive nature of Zn. Hence, the aldehyde and dimedone carbonyl groups may link with the nanocatalyst via coordination, catalyzing the transformation. ZnO@SiO$_2$ NPs also function as bases due to their zeta potential (ζ) value (over -30 mV), which corroborates well with oxygen deficiency. Basic ZnO@SiO$_2$ NPs abstract protons from dimedone to form carbanion. In the first step, aldehyde undergoes Knoevenagel condensation with dimedone to provide Michael acceptor heterodiene (adduct I). A Michael type reaction occurs between adduct I and 4-phenylurazole to form an open chain intermediate II. Finally, the desired triazolo[1,2-*a*]indazole-triones are obtained by intramolecular cyclisation followed by dehydration. Water removal occurs easily due to the bipolar nature of ZnO@SiO$_2$ NPs (Verma et al. 2017).

Panahi et al., accomplished the synthesis of 1,4-disubstituted-1,2,3-triazoles as well as new 1,2,3-triazoles including thiazolidin-2-thione groups in the presence of Cu/TiO$_2$ nanocatalyst in water by the multi component reaction of phenylacetylene, sodium azide, and 5-iodomethylthiazolidine-2-thiones under ultrasonic irradiation and conventional heating conditions (Alyari et al. 2017). All the resulting molecules were evaluated for anti bacterial activity. Some of them exhibited varying anti bacterial properties against *Escherichia coli* and *Staphylococcus aureus*. US irradiation provided the best results, dramatically reducing the temperature and reaction time.

2.12 Synthesis of Promising Organic Compounds based on Pyrrolidone Scaffold under Ultrasonic Irradiation

Pyrrolidone or pyrrolidinone derivatives have attracted the attention of scientists as various pharmaceutical drugs comprise pyrrolidone derivatives such as doxapram, cotinine, povidone, etc. (Lim et al. 2015).

2.12.1 Synthesis of 2-Pyrrolidinon-3-Olates

Ghasemzadeh et al., reported an efficient route to obtain 2-pyrrolidinon-3-olate derivatives through a one pot four component reaction of various aryl aldehydes, dimethyl acetylenedicarboxylate, 2-aminobenzothiazole, and morpholine or piperidine using reusable Co_3O_4@SiO_2 NPs in water: ethanol under ultrasonic irradiation (Equation 19) (Ghasemzadeh and Abdollahi-Basir 2006). Sonication was effective in preparing the core shell nanocomposite as well (Lin et al. 2015).

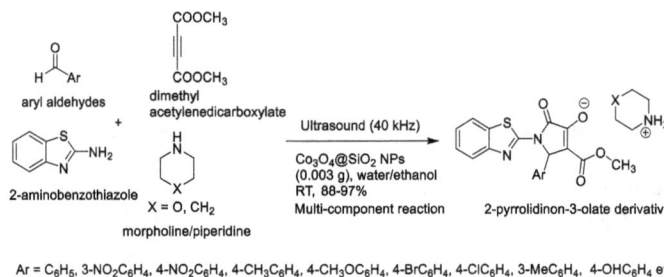

Ar = C_6H_5, 3-$NO_2C_6H_4$, 4-$NO_2C_6H_4$, 4-$CH_3C_6H_4$, 4-$CH_3OC_6H_4$, 4-BrC_6H_4, 4-ClC_6H_4, 3-MeC_6H_4, 4-OHC_6H_4 e

Representative examples

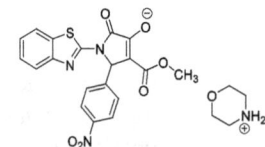

97%, sonication (23 min)
81%, conventional method (48 h)

1-(benzo[*d*]thiazol-2-yl)-
4-(methoxycarbonyl)-5-
(4-nitrophenyl)-2-oxo-
2,5-dihydro-1*H*-pyrrol-3-olate

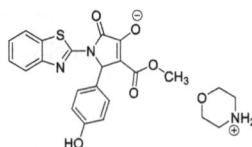

88%, sonication (40 min)
63%, conventional method (48 h)

1-(benzo[*d*]thiazol-2-yl)-
5-(4-hydroxyphenyl)-4-
(methoxycarbonyl)-2-oxo-
2,5-dihydro-1*H*-pyrrol-3-olate

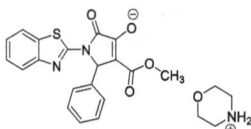

96%, sonication (30 min)
82%, conventional method (48 h)

1-(benzo[*d*]thiazol-2-yl)-
4-(methoxycarbonyl)-
2-oxo-5-phenyl-
2,5-dihydro-1*H*-pyrrol-3-olate

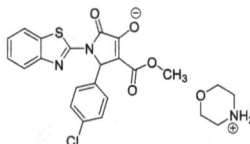

97%, sonication (23 min)
83%, conventional method (48 h)

1-(benzo[*d*]thiazol-2-yl)-
5-(4-chlorophenyl)-4-
(methoxycarbonyl)-2-oxo-
2,5-dihydro-1*H*-pyrrol-3-olate

Equation (19)

The synergistic effect of US irradiation offered an ecologically sound method for synthesizing the desired product with incredibly good yields and short reaction times.

Sonication was found to have an amazing effect on preparing the targeted products 2-pyrrolidinon-3-olates. The conventional method reduced reaction time to less than 40 min compared to 48 h, and the product yields were also enhanced by 10 to 25 percent under sonication (Ghasemzadeh and Abdollahi-Basir 2006).

2.13 Synthesis of Promising Organic Compounds based on Chromene Scaffold under Ultrasonic Irradiation

Chromene and its derivatives are the most efficient molecules in several biology, chemistry, and pharmacology areas. Among chromenes, 2-amino-4H-chromenes are one of the very potent family members due to their functionality in medicinal scaffolds, especially high spasmolytic, diuretic, anticoagulant, and anti anaphylactic activities (Safari et al. 2012).

2.13.1 Synthesis of 2-Amino-7-Hydroxy-4H-Chromenes

Rouhani et al., synthesized chromene derivatives in the presence of $MgFe_2O_4$ NPs through the one pot multi component reaction of various aldehydes, resorcinol and malononitrile in ethanol at 65°C under US irradiation as a promising tool (Equation 20) (Eshtehardian et al. 2020). The beneficial catalyst $MgFe_2O_4$ was prepared to follow a green protocol by tragacanth gum. The synthesized nanocatalyst was recovered simply with the help of an external magnet and smart recyclability up to four runs without any remarkable loss of catalytic activity.

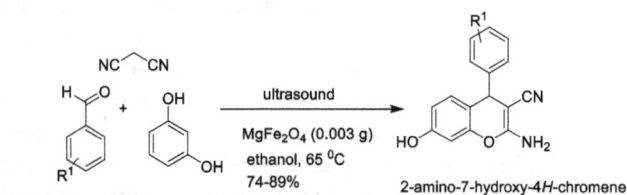

R^1 = H, 2-OCH$_3$, 4-OCH$_3$, 2-F, 3-Cl, 2-CH$_3$, 4-CH$_3$

Representative examples

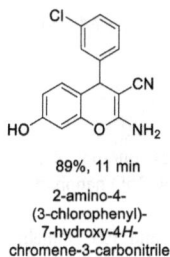

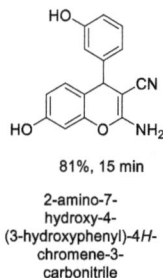

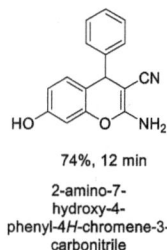

89%, 11 min

2-amino-4-
(3-chlorophenyl)-
7-hydroxy-4H-
chromene-3-carbonitrile

85%, 11 min

2-amino-4-
(2-fluorophenyl)-7-
hydroxy-4H-
chromene-3-
carbonitrile

81%, 15 min

2-amino-7-
hydroxy-4-
(3-hydroxyphenyl)-4H-
chromene-3-
carbonitrile

74%, 12 min

2-amino-7-
hydroxy-4-
phenyl-4H-chromene-3-
carbonitrile

Equation (20)

Also, the synthesis of chromene derivatives was executed under ultrasonic irradiation with a greener approach by several NPs in good yields, some examples were presented below:

1. nano-Fe_3O_4/PEG was used as nanocatalyst at room temperature and with C_2H_5OH for 15 to 20 min, and 80 to 95 percent yield was obtained (Safaei-Ghomi et al. 2016).

2. Fe_3O_4@SiO_2-imid-PMA[n] nanocatalyst was used as nanocatalyst; at room temperature, and in water for five to 12 min, and 86 to 96 percent yield was obtained (Esmaeilpour et al. 2015).

3. Fe_3O_4@L-arginine was used as nanocatalyst; at room temperature in solvent free for 60 min, and 70 to 98 percent yield was obtained (Azizi et al. 2014).

4. MNPs-NH_2 NPs was used as nanocatalyst; at 30°C in C_2H_5OH:H_2O (5:2) for 15 to 40 min, and 80 to 96 percent yield was obtained (Safari and Zarnegar 2014).

5. The Fe_3O_4-chitosan NPs was used as nanocatalyst, in water at 50°C for 15 to 27 min; and a yield of 94 to 99 percent was obtained (Safari and Javadian 2015).

3. Conclusion

US mediated chemical transformations have received more attention since they significantly enhance reaction rates, improve yields, utilize mild reaction conditions, and have low energy consumption. Catalyst free organic synthesis agrees with the concept and philosophy of green chemistry as it is a simple, low cost, convenient protocol for easy separation and purification. Many chemical reactions have been conducted smoothly using nanoparticles (NPs) under US irradiation to deliver better yields, shorter reaction times, and enhanced selectivities. Hence, this chapter dealt with catalyst free organic reactions and nanocatalysts in organic synthesis under US irradiation as a source of clean energy. The present chapter aims to provide a good contribution to the synthetic organic community by preparing demanding scaffolds for medicine and biology using NPs under catalyst free conditions through US irradiation. The spectrum of uses of synthetic methodologies may promote the development of more eco friendly organic synthesis so that the next generation can live on this planet with a minimum energy requirement for chemical reactions with the least pollution.

Acknowledgements

S Majhi is thankful to his father, Tarani Majhi and mother, Sadeswari Majhi.

References

Abaszadeh, M., M. Seifi and A. Asadipour. 2015. Ultrasound promotes one-pot synthesis of 1,4-dihydropyridine and imidazo[1,2-a]quinolone derivatives, catalyzed by ZnO nanoparticles. Res Chem Intermed. 41: 5229–5238.

Alyari, M., M.G. Mehrabani, M. Allahvirdinesbat, K.D. Safa, H.S. Kafil and P.N. Panahi. 2017. Ultrasound assisted synthesis of thiazolidine thiones containing1,2,3-triazoles using Cu/TiO_2. arkivoc. IV: 145–157.

Anastas, P.T. and J.C. Warner. 1998. Green chemistry: theory and practice. Oxford University Press, New York.

Anastas, P.T., L.B. Bartlett, M.M. Kirchhoff and T.C. Williamson. 2000. The role of catalysis in the design, development, and implementation of green chemistry. Catal. Today. 22: 11–22.

Armaly, A.M., Y.C. DePorre, E.J. Groso, P.S. Riehl and C.S. Schindler. 2015. Discovery of novel synthetic methodologies and reagents during natural product synthesis in the post-palytoxin era. Chem. Rev. 115: 9232–9276.

Azizi, K., M. Karimi, H.R. Shaterianb and A. Heydari. 2014. Ultrasound irradiation for the green synthesis of chromenes using L-arginine-functionalized magnetic nanoparticles as a recyclable organocatalyst. RSC Adv. 4: 42220–42225.

Bailly, C.S. Echepare, F. Gago and M. Waring. 1999. Recognition elements that determine affinity and sequence-specific binding to DNA of 2QN, a biosynthetic bis-quinoline analogue of echinomycin. Anti-Cancer Drug Des. 15: 291–303.

Banerjee, B. 2019. Ultrasound and nano-catalysts: an ideal and sustainable combination to carry out diverse organic transformations. Chemistry Select. 4: 2484 –2500.

Barghi-Lish, A., S. Farzaneh and M. Mamaghani. 2016. One-pot three-component catalyst-free synthesis of novel derivatives of pyrido-[2,3-d]pyrimidines under ultrasonic irradiations. Synth. Commun. 46: 1209–1214.

Barkhordarion-Mohammadi, S. and J. Safaei-Ghomi. 2018. Synthesis of 2,4-diamino-6-aryl-5-pyrimidinecarbonitrile promoted by amino-functionalised $CoFe_2O_4@SiO_2$ nanoparticles under conventional heating, microwave and ultrasound irradiations. Z. Naturforsch. 73: 17.

Baruah, B. and M.L. Deb. 2021. Catalyst-free and additive-free reactions enabling C–C bond formation: a journey towards a sustainable future. Org. Biomol. Chem. 19: 1191–1229.

Bhardwaj, D., A. Singh and R. Singh. 2019. Eco-compatible sonochemical synthesis of 8-aryl-7,8-dihydro-[1,3]-dioxolo[4,5-g]quinolin-6(5H)-ones using green TiO_2. Heliyon. 5: e01256.

Bigdeli, M.A., S. Jafari, G.H. Mahdavinia and H. Hazarkhani. 2007. Trichloroisocyanuric acid, a new and efficient catalyst for the synthesis of dihydropyrimidinones. Catal. Commun. 8: 1641–1644.

Cravotto, G. and P. Cintas. 2006. Power ultrasound in organic synthesis: moving cavitational chemistry from academia to innovative and large-scale applications. Chem. Soc. Rev. 35: 180–196.

Czarnecka, K., M. Girek, P. Krecisz, R. Skibinski, K. Latka, J. Jonczyk et al. 2019. Discovery of new cyclopentaquinoline analogues as multifunctional agents for the treatment of alzheimer's disease. Int. J. Mol. Sci. 20: 498.

da Silva, D.L., S.A. Fernandes, A.A. Sabino and A. de Fatima. 2011. p-Sulfonic acid calixarenes as efficient and reusable organocatalysts for the synthesis of 3,4-dihydropyrimidin-2(1H)-ones/-thiones. Tetrahedron Lett. 52: 6328–6330.

Dadhania, A.N., V.K. Patel and D.K. Raval. 2012. Catalyst-free sonochemical synthesis of 1,8-dioxo-octahydroxanthene derivatives in carboxy functionalised ionic liquid. C. R. Chimie. 15: 378–383.

Devi, L., A.R. Robert, H. Ganja, S. Maddila and S.B. Jonnalagadda. 2020. A rapid, sustainable and environmental friendly protocol for the catalyst-free synthesis of 2-methyl-5-oxo-hexahydroquinoline-3-carboxylate via ultrasonic irradiation. Chem. Data Collect. 28: 100432.

Eshtehardian, B., M. Rouhani and Z. Mirjafary. 2020. Green protocol for synthesis of $MgFe_2O_4$ nanoparticles and study of their activity as an efficient catalyst for the synthesis of chromene and pyran derivatives under ultrasound irradiation. J. Iran. Chem. Soc. 17: 469–481.

Esmaeilpour, M., J. Javidi, F. Dehghania and F.N. Dodeji. 2015. A green one-pot three-component synthesis of tetrahydrobenzo [b]pyran and 3,4-dihydropyrano[c]chromene derivatives using $Fe_3O_4@SiO_2$-imid-PMA[n] magnetic nanocatalyst under ultrasonic irradiation or reflux conditions, RSC Adv. 5: 26625–26633.

Faisal, M., A. Saeed, S. Hussain, P. Dar and F.A. Larik. 2019. Recent developments in synthetic chemistry and biological activities of pyrazole derivatives. J. Chem. Sci. 131: 70.

Fathima, N. 2021. Nano NiO-an efficient and a reusable catalyst for the one-pot synthesis of novel tetrahydropyridine-3-carboxylates under sonication. Mater. Today: Proc. 45: 4063–4066.

Ghahsare, A.G., Z.S. Nazifi and S.M.R. Nazifi. 2019. Structure-bioactivity relationship study of xanthene derivatives: a brief review. Curr. Org. Synth. 16: 1071–1077.

Ghasemzadeh, M.A. and M.H. Abdollahi-Basir. 2016. Ultrasound-assisted one-pot multi-component synthesis of 2-pyrrolidinon-3-olates catalysed by $Co_3O_4@SiO_2$ core–shell nanocomposite. Green Chem. Lett. Rev. 9: 156–165.

Guo, W.X., H.L. Jin, J.X. Chen, F. Chen, J.C. Dinga and H.Y. Wu. 2009. An efficient catalyst-free protocol for the synthesis of quinoxaline derivatives under ultrasound irradiation. J. Braz. Chem. Soc. 20: 1674–1679.

Gupta, P.K., M.R. Nassiri, L.A. Coleman, L.L. Worting, J.C. Drach and L.B. Townsend. 1989. Synthesis, cytotoxicity, and anti-viral activity of certain 7-[(2 hydroxyethoxy)methyl]pyrrolo[2,3-d]pyrimidine nucleosides related to toyocamycin and sangivamycin. J. Med. Chem. 32: 1420–1425.

Hafez, H.N., M.I. Hegab, I.S. Ahmed-Farag and A.B.A. El-Gazzar. 2008. A facile regioselective synthesis of novel spiro-thioxanthene and spiro-xanthene-9 0,2-[1,3,4] thiadiazole derivatives as potential analgesic and anti-inflammatory agents. Bioorg. Med. Chem. Lett. 18: 4538–4543.

Hasaninejed, A., M.R. Kazerooni, A. Zare. 2013. Room-temperature, catalyst-free, one-pot pseudo-five-component synthesis of 4,4-(arylmethylene)bis(3-methyl-1-phenyl-1H-pyrazol-5-ols) under ultrasonic irradiation. ACS Sustainable Chem. Eng. 1: 679–684.

Heravi, M.M., F. Derikvand and F.F. Bamoharram. 2005. A catalytic method for synthesis of biginelli-type 3,4-dihydropyrimidin-2(1H)-one using 12-tungstophosphoric acid. J. Mol. Catal. A: Chem. 242: 173–175.

Hilmy, K.M.H., M.M.A. Khalifa, M.A.A. Hawata, R.M.A. Keshk and A.A. El-Torgman. 2010. Synthesis of new pyrrolo[2,3-d]pyrimidine derivatives as antibacterial and antifungal agents Eur. J. Med. Chem. 45(11): 5243–5250.

Hong, M., C. Cai and W.B. Yi. 2010. Hafnium (IV) bis(perfluorooctanesulfonyl)imide complex catalysed synthesis of polyhydroquinoline derivatives via unsymmetrical Hantzsch reaction in fluorous medium. J. Fluorine Chem. 131: 111–114.

Ibrahim, D.A. and A.M. El-Metwally. 2010. Design, synthesis, and biological evaluation of novel pyrimidine derivatives as CDK2 inhibitors. Eur. J. Med. Chem. 45: 1158–1166.

Kappe, C.O. 1997. A reexamination of the mechanism of the biginelli dihydropyrimidine synthesis. Support for an N-Acyliminium Ion Intermediate (1). J. Org. Chem. 62: 7201–7204.

Kitanoso, T., K. Masuda, P. Xu and S. Kobayashi. 2018. Catalytic organic reactions in water toward sustainable society. Chem. Rev. 118: 679–746.

Kim, Y.B., Y.H. Kim, J.Y. Park and S.K. Kim. 2004. Synthesis and biological activity of new quinoxaline antibiotics of echinomycin analogues. Bioorg. Med. Chem. Lett. 14: 541–544.

Konkala, K., N.M. Sabbavarapu, R. Katla, N.Y.V. Durga, T.V.K. Reddy, L.A.P.D. Bethala et al. 2012. Revisit to the biginelli reaction: a novel and recyclable bioglycerol-based sulfonic acid functionalised carbon catalyst for one-pot synthesis of substituted 3,4-dihydropyrimidin-2-(1h)-ones. Tetrahedron Lett. 53: 1968–1973.

Kotha, S. and P. Khedkar. 2012. Rongalite: a useful green reagent in organic synthesis. Chem. Rev. 112: 1650–1680.

Kumar, A. and R.A. Maurya. 2007. Synthesis of 3,4-dihydropyrimidin-2(1H)-ones using Ziegler–Natta catalyst system under solvent free conditions. J. Mol. Catal. A: Chem. 272: 53–56.

Lim, J.I., H. İm and W.K. Lee. 2015. Fabrication of porous chitosan-polyvinyl pyrrolidone scaffolds from a quaternary system via phase separation. J. Biomater. Sci. Polym. Ed. 26: 32–41.

Lin, C.C., Y. Guo and J. Vela. 2015. Microstructure effects on the water oxidation activity of Co_3O_4/porous silica nanocomposites. ACS Catal. 5: 1037–1044.

Maddila, S., S. Gorle and S.B. Jonnalagadda. 2020. Drug screening of rhodanine derivatives for antibacterial activity. expert opin. Drug Discovery. 15: 203–229.

Majhi, S. 2020. Diterpenoids: natural distribution, semisynthesis at room temperature and pharmacological aspects - a decade update. ChemistrySelect. 5: 12450–12464.

Majhi, S. 2021a. Applications of norrish type i and ii reactions in the total synthesis of natural products: a review, photochem. Photobiol. Sci. 20: 1357–1378.

Majhi, S. 2021b. Applications of yamaguchi method to esterification and macrolactonization in total synthesis of bioactive natural products. ChemistrySelect. 6: 4178–4206.

Majhi, S. 2021c. Discovery, development, and design of anthocyanins-inspired anticancer agents-a comprehensive review. Anti-cancer Agents Med Chem. doi: 10.2174/1871520621666211015142310.

Majhi, S. and D. Das. 2021. Chemical derivatisation of natural products: semisynthesis and pharmacological aspects- A decade update. Tetrahedron. 78: 131801.

Majhi, S. 2022. Recent developments in the synthesis and anti-cancer activity of acridine and xanthine-based molecules. Phys. Sci. Rev. 0: 1–35.

Mamaghani, M., K. Tabatabaeian, R. Araghi, A. Fallah and R.H. Nia. 2014. An efficient, clean, and catalyst-free synthesis of fused pyrimidines using sonochemistry. Org. Chem. Int. ID: 406869.

Mason, T.J. 1997. Ultrasound in synthetic organic chemistry. Chem. Soc. Rev. 26: 443–451.

McDonald, E., K. Jones, P.A. Brough, M.J. Drysdale and P. Workman. 2006. Discovery and development of pyrazole-scaffold hsp90 inhibitors. curr. Top. Med. Chem. 6: 1193–1203.

Meldal, M. and C.W. Tornøe. 2008. Cu-catalyzed azide–alkyne cycloaddition. Chem. Rev. 108: 2952–3015.

Mishra, A., S. Singh, M.A. Quraishi and V. Srivastava. 2019. A catalyst-free expeditious green synthesis of quinoxaline, oxazine, thiazine, and dioxin derivatives in water under ultrasound irradiation. Org. Prep. Proceed. Int. 51: 345–356.

Mittersteiner, M., F.F.S. Farias, H.G. Bonacorso, M.A.P. Martins and N. Zanatta. 2021. Ultrasound-assisted synthesis of pyrimidines and their fused derivatives: a review. Ultrason. Sonochem. 79: 105683.

Moradi, L. and P. Mahdipour. 2019. Green and rapid synthesis of dihydropyrimido [4,5-b] quinolinetrione derivatives using $CoFe_2O_4@PPA$ as high efficient solid acidic catalyst under ultrasonic irradiation. Appl. Organometal. Chem. 33: e4996.

Moradi, L. and M. Zare. 2018. Ultrasound-promoted green synthesis of 1,4-dihydropyridines using functionalised MWCNTs as a highly efficient heterogeneous catalyst. Green Chem. Lett. Rev. 11: 197–208.

Mulakayala, N., D. Rambabu, M.R. Raja, M. Chaitanya, C.S. Kumar, A.M. Kalle et al. 2012. Ultrasound mediated catalyst free synthesis of 6H-1-benzopyrano [4,3-b]quinolin-6-ones leading to novel quinolone derivatives: their evaluation as potential anti-cancer agents, bioorg. Med. Chem. 20: 759–768.

Naeimi, H. and A. Didar. 2017. Efficient sonochemical green reaction of aldehyde, thiobarbituric acid and ammonium acetate using magnetically recyclable nanocatalyst in water. Ultrason. Sonochem. 34: 889–895.

Nemati, F., S. Nikkhah and A. Elhampour. 2015. An environmental friendly approach for the catalyst-free synthesis of highly substituted pyrazoles promoted by ultrasonic radiation. Chin. Chem. Lett. 26: 1397–1399.

Nicolaou, K.C. 2014. Organic synthesis: the art and science of replicating the molecules of living nature and creating others like them in the laboratory. Proc. R. Soc. A. 470: 20130690.

Odagi, M. and K. Nagasawa. 2019. Recent advances in natural products synthesis using bifunctional organocatalysts bearing a hydrogen-bonding donor moiety. Asian J. Org. Chem. 8: 1766–1774.

Pagadala, R., S. Maddila and S.B. Jonnalagadda. 2014. Ultrasonic-mediated catalyst-free rapid protocol for the multi-component synthesis of dihydroquinoline derivatives in aqueous media. Green Chem. Lett. Rev. 7: 131–136.

Peyton, L.R., S. Gallagher and M. Hashemzadeh. 2015. Triazole anti-fungals: a review. Drugs Today (Barc). 51: 705–18.

Pourian, E., S. Javanshir, Z. Dolatkhah, S. Molaei and A. Maleki. 2018. Ultrasonic-assisted preparation, characterisation, and use of novel biocompatible core/shell $Fe_3O_4@GA@$isinglass in the synthesis of 1,4-dihydropyridine and 4H-pyran derivatives. ACS Omega. 3: 5012–5020.

Purkhosro, A., A. Khalili, A.C. Ho, S.M. Haghighi, S. Fakher and A. Khalafi-Nezhad. 2019. Highly efficient, one pot, solvent and catalyst, free synthesis of novel quinazoline derivatives under ultrasonic irradiation and their vasorelaxant activity isolated thoracic aorta of rat. Iran. J. Pharm. Sci. 18: 607–619.

Quan, Z.J., Y.X. Da, Z. Zhang and X.C. Wang. 2009. PS–PEG–SO_3H as an efficient catalyst for 3,4-dihydropyrimidones via biginelli reaction. Catal. Commun. 10: 1146–1148.

Rafiee, E. and F. Shahbazi. 2006. One-pot synthesis of dihydropyrimidones using silica-supported heteropoly acid as an efficient and reusable catalyst: improved protocol conditions for the biginelli reaction. J. Mol. Catal. A: Chem. 250: 57–61.

Safaei-Ghomi, J., F. Eshteghal and H. Shahbazi-Alavi. 2016. A facile one-pot ultrasound assisted for an efficient synthesis of benzo[g] chromenes using Fe_3O_4/polyethylene glycol (PEG) core/shell nanoparticles. Ultrason. Sonochem. 33: 99–105.

Safaei-Ghomi, J., M. Tavazo and G.H. Mahdavinia. 2018. Ultrasound promoted one-pot synthesis of 3,4-dihydropyrimidin-2(1H)-ones/thiones using dendrimer-attached phosphotungstic acid nanoparticles immobilised on nanosilica. Ultrason. Sonochem. 40: 230–237.

Safari, J., Z. Zarnegar and M. Heydarian. 2012. Magnetic Fe_3O_4 nanoparticles as efficient and reusable catalyst for the green synthesis of 2-amino-4H-chromene in aqueous media. Bull. Chem. Soc. Jpn. 85: 1332–1338.

Safari, J. and Z. Zarnegar. 2014. Ultrasonic activated efficient synthesis of chromenes using amino-silane modified Fe_3O_4 nanoparticles: A versatile integration of high catalytic activity and facile recovery. J. Mol. Struct. 1072: 53–60.

Safari, J. and L. Javadian. 2015. Ultrasound assisted the green synthesis of 2-amino-4H-chromene derivatives catalysed by Fe_3O4-functionalized nanoparticles with chitosan as a novel and reusable magnetic catalyst. Ultrason. Sonochem. 22: 341–348.

Salehi, P., M. Dabiri, M.A. Zolfigol and M.A.B. Fard. 2003. Silica sulfuric acid: an efficient and reusable catalyst for the one-pot synthesis of 3,4-dihydropyrimidin-2(1H)-ones. Tetrahedron Lett. 44: 2889–2891.

Sarmah, M.M. and D. Prajapati. 2015. Ultrasound-mediated one-pot synthesis of 7-methyl-substituted pyrido[4,3-d]pyrimidine scaffolds by a catalyst-free protocol. Synlett 26: 91–94.

Schweitzer, N.M., B. Hu, U. Das, H. Kim, J. Greeley, L.A. Curtiss et al. 2014. Propylene hydrogenation and propane dehydrogenation by a single-site Zn^{2+} on silica catalyst. ACS Catal. 4: 1091–1098.

Shabalala, N.G., N. Kerru, S. Maddila, W. van Zyl and S.B. Jonnalagadda. 2020. Ultrasound-mediated catalyst-free protocol for the synthesis of bis-3-methyl-1-phenyl-1 H -pyrazol-5-ols in aqueous ethanol. Chem. Data Collect. 28: 100467.

Shabalala, N.G., R. Pagadala and S.B. Jonnalagadda. 2015. Ultrasonic-accelerated rapid protocol for the improved synthesis of pyrazoles. Ultrason Sonochem. 27: 423–429.

Sharma, R.K., S. Dutta, S. Sharma, R. Zboril, R.S. Varma and M.B. Gawande. 2016. Fe_3O_4 (iron oxide)-supported nanocatalysts: synthesis, characterisation and applications in coupling reactions. Green Chem. 18: 3184–3209.

Sharma, V., R. Das, D.K. Mehta, D. Sharma and R.K. Sahu. 2021. Exploring quinolone scaffold: unravelling the chemistry of anti-cancer drug design. Mini Rev Med Chem. 22(1):69–88.

Tabarsaei, N., N.F. Hamedani, S. Shafiee, S. Khandan and Z. Hossaini. 2020. Catalyst-free green synthesis and study of antioxidant activity of new pyrazole derivatives. J Heterocyclic Chem. 57: 2945–2954.

Taheri-Ledari, R., J. Rahimi and A. Maleki. 2019. Synergistic catalytic effect between ultrasound waves and pyrimidine-2,4-diamine-functionalized magnetic nanoparticles: applied for synthesis of 1,4-dihydropyridine pharmaceutical derivatives. Ultrason. Sonochem. 59: 104737.

Tang, S.Y., R.A. Bourne, R.L. Smith and M. Poliakoff. 2008. The 24 principles of green engineering and green chemistry: "IMPROVEMENTS PRODUCTIVELY". Green. Chem. 10: 268–269.

Tiwari, A.K., V.K. Singh, A. Bajpai, G. Shukla, S. Singh and A.K. Mishra. 2007. Synthesis and biological properties of 4-(3H)-quinazolone derivatives. Eur. J. Med. Chem. 42: 1234–1238.

Upadhyay, A., P. Chandrakar, S. Gupta, N. Parmar, S.K. Singh, M. Rashid et al. 2019. Synthesis, biological evaluation, structure-activity relationship, and mechanism of action studies of quinoline-metronidazole derivatives against experimental visceral leishmaniasis. J. Med. Chem. 62: 5655–5671.

Verma, D., V. Sharma, Okram, G.S. and S. Jain. 2017. High-yield multi-component synthesis of triazolo[1,2-a] indazoletriones using silica-coated ZnO nanoparticles as heterogeneous catalyst in the presence of ultrasound. Green Chem. 19: 5885–5899.

Wang, L.W., J.J. Kang, I.J. Chen, C.M. Teng and C.N. Lin. 2002. Antihypertensive and vasorelaxing activities of synthetic xanthone derivatives. Bioorg. Med. Chem. 10: 567–572.

Zarnegar, Z. and J. Safari. 2014. Ultrasonic activated efficient synthesis of chromenes using aminosilane modified Fe_3O_4 nanoparticles: a versatile integration of high catalytic activity and facile recovery. J. Mol. Struct. 1072: 53–60.

Zelefack, F., D. Guilet, N. Fabre, C. Bayet, S.V. Chevalley, S.R. Ngouela et al. 2009. Cytotoxic and antiplasmodial xanthones from pentadesma butyracea. J. Nat. Prod. 72: 954–957.

Ziarani, G.M., Z. Kheilkordi, F. Mohajer, A. Badiei and R. Luque. 2021. Magnetically recoverable catalysts for the preparation of pyridine derivatives: an overview. RSC Adv. 11: 17456–17477.

CHAPTER 5

Solvent-free and Aqueous Organic Synthesis Using Ultrasound toward Sustainable Society

Sasadhar Majhi,[1,]* *Sivakumar Manickam*[2,]* and *Giancarlo Cravotto*[3]

1. Introduction

The chemical synthesis needs energy often; ultrasound (US) can contribute this energy as a non contaminating source (Mason 1997, Cravotto and Cintas 2006). US irradiation has been positively employed as a source of clean energy to produce promising organic molecules and bioactive natural products in good yields with better selectivity (Javed et al. 1995, Li et al. 2005, Machado et al. 2021, Majhi 2020, 2021a, 2021b, 2021c, 2021d, 2021e, 2022a, 2022b, Majhi and Das 2021). Ultrasonic waves in organic synthesis provide advantages over traditional heating. Many chemical reactions need extreme temperature and pressure under classical thermal protocols (Penteado et al. 2018). On the other hand, several merits of US mediated methodology consist of higher yield, easy work up, environmentally benign conditions, greater selectivity, short reaction times, and purity of the products (Puri et al. 2013, Ellstrom and Török 2018).

Many chemists still perform chemical reactions in solutions or different organic solvents. However, some of these toxic solvents create serious problems for the health and environment (Metzger 1998). It has been observed that various chemical reactions proceed effectively in solid state or solvent free conditions (Metzger 1998, Sainath and Pravinkumar 2019). Solvent free organic transformation occurs more efficiently and selectively than its solution counterpart, as molecules in a crystal are arranged compactly and regularly (Tanaka and Toda 2000). In addition, solvent free reactions comprise several advantages: decreased pollution, low costs, simplicity in

[1] Department of Chemistry (UG & PG Dept.), Triveni Devi Bhalotia College, Raniganj, Kazi Nazrul University, West Bengal 713347, India.
[2] Petroleum and Chemical Engineering, Faculty of Engineering, Universiti Teknologi Brunei, Bandar Seri Begawan, BE1410, Brunei Darussalam.
[3] Department of Drug Science and Technology, University of Turin, Via P. Giuria 9, 10125 Turin, Italy.
* Corresponding authors: sasadharmajhi@gmail.com; manickam.sivakumar@utb.edu.bn

the protocol, and easy handling of the methodology (Metzger 1998, Sainath and Pravinkumar 2019, Tanaka and Toda 2000).

The substitution of many harmful volatile organic solvents with water in organic synthesis is one of the main aims of green chemistry. Water is easily available, cheap, safe, nontoxic, and eco-friendly green medium (Chanda and Fokin 2009). Many organic transformations occur in aqueous media with higher reaction rates and greater selectivities than results obtained employing classical organic solvent-based systems (Cortes-Clerget et al. 2021). The present chapter aims to deal with sonochemical organic synthesis under solvent free conditions and in an aqueous medium as a green solvent.

2.   Sonochemical Organic Synthesis under Solvent Free Conditions and in an Aqueous Medium toward Sustainable Society

2.1 *Applications of Sonochemistry in Organic Synthesis under Solvent free Conditions*

A key factor in sustainable chemistry is the need to restrict the application of ecologically harmful organic solvents. Solvent free chemical transformations are not only of interest from an eco-friendly viewpoint but also offer considerable merits in terms of selectivity, yield, and simplicity of the reaction methodologies (Tanaka and Toda 2000). Hence, this section covers applications of sonochemistry in organic synthesis under solvent free conditions.

2.1.1 *Cross Dehydrogenative Coupling (CDC) Reaction*

In organic synthesis, the coupling reaction plays a crucial role in constructing a new C-C and C-heteroatom bond. Cross dehydrogenative coupling (CDC), also known as oxidative coupling, is an effective novel technique for forming C-C and C-heteroatom bonds in organic chemistry. The CDC permits the direct coupling of two inactivated C–H bonds of substrates, or a C–H bond along with an X–H bond (X = O, N, S, P, B, or Si) of substrates to generate different beneficial bonds for complex organic synthesis (Phillips et al. 2020). An oxidant is needed for the oxidative coupling, which behaves as the terminal acceptor of the two hydrogen atoms. This coupling does not need reaction substrates having functional groups. So, it eliminates redundant steps and gains smaller synthetic routes as well as better efficiency (Jiang et al. 2019).

2.1.1.1 Synthesis of *O*-alkylated Hydroximides

Gui et al., achieved a novel, high substrate scope, eco-friendly methodology for the green synthesis of various *O*-alkylated hydroximides by the cross dehydrogenative coupling (CDC) reaction of *N*-hydroxyphthalimide and benzyl/ether molecules under US irradiation as a green tool without the need of any solvent in good to high yields (30 examples, 69 to 91 percent yield) (Equation 1) (Jiang et al. 2019). Applying US

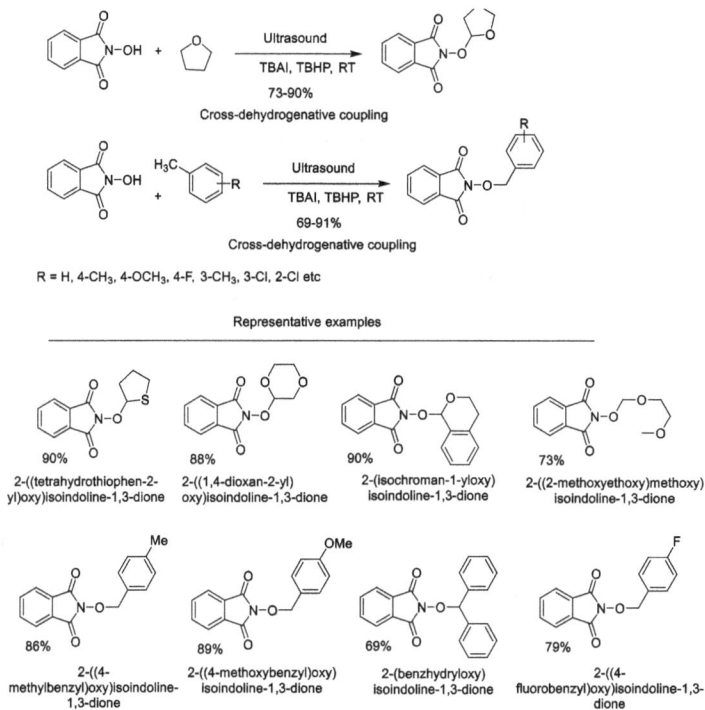

Equation (1)

techniques without any solvent and metal catalyst enhances the reaction efficiency and rate of transformation. It reduced the side reactions compared to conventional heating conditions. This conversion includes a good functional group tolerance and proceeds through a highly selective manner to deliver the desired products.

2.1.2 Baeyer-Villiger Oxidation

Lactones are very suitable synthetic products in the polymer, agrochemical, and pharmaceutical industries (Targel and Portnoy 2020). The Baeyer-Villiger oxidation is a popular synthetic tool for quickly creating esters or lactones from the corresponding ketones by inserting an oxygen atom. It was generally carried out in organic peracids such as peracetic acid. Green primary oxidants such as aqueous hydrogen peroxide are required to produce lactone or ester, as organic peracids are expensive and/or hazardous (because of shock sensitivity). Hydrogen peroxide, *m*-chloroperbenzoic acid, hydroperoxides, perfluoroacetic acid, peroxyacetic acid, peroxy acid (RCO_3H), etc., are employed as the peroxide derivatives to prepare lactones or esters in this oxidation. The Baeyer-Villiger oxidation is still one of the most powerful transformations in organic chemistry involving the synthesis of steroids, antibiotics, pheromones, etc. (Strukul 1998).

2.1.2.1 Synthesis of Esters or Lactones

Pombeiro et al., applied an organometallic coordination polymer as the heterogeneous catalyst to synthesize esters or lactones from linear or cyclic ketones under the US or microwave irradiation without solvent (Equation 2) (Martins et al. 2016). Aqueous hydrogen peroxide was used as an oxidant under additive free conditions through the Baeyer-Villiger oxidation for a short reaction period. The attractive features were low catalyst loading (0.1 mol percentage) in these transformations and smart recyclability up to five runs without any remarkable loss of catalytic activity. Interestingly, sonication was the best method compared to microwave irradiation.

2.1.3 Passerini Reaction

In 1921, Mario Passerini invented isocyanide based multi component reactions (IMCRs) comprising an aldehyde (or ketone), an isocyanide, and a carboxylic acid (acetic acid) to construct an α-acyloxy amide without any side product (Ramozzi and Morokuma 2015). This reaction is kinetically favored in aprotic solvents such as DCM (dichloromethane); however, the closely linked Ugi reaction includes an opposite trend (Iacobucci et al. 2014). This reaction contributes a vital role in combinatorial chemistry.

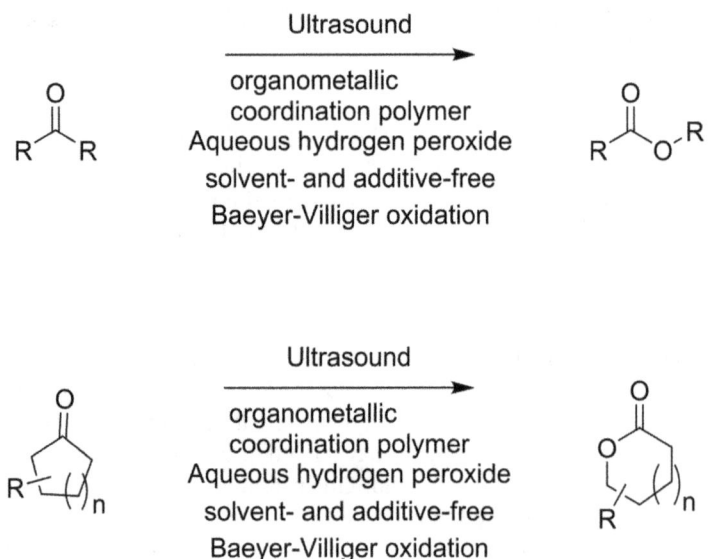

Equation (2)

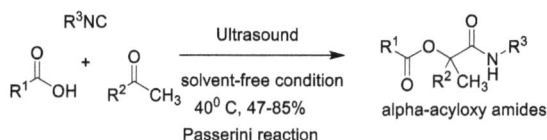

R^1 = H, 4-OCH₃, 4-Ph, 4-PhO

R² = 4-ClPh, 4-OCH₃Ph, 4-MePh, 2-NO₂-4-MeOPh, PhCH₂ etc.

Representative examples

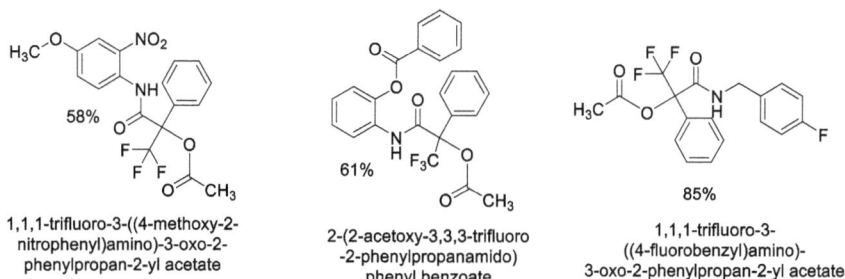

1,1,1-trifluoro-3-((4-methoxy-2-nitrophenyl)amino)-3-oxo-2-phenylpropan-2-yl acetate

58%

2-(2-acetoxy-3,3,3-trifluoro-2-phenylpropanamido) phenyl benzoate

61%

1,1,1-trifluoro-3-((4-fluorobenzyl)amino)-3-oxo-2-phenylpropan-2-yl acetate

85%

Equation (3)

2.1.3.1 Synthesis of α-Acyloxy Amides

Zhao et al., developed a sterically congested Passerini reaction to prepare α-acyloxy amides under solvent free conditions at 40°C (Equation 3) (Cui et al. 2012). This mild and time saving US assisted methodology accomplished various merits, such as being eco-friendly, a simpler work up methodology, shorter reaction times, and better yields over current reaction procedures.

2.1.4 Paal-Knorr Synthesis

Five membered heterocyclic systems, namely furan, pyrrole, and thiophene ring comprising species, are beneficial in generating pharmacologically important species employing bioisosteric replacement. The pharmacophoric and toxicophoric features obtained from these heterocyclic systems are widely explored in drug design and metabolism. The Paal Knorr reaction is important to efficiently create five membered polysubstituted heterocycles (furan, pyrrole, and thiophene). Catalyst and solvent free conditions have been applied to prepare underivatized with *N*-substituted pyrroles. The significance of this synthetic transformation is highlighted in the manufacture of many drugs such as Aloracetam (anti Alzheimer), Roseophilin (anti tumor), Atorvastatin (antitriglyceride), and Prodiogiosin (antifungal, antibacterial, and antimalarial) (Abbat et al. 2015).

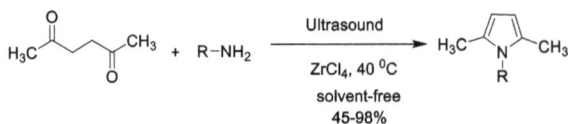

R = C_6H_5, $C_6H_5CH_2$, 4-$CH_3C_6H_4$, 4-$OCH_3C_6H_4$, 3-MeC_6H_4, 2-BrC_6H_4 etc.

Representative examples

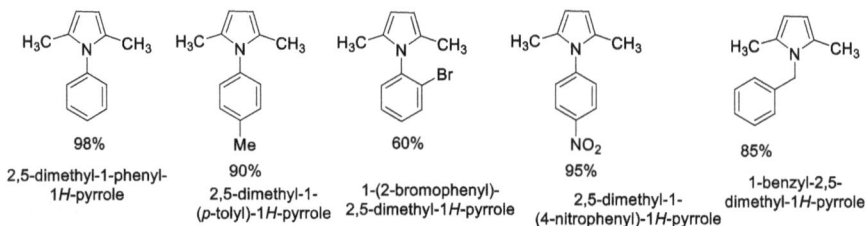

98%	Me	60%	NO₂	85%
2,5-dimethyl-1-phenyl-1H-pyrrole	90% 2,5-dimethyl-1-(p-tolyl)-1H-pyrrole	1-(2-bromophenyl)-2,5-dimethyl-1H-pyrrole	95% 2,5-dimethyl-1-(4-nitrophenyl)-1H-pyrrole	1-benzyl-2,5-dimethyl-1H-pyrrole

Equation (4)

2.1.4.1 Synthesis of Substituted Pyrroles

Li et al., demonstrated a modified Paal-Knorr reaction for the straightforward preparation of important substituted pyrroles from acetonylacetone and various amines utilizing zirconium chloride as an inexpensive and easily accessible catalyst under US assisted solvent free conditions (Equation 4) (Zhang et al. 2008). The green procedure delivered desired substituted pyrrole derivatives at 40°C in high yield under mild reaction conditions.

2.1.5 Condensation Reactions

Condensation reactions have a profound role in our lives since these reactions are essential for constructing peptide bonds between amino bonds and the biosynthesis of fatty acids (Fakirov 2019). The condensation reaction is a multifaceted class of reactions in which two compounds are combined to provide a single compound; it includes the loss of a small molecule, frequently water. It often needs heat, and it can proceed under basic or acidic conditions or requires a catalyst.

2.1.5.1 Synthesis of 6,6'-((1E,1'E)-(1,2-Phenylenebis(Azanylylidene)) bis(Methanylylidene)) bis(2-Methoxyphenol) and 2-(4,4-Dimethyl-2,6-Dioxocyclohexyl)-2-Hydroxy-1H-Indene-1,3(2H)-Dione

Crawford reported the first examples of condensation reactions for preparing a Schiff base and a 1,3-indandione under US irradiation without solvent as a green methodology (Figure 1) (Crawford 2017). In these transformations, the crucial parameter was the particle size of the starting materials. The particle size of both starting materials was decreased to <200 μm in a homogeneous mixture to obtain the complete conversion to the desired products.

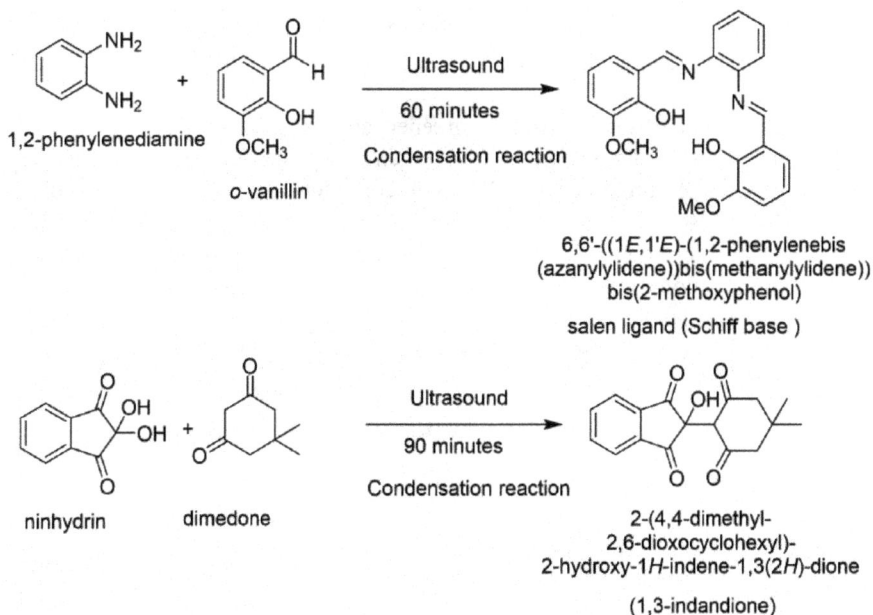

Figure 1. US promoted synthesis of salen ligand (Schiff base) and 1,3-indandione under solvent free conditions.

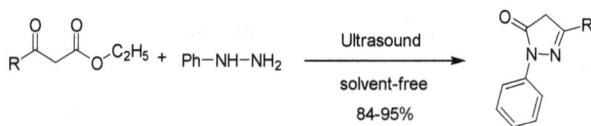

$R = CH_3CH_2CH_2, CH_3CO, CH_3, CF_3COCH_2, CH_3CH_2CH_2COCH_2$ etc.

Representative examples

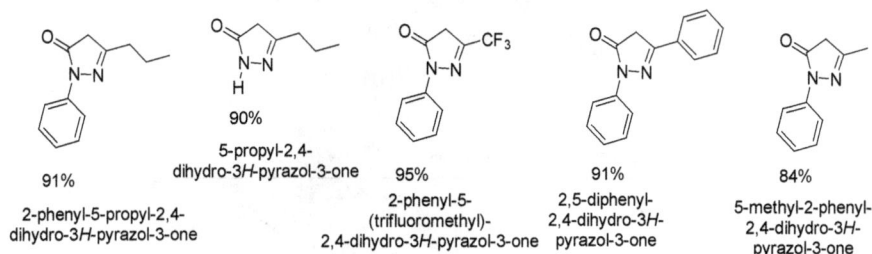

Equation (5)

Abaee et al., disclosed a condensation reaction between phenyl hydrazine or hydrazine and β-keto ester under US assisted solvent free conditions for the synthesis of valuable pyrazolone derivatives in good to excellent yields (84 to 95 percent) for short reaction time (Equation 5) (Mojtahedi et al. 2008). The eco-friendly protocol produces the sole product at ambient temperature without any additives.

2.1.6 Miscellaneous

2.1.6.1 Synthesis of α-Hydroxy Phosphonates

Gill et al., accomplished a novel, greener approach to preparing α-hydroxy phosphonates employing potassium dihydrogen phosphate (KH_2PO_4) as an efficient catalyst under US promoted solvent free conditions (Equation 6) (Mandhane et al. 2010). A series of phosphonate compounds were prepared from aromatic or heteroaromatic aldehydes as starting materials with improved yields through a facile methodology at ambient temperature for 5 to 45 min; however, aliphatic

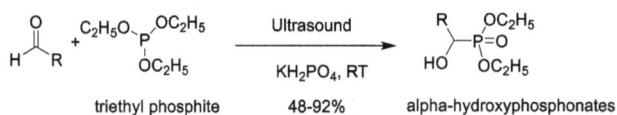

triethyl phosphite   48-92%   alpha-hydroxyphosphonates

R = C_6H_5, 4-$CH_3C_6H_5$, 4-$OCH_3C_6H_5$, 3-OHC_6H_5, 2-ClC_6H_5 etc.

Representative examples

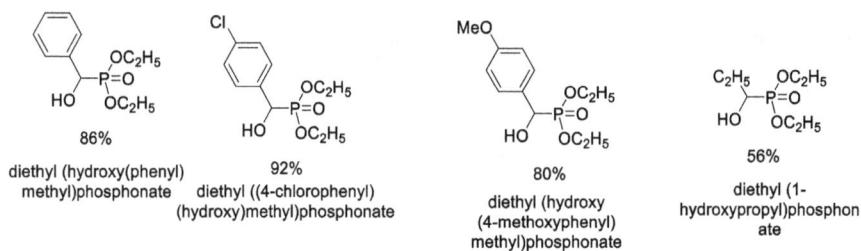

86%

diethyl (hydroxy(phenyl) methyl)phosphonate

92%

diethyl ((4-chlorophenyl) (hydroxy)methyl)phosphonate

80%

diethyl (hydroxy (4-methoxyphenyl) methyl)phosphonate

56%

diethyl (1-hydroxypropyl)phosphonate

Equation (6)

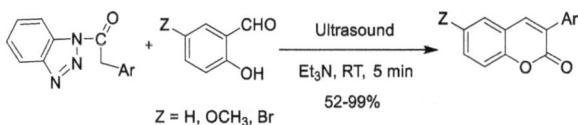

Ultrasound

Et$_3$N, RT, 5 min

52-99%

Z = H, OCH$_3$, Br

Representative examples

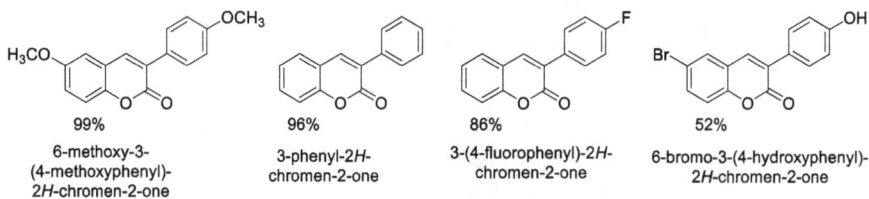

99%

6-methoxy-3-(4-methoxyphenyl)-2H-chromen-2-one

96%

3-phenyl-2H-chromen-2-one

86%

3-(4-fluorophenyl)-2H-chromen-2-one

52%

6-bromo-3-(4-hydroxyphenyl)-2H-chromen-2-one

Equation (7)

aldehydes provided the targeted products in low yields after prolonged reaction time. No conversion occurred applying ketones as starting material even after prolonged reactiong time.

2.1.6.2 Synthesis of Substituted 3-Arylcoumarins

Pattarawarapan et al., disclosed a US mediated one pot reaction to form substituted 3-arylcoumarins from 2-hydroxybenzaldehydes and *N*-acylbenzotriazoles under solvent free together with chromatography free conditions using triethylamine as a base (Equation 7) (Wet-osot et al. 2016). The rapid synthesis of a small library of substituted 3-arylcoumarins (24 examples) occurred under mild conditions in improved yields (up to 99 percent) in the presence of *N*-acylbenzotriazoles as the acylating agents at room temperature for five min.

2.2 Applications of Sonochemistry in Organic Synthesis in Aqueous Medium

Traditionally, the solvents employed in chemical synthesis are volatile organic compounds (VOCs); these solvent vapors are harmful and toxic to living organisms, including humans, damaging most organs, mainly the respiratory and nervous systems (Cortes-Clerget et al. 2021). Recently, water has appeared as an environmentally benign solvent in organic synthesis due to its unique physical and chemical properties, including large heat capacity, high dielectric constant, wide hydrogen bonding, etc., (Chanda and Fokin 2009). Hence, this section highlights the uses of sonochemistry in the organic synthesis using an aqueous medium.

2.2.1 US Promoted C-C Bond Forming Reactions

Carbon-carbon bond forming transformations are efficient tools to build the carbon backbone of organic compounds, the most effective and fundamental reactions in the evolution of organic synthesis. Organic molecules containing the carbon skeleton find various applications as medicinal and pharmaceutical agents, agrochemicals, fine chemicals, etc. (Brahmachari 2016).

2.2.1.1 Synthesis of 2,2'-Arylmethylene bis(3-Hydroxy-5,5-Dimethyl-2-Cyclohexene-1-one) Derivatives

In preparing biologically potent xanthenes, 2,2'-arylmethylene bis(3-hydroxy-5,5-dimethyl-2-cyclohexene-1-one) derivatives can serve as multifaceted precursors along with they can provide important synthetic intermediates and can be estimated as tyrosinase inhibitors (Khan et al. 2006). Chen et al., synthesized these derivatives through the condensation reaction of 5,5-dimethyl-1,3-cyclohexanedione and several aromatic aldehydes in aqueous media under power US in good to excellent yields (80 to 98 percent) at 50°C (Equation 8) (Li et al. 2012).

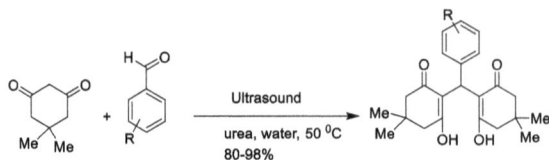

R = H, 2-Cl, 3-Cl, 4-CH₃, 3-NO₂, 2-NO₂, etc.

Representative examples

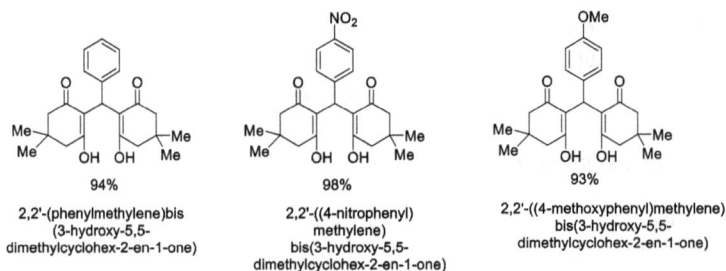

94%

2,2'-(phenylmethylene)bis
(3-hydroxy-5,5-
dimethylcyclohex-2-en-1-one)

98%

2,2'-((4-nitrophenyl)
methylene)
bis(3-hydroxy-5,5-
dimethylcyclohex-2-en-1-one)

93%

2,2'-((4-methoxyphenyl)methylene)
bis(3-hydroxy-5,5-
dimethylcyclohex-2-en-1-one)

Equation (8)

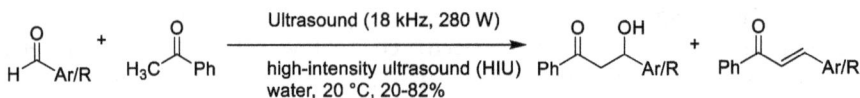

Representative examples

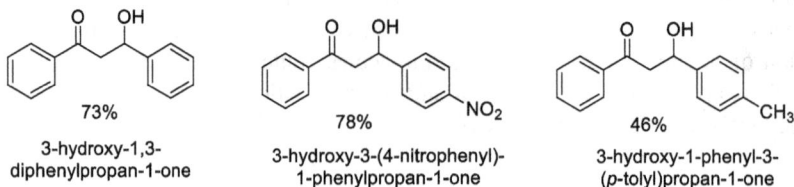

73%

3-hydroxy-1,3-
diphenylpropan-1-one

78%

3-hydroxy-3-(4-nitrophenyl)-
1-phenylpropan-1-one

46%

3-hydroxy-1-phenyl-3-
(p-tolyl)propan-1-one

Equation (9)

2.2.1.2 Synthesis of Aldols

In 1838, Kane invented the self-condensation of acetone to produce mesityl oxide through the aldol reaction, the most prominent transformation of enolate ions. It remains a common synthetic protocol (Equation 9) (Heathcock 1981). Cravotto et al., applied high intensity US (HIU) to synthesize several aldols in aqueous suspension; normally, customary conditions could not afford the same products since elimination occurred immediately (Cravotto et al. 2003). The green protocol provides an entrance for the generation of polyols and bis-benzylidene adducts.

2.2.1.3 Synthesis of 3-(Aryl)Methyl-4-Hydroxycoumarins at 30 to 40°C without Isolating any Intermediate

The coumarin ring system is the key building block of natural and non-natural analogues exhibiting interesting anticoagulant, anti-neoplastic, and photosensitizing activities (Petersen 2009). Recently, several 3(aryl)methyl-4-hydroxycoumarins display HIV integrase or HIV protease inhibitors together with potent non-nucleoside RT inhibitors (Dubey et al. 2007). Hence, the sonochemical mediated methodology was developed by Cravotto et al., for the synthesis of 3-(aryl)methyl-4-hydroxycoumarin derivatives from (hetero)aromatic aldehydes and 4-hydroxycoumarin applying Hantzsch 1,4-dihydropyridine (HEH) as a hydride donor in water through a tandem Knoevenagel reductive Michael addition under power US (Equation 10) (Palmisano et al. 2011). This sonochemical transformation was performed by *p*-dodecylbenzenesulphonic acid (DBSA) because of its surfactant activity and the strong Bronsted acidity at 30 to 40°C without isolating any intermediate.

2.2.1.4 Synthesis of Bis(Indolyl)Methanes

Due to their wide pharmacological properties, indole derivatives are important intermediates in organic synthesis (Sirisoma et al. 2009). Bis (indolyl) methane derivatives affect the central nervous system and are employed as tranquillizers (Michnovicz and Bradlow 1991). In addition, they exhibit anti tumor together with antibacterial activities (Michnovicz and Bradlow 1991). Li et al., performed a US mediated green, facile and efficient synthesis of bis(indolyl)methanes in aqueous media from various aromatic aldehydes and indole or *N*-methylindole in the presence

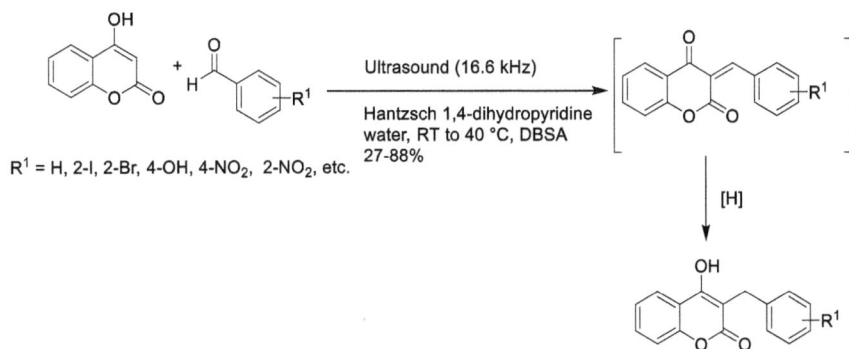

Representative examples

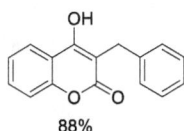

88%

3-benzyl-4-hydroxy-
2*H*-chromen-2-one

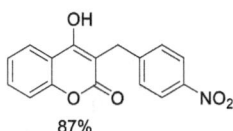

87%

4-hydroxy-3-
(4-nitrobenzyl)-2*H*-chromen-2-one

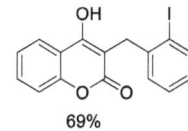

69%

4-hydroxy-3-(2-iodobenzyl)-
2*H*-chromen-2-one

Equation (10)

of dodecylbenzenesulfonic acid (ABS) catalyst in excellent yields at 23 to 25°C (Li et al. 2011).

2.2.2 US Promoted C-N Bond Forming Reactions

Organic compounds carrying C-N bonds play a crucial role in synthetic and biological science due to their attractive and diverse biological properties. Nitrogen containing organic molecules are the most significant ones in drugs and agricultural chemicals. Nowadays, more than 80 percent of used drugs possess at least a C-N bond (Shin et al. 2015). So, we need a green technique such as US irradiation to synthesize diverse C-N bonds to eliminate conventional harsh preparation methodologies.

2.2.2.1 Synthesis of β-amino Carbonyl Compounds

In heterocyclic synthesis, a powerful preparative protocol is the conjugate addition of electron rich nucleophiles (nitrogen centered heterocyclic) to electron deficient conjugated olefins. In this context, the aza-Michael addition of nucleophiles to α,β-unsaturated esters, ketones, and nitriles is a potent tool for constructing carbon-nitrogen bonds. Banik et al., accomplished an aza-Michael reaction for the US assisted addition of various amines to α,β-unsaturated ketones, nitriles and esters in water as a green solvent (Equation 11) (Bandyopadhyay et al. 2012). There was no need for catalysts or solid supports, which were the attractive features of this environmentally benign methodology. Significant improvement in reaction rate has been found in water under US mediated methodology. This eco-friendly aza-Michael transformation comprises notable merits such as better yields of products, shorter time, and no requirement for metallic or corrosive catalysts.

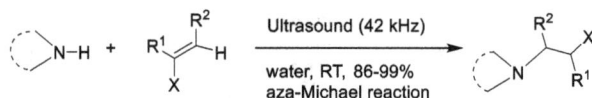

$R^1 = R^2$ = alkyl, H

X = CN, COOEt, COMe, COOMe

Representative examples

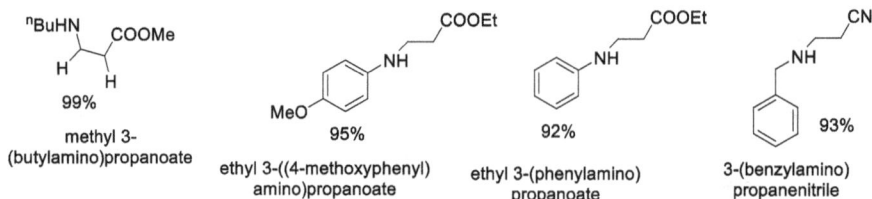

99%

methyl 3-(butylamino)propanoate

95%

ethyl 3-((4-methoxyphenyl)amino)propanoate

92%

ethyl 3-(phenylamino)propanoate

93%

3-(benzylamino)propanenitrile

Equation (11)

2.2.3 US Promoted C-C and C-N Bond Forming Reactions

2.2.3.1 Synthesis of 4,4-(Arylmethylene)-Bis-(3-Methyl-1-Phenyl-1H-Pyrazol-5-ols)Pyrazoles

Pyrazoles include di-aza heterocyclic molecules with extensive applications in biology, dyes, catalysis, and extraction metallurgy. The preparation of 4,4-(arylmethylene)-bis-(3-methyl-1-phenyl-1H-pyrazol-5-ols) pyrazole derivatives comprise a wide spectrum of biological properties including anti tumor, anti-inflammatory, antioxidant, etc. (Sujatha et al. 2009). Shankarling et al., demonstrated Knorr pyrazole synthesis as well as tandem Knoevenagel-Michael addition transformation for the preparation of 4,4-(arylmethylene)-bis-(3-methyl-1-phenyl-1H-pyrazol-5-ol)s from

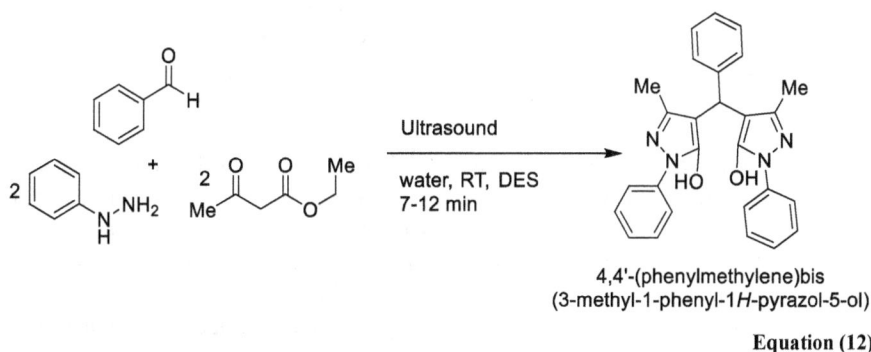

4,4'-(phenylmethylene)bis
(3-methyl-1-phenyl-1H-pyrazol-5-ol)

Equation (12)

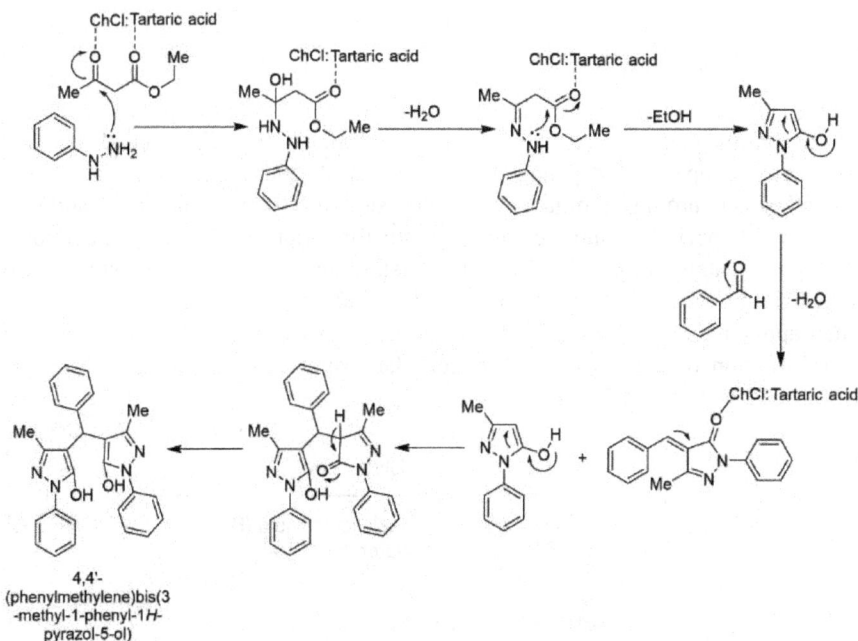

4,4'-
(phenylmethylene)bis(3
-methyl-1-phenyl-1H-
pyrazol-5-ol)

Figure 2. A plausible mechanism for synthesizing 4,4-(arylmethylene)-bis-(3-methyl-1-phenyl-1H-pyrazol-5-ol)s.

phenylhydrazine, ethyl acetoacetate, and aldehyde under US irradiation in water as a green solvent using Deep Eutectic Solvent (choline chloride, tartaric acid DES, 10 mol percentage) through the formation of 3-methyl-1-phenyl-5-pyrazolone (Equation 12) (Kamble and Shankarling 2018). One pot reaction and water as a cheap, eco friendly solvent make this protocol attractive for synthesizing pyrazole derivatives. As a greener technology, US decreased the required time and temperature for the transformation and improved the yield. The DES and water employed were recycled five times without performance loss significantly.

Figure 2 shows a plausible mechanism for this reaction.

2.2.3.2 Synthesis of Quinolone Derivatives

Sustainable chemistry needs less polluting neoteric solvents, as well as it avoids the need for any added catalyst under the green technique, namely ultrasonic irradiation. Chemical transformations in aqueous media offer merits, such as simple operation and better efficiency in several organic reactions, including water soluble substrates and reagents. These merits become even more attractive if such chemical transformations can be performed by applying ionic liquids in aqueous media. Kowsari et al., achieved an effective two component quinolone synthesis from isatin and ketones employing a basic ionic liquid (imidazolium cation) catalyst in aqueous media under ultrasonic irradiation (Equation 13) (Kowsari and Mallakmohammadi 2011). A mixture of two quinolines is generated if the ketone contains two different α-protons. However, applying basic ionic liquid (BIL) provides quinolones with improved yields.

2.2.3.3 Synthesis of Pyrazolone Derivatives

Copper iodide has gained much interest due to uncommon behaviors, including unusual diamagnetism, negative spin orbit splitting, large direct band gap, and rarely high temperature dependency (Feraoun et al. 2003). Moreover, it is useful in synthesizing promising organic molecules, superionic conductors, and solid-state solar cells. Especially in multi component transformations, CuI NPs were employed as potential catalysts for reactivity, higher yields, and selectivity (Bock et al. 2006). Safaei-Ghomi et al., prepared copper iodide NPs through an efficient protocol. They applied for the synthesis of pyrazolone derivatives by the four component transformation of aldehyde, ethyl acetoacetate, hydrazine, and β-naphthol under

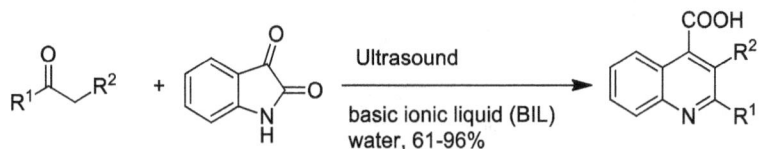

R^1 = H, C_6H_5, CH_3, 4-$NO_2C_6H_4$, 2-MeC_6H_4, 4-BrC_6H_4 etc.
R^2 = H, CH_3, 4-$NO_2C_6H_4$, 3-MeC_6H_4, 4-BrC_6H_4 etc.

Equation (13)

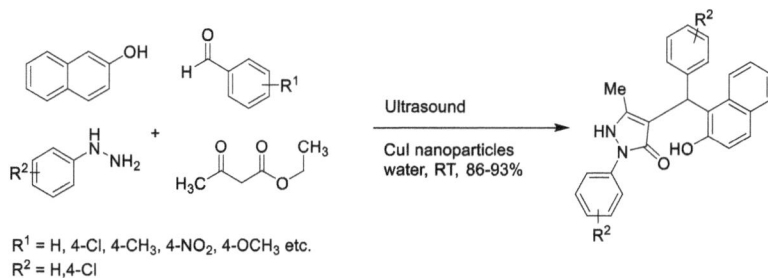

R[1] = H, 4-Cl, 4-CH₃, 4-NO₂, 4-OCH₃ etc.
R² = H,4-Cl

Representative examples

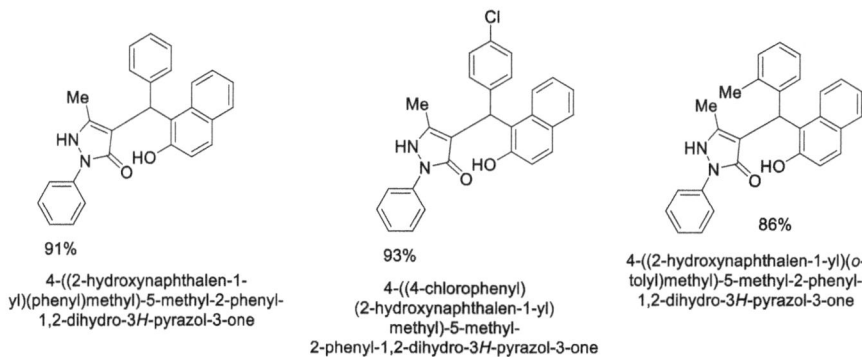

91%

4-((2-hydroxynaphthalen-1-
yl)(phenyl)methyl)-5-methyl-2-phenyl-
1,2-dihydro-3*H*-pyrazol-3-one

93%

4-((4-chlorophenyl)
(2-hydroxynaphthalen-1-yl)
methyl)-5-methyl-
2-phenyl-1,2-dihydro-3*H*-pyrazol-3-one

86%

4-((2-hydroxynaphthalen-1-yl)(o-
tolyl)methyl)-5-methyl-2-phenyl-
1,2-dihydro-3*H*-pyrazol-3-one

Equation (14)

US irradiation as a non-conventional source of energy using green solvent water (Equation 14) (Ziarati et al. 2013). The main merits of this methodology comprise short reaction times, a better yield of products, easy work up, the smallest size and the largest surface of the catalyst.

2.2.4 US Promoted C-O Bond Forming Reactions

Generating a C-O bond in an organic compound is one of the most important methodologies in organic synthesis. Metal catalyzed transformations for constructing C-O bonds significantly impact the syntheses of natural products. Aryl or alkenyl halides or synthetic equivalents and aliphatic alcohols, phenols, and water provided valuable various C-O bonds through cross coupling reactions (Evano et al. 2017).

2.2.4.1 Synthesis of Ketone and Aldehyde through Deprotection

The recovery of carbonyl compounds from oximes is the key transformation as oxime provides an effective protective group for ketone and aldehyde; it was widely employed to purify ketone and aldehyde. Li et al., accomplished a practical deprotection of oximes to carbonyls in the presence of silica sulfuric acid as an efficient catalyst in green solvent water under US irradiation as a non-polluting source of energy (Equation 15) (Li et al. 2010). The green methodology provided excellent yields at 50°C without any organic solvent, and the catalyst silica sulfuric acid was recycled successfully.

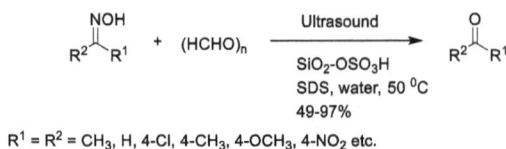

R^1 = R^2 = CH$_3$, H, 4-Cl, 4-CH$_3$, 4-OCH$_3$, 4-NO$_2$ etc.

Representative examples

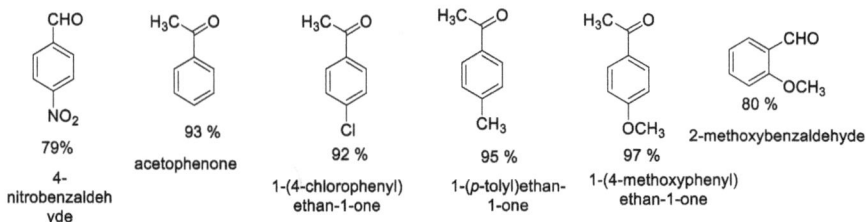

Equation (15)

2.2.5 *Ultrasound (US) Promoted C-C and C-O Bond Forming Reactions*

2.2.5.1 Synthesis of 2,3-Dihy Drofuranediones

In 2018, Fekri et al., demonstrated a multi component synthesis of 2,3-dihydrofuranediones from pyrazolecarbaldehydes or indole-3-carbaldehyde (aryl aldehydes), bromine, and ethyl pyruvate in the presence of aqueous NaOH under ultrasonic irradiation as an efficient green procedure (Figure 3) (Fekri et al. 2018). The eco-friendly protocol was performed at room temperature from inexpensive materials.

2.2.5.2 Synthesis of 2-Amino-4,8-Dihydropyrano[3,2-b]Pyran-3-Carbonitriles

Derivatives of polysubstituted 2-amino-4*H*-pyran-3-carbonitrile display diverse biological properties (Kemnitzer et al. 2008). Conventionally, they are synthesized by various methodologies; however, traditional methods suffer from several demerits, including prolonged reaction times, poor yields, an additional application of catalysts and reagents, and harsh reaction conditions (Reddy et al. 2010). Kojic acid is a fascinating compound in pharmaceutical chemistry and multi component transformations because of its easy availability, better reactivity, and significant biological property (Xiong 2008). Safari et al. demonstrated a US assisted method for the synthesis of 2-amino-4,8-dihydropyrano[3,2-b]pyran-3-carbonitrile scaffolds through the multi component reaction of kojic acid, various aromatic aldehydes, and malononitrile in aqueous media as a green protocol at 50°C (Equation 16) (Banitaba et al. 2013). This environmentally benign methodology provides better results than traditional methods, including short reaction time, high yields, better functional group tolerance, experimental simplicity, and selectivity in the absence of transition metal catalysts.

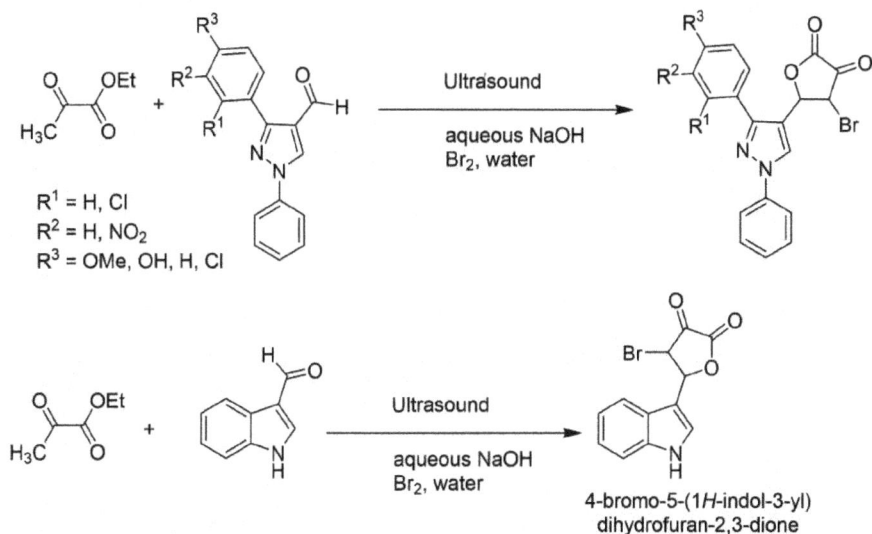

Figure 3. US mediated synthesis of 2,3-dihydrofuranediones in an aqueous medium.

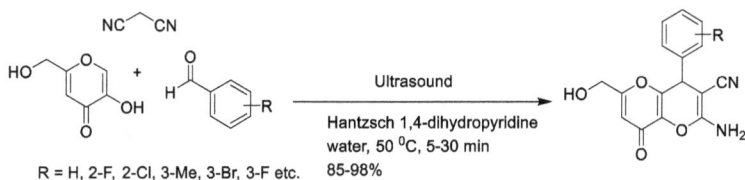

R = H, 2-F, 2-Cl, 3-Me, 3-Br, 3-F etc.   85-98%

Representative examples

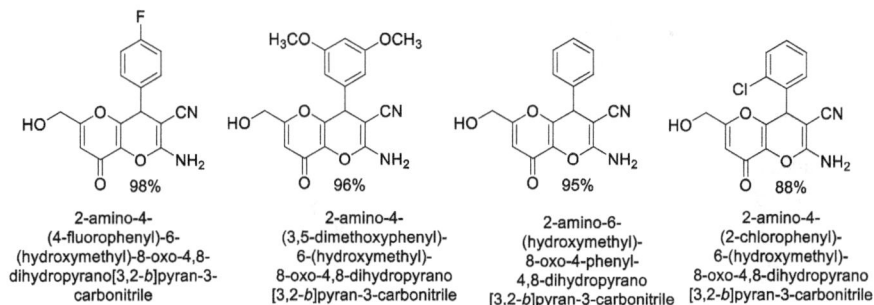

| 98% | 96% | 95% | 88% |

2-amino-4-(4-fluorophenyl)-6-(hydroxymethyl)-8-oxo-4,8-dihydropyrano[3,2-*b*]pyran-3-carbonitrile

2-amino-4-(3,5-dimethoxyphenyl)-6-(hydroxymethyl)-8-oxo-4,8-dihydropyrano[3,2-*b*]pyran-3-carbonitrile

2-amino-6-(hydroxymethyl)-8-oxo-4-phenyl-4,8-dihydropyrano[3,2-*b*]pyran-3-carbonitrile

2-amino-4-(2-chlorophenyl)-6-(hydroxymethyl)-8-oxo-4,8-dihydropyrano[3,2-*b*]pyran-3-carbonitrile

Equation (16)

2.2.5.3 Synthesis of 1,8-Dioxo-Octahydroxanthens

The organic synthetic community is motivated to prepare 1,8-dioxo-octahydroxanthens because of their important biological activity (Poupelin et al. 1978); several benzoxanthene derivatives are applied in industries as dyes in laser

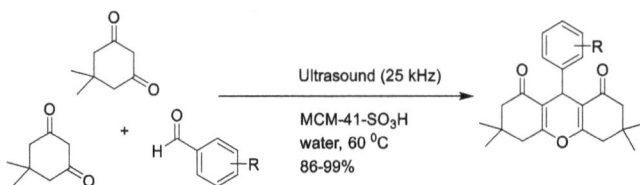

R = 4-CN, 2-OMe, 4-Cl, 4-Me, 4-NO$_2$ etc.

Representative examples

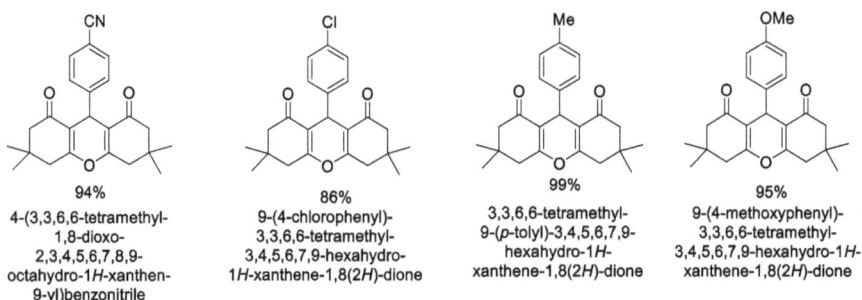

94%	86%	99%	95%
4-(3,3,6,6-tetramethyl-1,8-dioxo-2,3,4,5,6,7,8,9-octahydro-1H-xanthen-9-yl)benzonitrile	9-(4-chlorophenyl)-3,3,6,6-tetramethyl-3,4,5,6,7,9-hexahydro-1H-xanthene-1,8(2H)-dione	3,3,6,6-tetramethyl-9-(p-tolyl)-3,4,5,6,7,9-hexahydro-1H-xanthene-1,8(2H)-dione	9-(4-methoxyphenyl)-3,3,6,6-tetramethyl-3,4,5,6,7,9-hexahydro-1H-xanthene-1,8(2H)-dione

Equation (17)

technology. Moreover, these derivatives have also been employed in photodynamic therapy (Ion et al. 1998) and are fluorescent materials for sensing biomolecules (Sarma and Baruah 2005). Rostamizadeh et al. disclosed a green and effective method for the preparation of 1,8-dioxo-octahydroxanthenes derivatives from dimedone (two mmol) and various aromatic aldehydes (one mmol) comprising electron donating groups along with electron withdrawing groups in the presence of nanosized MCM-41-SO$_3$H catalyst in water under ultrasonic irradiation at 60°C in good to excellent yields for short reaction times (Equation 17) (Rostamizadeh et al. 2010).

3. Conclusion

Organic syntheses have the power to construct valuable organic compounds for potential uses in biology and medicine, cosmetics, dyes, and agricultural chemicals, as well as in advanced materials employed in cell phones, computers, and the space shuttle. Organic synthesis needs energy; ultrasound can contribute this energy as a clean source. Ultrasound (US) appears as an unconventional, powerful activation technique to reduce energy consumption in organic synthesis. US-mediated chemical transformations have received more attention since they significantly enhance reaction rates, improve yields, utilize mild reaction conditions, and have low energy consumption. Solvent-free transformations agree with the concept and philosophy of sustainable chemistry as they decrease pollutant production, cost, and application of harmful chemicals. Moreover, selecting suitable solvents in organic synthesis is also important from the green point of view as sustainability is a concept used to distinguish methods and procedures that can secure the long-term productivity of Mother Nature. In this context, the application of water as a solvent has attracted attention as it includes advantages such as being cheap, nontoxic, easily available, and non-flammable. It can be more selective compared to conventional organic solvents. Hence, the present chapter focuses on recent developments of promising organic syntheses in the absence of the solvent as well as the presence of water as a green solvent under the influence of ultrasonic irradiation as a green technique. It may stimulate the improvement of more eco-friendly organic methodologies so that the next generation can protect Mother Nature from pollution.

Acknowledgements

S Majhi is thankful to his father, Tarani Majhi and mother, Sadeswari Majhi.

References

Abbat, S., D. Dhaked, M. Arfeenb and P.V. Bharatam. 2015. Mechanism of the paal-Knorr reaction: the importance of water mediated hemialcohol pathway. RSC Adv. 5: 88353–88366.

Bandyopadhyay, D., S. Mukherjee, L.C. Turrubiartes and B.K. Banik. 2012. Ultrasound-assisted aza-michael reaction in water: a green procedure, Ultrason Sonochem. 19: 969–973.

Banitaba, S.H., J. Safari and S.D. Khalili. 2013. Ultrasound promoted one-pot synthesis of 2-amino-4,8-dihydropyrano [3,2-b]pyran-3-carbonitrile scaffolds in aqueous media: a complementary 'green chemistry' tool to organic synthesis. Ultrason Sonochem. 20: 401–407.

Bock, V.D., H. Heimstra and J.H. Van Maarseveen. 2006. Cu[I]-catalyzed alkyne–azide "click" cycloadditions from a mechanistic and synthetic perspective. Eur. J. Org. Chem. 51–68.

Brahmachari, G. 2016. Design for carbon-carbon bond forming reactions at ambient conditions. RSC Adv. 6: 64676–64725.

Chanda, A. and V.V. Fokin. 2009. Organic synthesis "on water". Chem. Rev. (109): 725–748.

Cortes-Clerget, M., J. Yu, J.R.A. Kincaid, P. Walde, F. Gallou and B.H Lipshutz. 2021. Water as the reaction medium in organic chemistry: from our worst enemy to our best friend. Chem. Sci. 12: 4237–4266.

Cravotto, G., A. Demetri, G.M. Nano, G. Palmisano, A. Penoni and S. Tagliapietra. 2003. The aldol reaction under high-intensity ultrasound: a novel approach to an old reaction. Eur. J. Org. Chem. 4438–4444.

Cravotto, G. and P. Cintas. 2006. Power ultrasound in organic synthesis: moving cavitational chemistry from academia to innovative and large-scale applications. Chem. Soc. Rev. 35: 180–196.

Crawford, D.E. 2017. Solvent-free sonochemistry: sonochemical organic synthesis in the absence of a liquid medium. Beilstein J. Org. Chem. 13: 1850–1856.

Cui, C., C. Zhu, X.J. Du, Z.P. Wang, Z.M. Li and W.G. Zhao. 2012. Ultrasound-promoted sterically congested passerini reactions under solvent-free conditions. Green Chem. 14: 3157–3163.

Dubey, S., Y.D. Satyanarayana and H. Lavania. 2007. Development of integrase inhibitors for treatment of AIDS: an overview. Eur. J. Med. Chem. 42: 1159–1168.

Ellstrom, C.J. and B. Török. 2018. Application of sonochemical activation in green synthesis. pp. 673–693. *In*: Török, B. and T. Dransfield [eds.]. Green Chem. An Incl. Approach. Elsevier Inc.

Evano, G., J. Wang and A. Nitelet. 2017. Metal-mediated C–O bond forming reactions in natural product synthesis. Org. Chem. Front. 4: 2480–2499.

Fakirov, S. 2019. Condensation polymers: their chemical peculiarities offer great opportunities. Prog. Polym. Sci. 89: 1–18.

Fekri, L.Z., M. Nikpassand and L. Imani-Darestani. 2018. Ultrasonochemical synthesis of 2,3-dihydrofuranediones in aqueous medium. Heterocycl. Commun. 24: 151–154.

Feraoun, H., H. Aourag and M. Certier. 2003. Theoretical studies of substoichiometric cuI. Mater. Chem. Phys. 82: 597–601.

Heathcock, C.H. 1981. Acyclic stereocontrol through the aldol condensation. science. 214: 395–400.

Iacobucci, C., S. Reale, J.F. Gal and F. De Angelis. 2014. Insight into the mechanisms of the multicomponent ugi and ugi–smiles reactions by ESI-MS(/MS). Eur. J. Org. Chem. 32: 7087–7090.

Ion, R.M., A. Planner, K. Wiktorowicz and D. Frackowiak. 1998. The incorporation of various porphyrins into blood cells measured via flow cytometry, absorption and emission spectroscopy. Acta Biochim. Pol. 45: 833–45.

Javed, T., T.J. Mason, S.S. Phull, N.R. Baker and A. Robertson. 1995. Influence of ultrasound on the diels-alder cyclization reaction: synthesis of some hydroquinone derivatives and lonapalene, an anti-psoriatic agent. Ultrason Sonochem. 2: S3–S4.

Jiang, H., X. Tang, S. Liu, L. Wang, H. Shen, J. Yang et al. 2019. Ultrasound accelerated synthesis of O-alkylated hydroximides under solvent- and metal-free conditions. Org. Biomol. Chem. 17: 10223.

Kamble, S.S. and G.S. Shankarling. 2018. A unique blend of water, des and ultrasound for one-pot knorr pyrazole synthesis and knoevenagel-michael addition reaction. Chemistry Select 3: 2032–2036.

Khan, K.M., G.M. Maharvi, M.T.H. Khan, A.J. Shaikh, S. Perveen, S.B. Mild et al. 2006. Tetraketones: a new class of tyrosinase inhibitors. Bioorg. Med. Chem. 14: 344–351.

Kemnitzer, W., J. Drewe, S. Jiang, H. Zhang, C. Crogan-Grundy, D. Labreque et al. 2008. Discovery of 4-Aryl-4*H*-chromenes as a new series of apoptosis inducers using a cell- and caspase-based high throughput screening assay. 4. Structure–Activity Relationships of *N*-Alkyl Substituted Pyrrole Fused at the 7,8-Positions. J. Med. Chem. 51(3): 417–423.

Kowsari, E. and M. Mallakmohammadi. 2011. Ultrasound promoted synthesis of quinolines using basic ionic liquids in aqueous media as a green procedure. Ultrason Sonochem. 18: 447–454.

Li, J.T., S.X. Wang, G.F. Chen and T.S. Li. 2005. Some applications of ultrasound irradiation in organic synthesis, current organic synthesis. Curr. Org. Synth. 2: 415–436.

Li, J.T., X.T. Meng, B. Bai and M.X. Sun. 2010. An efficient deprotection of oximes to carbonyls catalyzed by silica sulfuric acid in water under ultrasound irradiation. Ultrason Sonochem. 17: 14–16.

Li, J.T., M.X. Sun, G.Y. He and X.Y. Xu. 2011. Efficient and green synthesis of bis(indolyl)methanes catalyzed by ABS in aqueous media under ultrasound irradiation. Ultrason Sonochem. 18: 412–414.

Li, J.T., Y.W. Li, Y.L. Song and G.F. Chen. 2012. Improved synthesis of 2,20-arylmethylene bis(3-hydroxy-5,5-dimethyl-2-cyclohexene-1-one) derivatives catalyzed by urea under ultrasound. Ultrason Sonochem. 19: 1–4.

Machado, I.V., J.R.N. dos Santos, M.A.P. Januario and A.G. Correa. 2021. Greener organic synthetic methods: sonochemistry and heterogeneous catalysis promoted multi-component reactions. Ultrason Sonochem. 78: 105704

Majhi, S. 2020. Diterpenoids: natural distribution, semisynthesis at room temperature and pharmacological aspects-a decade update. ChemistrySelect 5: 12450–12464.

Majhi, S. 2021a. Applications of ultrasound in total synthesis of bioactive natural products: a promising green tool. Ultrason Sonochem. 77: 105665.

Majhi, S. 2021b. Applications of norrish type I and II reactions in the total synthesis of natural products: a review, Photochem. Photobiol. Sci. 2021a. 20: 1357–1378.

Majhi, S. 2021c. Applications of yamaguchi method to esterification and macrolactonization in total synthesis of bioactive natural products. ChemistrySelect. 6: 4178–4206.

Majhi, S. 2021d. Discovery, development, and design of anthocyanins-inspired anticancer agents-a comprehensive review. Anti-cancer Agents Med Chem. 22(19): 3219–3238.

Majhi, S. 2021e. The art of total synthesis of bioactive natural products via microwaves. Curr Org Chem. 25: 1047–1069.

Majhi, S. 2022a. Recent developments in the synthesis and anti-cancer activity of acridine and xanthine-based molecules. Phys. Sci. Rev. 0: 1–35. doi.org/10.1515/psr-2021-0216.

Majhi, S. 2022b. Synthesis of bioactive natural products and their analogs at room temperature – an update. Phys. Sci. Rev. 1–27. doi.org/10.1515/psr-2021-0094

Majhi, S. and D. Das. 2021. Chemical derivatization of natural products: semisynthesis and pharmacological aspects- a decade update. Tetrahedron. 78: 131801.

Mandhane, P.G., R.S. Joshi, D.R. Nagargoje and C.H. Gill. 2010. Ultrasound-promoted greener approach to synthesize a-hydroxy phosphonates catalyzed by potassium dihydrogen phosphate under solvent-free condition. Tetrahedron Lett. 51: 1490–1492.

Martins, L.M.D.R.S., S. Hazra, M.F.C. Guedes da Silvaa and A.J.L. Pombeiro. 2016. A sulfonated schiff base dimethyltin (IV) coordination polymer: synthesis, characterization and application as a catalyst for ultrasound- or microwave-assisted baeyer-villiger oxidation under solvent-free conditions. RSC Adv. 6: 78225–78233.

Mason, T.J. 1997. Ultrasound in synthetic organic chemistry. Chem. Soc. Rev. 26: 443–451.

Metzger, J.O. 1998. Solvent-free organic syntheses. angew. Chem. Int. Ed. 37: 2975–2978.

Michnovicz, J.J. and H.L. Bradlow. 1991. Altered estrogen metabolism and excretion in humans following consumption of indole-3-carbinol. Nutr. Cancer 16: 59–66.

Mojtahedi, M.M., M. Javadpour and M.S. Abaee. 2008. Convenient ultrasound mediated synthesis of substituted pyrazolones under solvent-free conditions. Ultrason Sonochem. 15: 828–32.

Palmisano, G., F. Tibiletti, A. Penoni, F. Colombo, S. Tollari, D. Garella et al. 2011. Ultrasound-enhanced one-pot synthesis of 3-(Het)arylmethyl-4-hydroxycoumarins in water. Ultrason Sonochem. 18: 652–660.

Penteado, F., B. Monti, L. Sancineto, G. Perin, R.G. Jacob, C. Santi et al. 2018. Ultrasound-assisted multicomponent reactions, organometallic and organochalcogen chemistry. Asian J. Org. Chem. 7: 2368–2385.

Petersen, L.J. 2009. Anticoagulation therapy for prevention and treatment of venous thromboembolic events in cancer patients: a review of current guidelines. Cancer Treat. Rev. 35: 754–764.

Phillips, A.M.F., M.F.C. Guedes da Silva and A.J.L. Pombeiro. 2020. New trends in enantioselective cross-dehydrogenative coupling. Catalysts, 10(5): 529.

Poupelin, J.P., G. Saint-Rut, O. Fussard-Blanpin, G. Narcisse, G. Uchida-Ernouf and R. Lakroix. 1978. Synthesis and antiinflammatory properties of bis(2-hy- droxy, 1-naphthyl) me thane derivatives. Eur. J. Med. Chem. 13: 67–71.

Puri, S., B. Kaur, A. Parmar and H. Kumar. 2013. Applications of ultrasound in organic synthesis - a green approach. Curr. Org. Chem. 17: 1790–1828.

Ramozzi, R. and K. Morokuma. 2015. Revisiting the passerini reaction mechanism: existence of the nitrilium, organocatalysis of its formation and solvent effect. J. Org. Chem. 80: 5652–5657.

Reddy, B.V.S., M.R. Reddy, G. Narasimhulu and J.S. Yadav. 2010. InCl$_3$-catalyzed three-component reaction: a novel synthesis of dihydropyrano[3,2-b]chromenediones under solvent-free conditions. Tetrahedron Lett. 51: 5677–5679.

Rostamizadeh, S., A.M. Amani, G.H. Mahdavinia, G. Amiri and H. Sepehrian. 2010. Ultrasound promoted rapid and green synthesis of 1,8-dioxo-octahydroxanthenes derivatives using nanosized MCM-41-SO$_3$H as a nanoreactor, nanocatalyst in aqueous media. Ultrason Sonochem. 17: 306–309.

Sainath, Z. and P. Pravinkumar. 2019. A review on solvent-free methods in organic synthesis. Curr. Org. Chem. 23: 2295–2318.

Sarma, R.J. and J.B. Baruah. 2005. One step synthesis of dibenzoxanthenes. Dyes Pigm. 64: 91–92.

Shin, K., H. Kim and S. Chang. 2015. Transition-metal-catalyzed C-N bond forming reactions using organic azides as the nitrogen source: a journey for the mild and versatile C-H amination. Acc. Chem. Res. 48: 1040–1052.

Sirisoma, N., A. Pervin, J. Drewe, B. Tseng and S.X. Cai. 2009. Discovery of substituted N'-(2-oxoindol in-3-ylidene)benzohydrazides as new apoptosis inducers using a cell- and caspase-based HTS assay. Bioorg. Med. Chem. Lett. 19: 2710–3.

Strukul, G. 1998. Transition metal catalysis in the baeyer–villiger oxidation of ketones. Angew. Chem. Int. Ed. 37: 1198–1209.

Sujatha, K., G. Shanthi, N.P. Selvam, S. Manoharan, P.T. Perumal and M. Rajendran. 2009. Synthesis and antiviral activity of 4,4'-(arylmethylene)bis(1H-pyrazol-5-ols) against peste des petits ruminant virus (PPRV). Bioorg. Med. Chem. Lett. 19: 4501–4503.

Tanaka, K. and F. Toda. 2000. Solvent-free organic synthesis. Chem. Rev. 100: 1025–1074.

Targel, T. and M. Portnoy. 2020. Baeyer-villiger-including domino two-step oxidations of β-o-substituted primary alcohols: reflection of the migratory aptitudes of o-substituted alkyl group in the outcome of the reaction. Catalysts 10: 1275.

Wet-osot, S., C. Duangkamol, W. Phakhodee and M. Pattarawarapan. 2016. Ultrasound-assisted solvent-free parallel synthesis of 3-arylcoumarins using n-acylbenzotriazoles. ACS Comb. Sci. 18: 6, 279–282.

Xiong, X. and M.C. Pirrung. 2008. Modular synthesis of candidate indole-based insulin mimics by claisen rearrangement. Org. Lett. 10: 1151–1155.

Zhang, Z.H., J.J. Li and T.S. Li. 2008. Ultrasound-assisted synthesis of pyrroles catalyzed by zirconium chloride under solvent-free conditions. Ultrason. Sonochem. 15: 673–676.

Ziarati, A., J. Safaei-Ghomi and S. Rohani. 2013. Sonochemically synthesis of pyrazolones using reusable catalyst CuI nanoparticles that was prepared by sonication. Ultrason Sonochem. 20: 1069–1075.

CHAPTER 6

Nanomaterials and Nanocomposites with Sonochemistry

P.R. Bhilkar,[1] *A.K. Potbhare,*[1] *M.S. Nagmote,*[1] *T.S. Shrirame,*[1]
R.B. Jotania,[2] *M.F. Desimone*[3,*] and *R.G. Chaudhary*[1,*]

1. Introduction

Sonochemistry is a branch of chemical science dealing with the physicochemical process occurring in solution by means of ultrasonic waves. The ultrasonic sound wave discusses the portion of the acoustic spectrum that is beyond the hearing limit of the human, typically within the range from 18 to 20 kHz (Mason and Peter 2002b, Mason and Lorimer 2002a, Suslick et al. 1986). Generally, sonochemistry practises in between the frequencies of 20 to 40 kHz, as it is significant for designing common laboratory equipment, while the frequencies between one to10 MHz have many therapeutic and diagnostic applications (Cravotto and Cintas 2007, Mason 2003). Sonochemistry deals with sustainable chemistry with aim of consumption of less harmful chemicals, minimal energy consumption and lesser time for completion of reaction with better product yield. With an eye towards the green chemistry goals, such as minimising hazardous wastes, the use of catalysts; so, synthesis using ultrasonic waves is remarkable. The use of ultrasound to accelerate classical chemical reactions has been significant (Chatel 2018a, Machado 2021).

Sonochemical low frequencies ~ (20 to 80 kHz) results in changes by means of shockwaves, microjets, microconvection, etc. Whereas the large frequencies of ultrasonic waves range between 150 to 2,000 kHz favours the generation of highly reactive species (such as HO˙ radicals) achieved via local hotspots made by acoustic

---

[1] Post Graduate Department of Chemistry, Seth Kesarimal Porwal College of Arts and Science and Commerce, Kamptee-441001, India.
[2] Department of Physics, Electronics and Space Science, School of Sciences, Gujarat University, Ahmedabad-380 009, India.
[3] Universidad de Buenos Aires, Facultad de Farmacia y Bioquimica, Junin 956 Piso 3, (1113) Ciudad Autonoma de Buenos Aires, Argentina.
* Corresponding authors: chaudhary_rati@yahoo.com; martinfdesimone@gmail.com

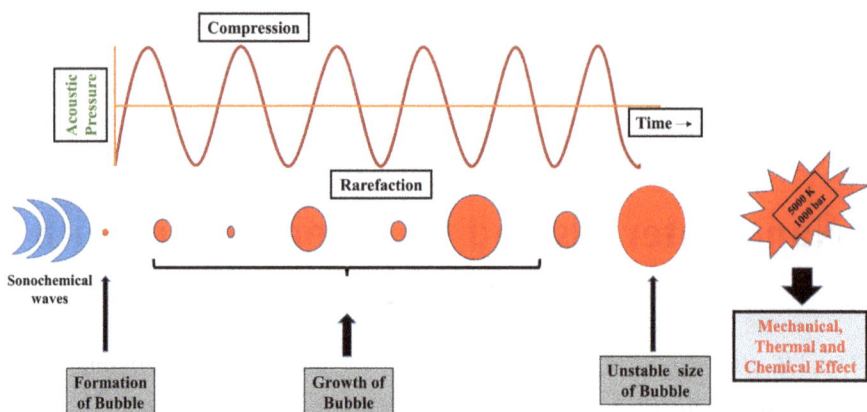

Figure 1. Schematic representation of the method of acoustic cavitations: generations, growth, and collapse of bubbles in an aqueous phase exposed to high intensity ultrasonic waves.

cavitations, that result in such chemical effects. The ultrasonic sound waves produce the acoustic cavitations phenomenon via the development, growth, and collapsing of microbubbles, elevating the temperature around 5,000 K and reaching the pressure up to 1,000 bar as shown in Figure 1 (Gong and Hart 1998, Masson and Peter 2002b, Wood et al. 2017).

The periodic cycles of compression and rarefaction phenomena generated in the liquid medium bring on perpetuating of ultrasound waves (Cobley et al. 2008). When supplied acoustic power is at threshold value, the rarefaction cycle overcomes existing attractive forces in the molecules of a liquid. However, the cavitation microbubbles are formed concertedly. During its expansion phase, a lesser amount of gas gets to enter the microbubbles from the surrounding medium whereas, in the compression phase, the gas is partially expelled. Furthermore, the bubbles developed gradually to attain an equilibrium size throughout cycles under the applied frequency. The hasty collapse of microbubbles generated previously results in strong local effects in terms of mechanical, thermal, and chemical. This significant effect leads to the potential application of sonochemistry in various fields (Cravotto and Cintas 2006, Mason 2003).

Owing to the salient features of sonochemistry, it can be employed for the preparation of certain organic compounds, organometallics, and materials. Especially, the sonochemistry methodology is adopted for the synthesis of various nanomaterials (NMs), and nanocomposites (NCs) from their respective precursors. The important parameters to be considered while synthesizing NMs and NCs is frequency and power of ultrasonic waves (Gumeci et al. 2013). There are two basic methods reported in sonochemistry: namely primary and secondary sonochemistry. Both these methods are adopted for the synthesis of NMs and NCs. In primary sonochemistry, the collapse of microbubbles raises the temperature causing the dissociation of bond and generates free metal atoms. This atom perchance injected into the solution and further undergoes nucleation to yield NMs and NCs according to the availability of relevant templates or stabilizers existing in the solution, (reaction occur inside the microbubble generating product). Whereas, in secondary sonochemistry, the precursors may further undergo sonochemical reactions, via collapsing of the microbubbles externally with high energy species (radicals) generated from the sonolysis of vapour molecules inside the collapsing microbubbles (reaction diffuse in liquid phase and react to give products). Thus, the entire reaction mechanism ends up with series of reactions and finally produces NMs or NCs (Xu et al. 2013).

As previously mentioned, the phenomena of microjets, and shockwaves plays an invaluable role in the synthesis of NMs and NCs. With increase in the radius of the bubbles, it increases the surface area up to certain extent. This recently generated bubble will collapse after having a larger diameter and a shorter lifespan. During this collapse, the bubble self-reinforces and induces in homogeneity and a sphericity in the surrounding bubbles. Such disturbances with the structure of microbubbles result in high speed microjets impact on the surfaces of NMs and NCs (Blake and Gibson 1987). However, if the bubbles remain unperturbed with respect to its surface there may be rapid rebound from the minimum radius of the spherical microbubbles. Because of reduced diameter of the bubble, surrounding liquid get compressed and propagates outward in the form of shock wave (Ohl et al. 1999). These shockwaves may experience pressure of 60 kbar and four km/s in water. In addition to this, the shockwave also generates physical effects like mechanical or vibrational, while chemical effects like radical formation. Interestingly the shockwaves accelerate suspended particles present in the liquid subsequently changing shapes of particle, size distributions, surface morphologies of NMs and NCs (Doktycz and Suslick 1990, Prozorov et al. 2004). Beside these parameters, particle agglomeration and fragmentation are also evident from various scientific studies. Similarly, it was also observed that layered material exfoliated into 2D layers under the effect of shockwaves (Bang and Suslick 2010, Shchukin et al. 2010).

In this chapter, we tried to explore the importance of sonochemistry with special emphasis on sonochemical mediated synthesis of NMs and NCs. It also describes the present and future aspects of sonochemistry.

2. Nanomaterials and Nanocomposites with Sonochemistry

2.1 Historical Background

In 1794, Italian biologist Lazzaro Spallanzani discovered ultrasounds from the movements of bats. Later, ultrasound was discovered in 1883 by the English physiologist Sir Francis Galton, who created the "Galton's whistle", emitting ultrasound in a range that could only be heard by dogs, not by humans. Further, the French physicists Jacques Curie and Pierre Curie (1859 to 1906) discovered the piezoelectric material and some powerful electronic devices which generated ultrasonic waves in water. Subsequently, in 1917, Paul Langevin offered the first industrial application of ultrasound by developing the SONAR system, using ultrasonic vibrations to detect the submarines. In submarine propellers, Sir John Isaac Thornycroft and Sydney Walker Barnaby first observed cavitations as propellers became pitted and eroded over a relatively short period of time. Alfred Lee Loomis, Robert William Wood, and Theodore William Richards reported for the first time in 1927 the biochemical effects of ultrasounds which show that it can be a useful tool in chemistry. Especially, sonochemistry can be considered as an important tool in green chemistry. Although, it took until the 1980s, and the advent of reliable and commercially available ultrasonic generators for researchers to demonstrate that ultrasonic waves offer undeniable opportunities in chemistry, particularly in the synthesis of organic compounds and materials (Chatel 2016).

In 1992, Suslick et al., for the first time demonstrated the synthesis of metal carbonyl using a horn type sonochemical reactor, for the synthesis nanomaterials. The fabrication of NMs through ultrasound irradiation was reported by the author for the first time (Suslick et al. 1992). In another study, Nagata et al., examined how sonochemical dispersion of silver nanoparticles can be prepared by using ultrasound irradiation (Nagata et al. 1992). Similarly, Yeung et al., investigated the formation of gold and other metal particles by sonolysis of their aqueous metal ion solutions (Yeung et al. 1993). Afterward 2000, there is exponential growth observed in the study of sonochemistry in the field of nanotechnology, especially for the synthesis of NMs and NCs. Initially, the synthesis of metal oxides and metal chalcogenides was reported by Qian et al., (ZnO), Liang et al., (ZrO_2), Wang et al. (ZnS), Qiu et al., (CuS), etc., via sonochemical irradiation technique (Liang et al. 2003, Qian et al. 2003, Qiu et al. 2003, Wang et al. 2003). These metal oxides and metal chalcogenides have great potential applications, especially in electrochemical and photo degradation. Moreover, the research describes the sonication assisted method in the fabrication of metallic nanoparticles with a green reducing agent such as ascorbic acid (Qiu et al. 2003), zinc powder (Yu et al. 2003), and hydrazine (Wu and Chen 2003). Similar to this, Katoh et al. generated nanostructured materials in 1999 using this green technology with varied dimensionality. They explain the carbon nanotubes by applying ultrasonic irradiation in liquid chlorobenzene with $ZnCl_2$ particles (Katoh et al. 1999). In 2004, Gedanken discussed the fabrication of NMs and NCs by using the sonochemical process in detail (Gedanken 2004).

2.2 Recent Studies on Sonochemistry

Baladi et al., prepared Dy_2ZnMnO_6 NCs by co precipitation method aided with sonication technique for examination of photocatalytic activity on organic dyes contaminants in water (Baladi et al. 2019). Dheyab et al., synthesized gold nanoparticles for photocatalytic degradation of organic dyes with a vibra-cell™ ultrasonic solid horn with a frequency of 20 kHz and different ultrasound output powers (12, 20, and 36 W) (Dheyab et al. 2021) Monsef et al., sonochemically fabricated the $PrVO_4$ NCs with the 60 W power and frequency of 18 kHz applied for degradation of organic dyes via photocatalytic route for wastewater treatment (Monsef et al. 2019) Chatel reviewed the use of sonochemistry in NMs as a nanocatalyst for the catalytic reaction (Chatel 2018b).

Using ultrasonic cavitations, Palomino et al., investigated the effect of sonication time and thermal energy on the crystal structure and magnetic properties of $SrFe_{12}O_{19}$ NCs (Palomino et al. 2016). Esran et al., modified the synthesis of zinc borate ($Zn_3B_6O_{12}\cdot3.5H_2O$) NCs using ultrasonic probe. Here, in this they reported the accelerated way of synthesis with help compared to other conventional literature methods (Ersan et al. 2020). Dheyab et al., reported Fe_3O_4@Au core shell NCs by reviewing the sonochemical method. As acoustic cavitations help to prevent the clusters agglomeration, forming more stable dispersion which affects the structure and surface of the Fe_3O_4@Au NPs (Dheyab et al. 2020). Park et al., studied the step by step, the mechanism of formation of Fe@Pt NCs through sonochemistry (Park et al. 2021). Afreen et al., reviewed bulk scaled synthesis of polyhydroxylated carbon NMs (graphene and Buckminsterfullerene) by the sonochemical process. Moreover, they also inspected the acute aspects of sononanochemistry, and the statistical tactics to improve large-scale production of carbon NMs by sonochemical process (Afreen et al. 2018)

Nobre and coworker utilized a combination of sonochemical and hydrothermal techniques to synthesize α-Ag_2WO_4 NCs, which exhibited notable efficacy against both bacterial and fungal cells (Nobre et al. 2019). Bayrami et al., biosynthesized ZnO NMs using the leaf extract of whortleberry (*Vaccinium arctostaphylos* L.) through an ultrasonic method. These phytosynthesized ZnO NMs employed for antidiabetic, antibacterial, as well as their sonophotocatalytic performances (Bayrami et al. 2019) Islam et al., reviewed development of sonoelctrochemicals, a hybrid technique for fabrication of controlled size and shapes with high scaled yield of NMs (Islam et al. 2019).

2.3 Role of Sonochemistry in NMs and NCs Synthesis

2.3.1 Synthesis of Graphene Based NMs/NCs

The ultrasound technique is the most important and operative technique for the synthesis of nanoparticles and nanocomposites. In this ultrasound or sonochemical techniques, the chemical effects and physical effects of ultrasound arise from acoustic cavitations, which is, the quick development and formation of bubbles and finally affricate collapse of bubbles in a liquid due to the irradiation with high intensity of ultrasound. This crumpling of bubbles generates extreme heat up to 5,000 K and pressure greater than a hundred bar within a very short period (Potbhare et al. 2020b, Umekar et al. 2021b). Ultrasound based sonochemical approach is comprehensively reflected in the design of a novel fabrication method, biochemical analysis, or biochemical routes. The utilization of non-hazardous compounds, eco-friendly solvents, and sustainable resources are some of the major concerns in green synthesis methods. These techniques being fast, operative, and inexpensive, the green sonochemical method may be developed as an encouraging technique for the synthesis of graphene-based nanomaterials because of their widespread futuristic applications (Chaudhary et al. 2021, Umekar et al. 2021a).

The present chapter emphasizes on ultrasound mediated synthesis of graphene-based nanomaterials, and nanocomposites which are shown in Figure 2. The production of graphene-based nanocomposites through exfoliation from graphitic oxide and the synthesis of metal NPs through reduction process. In an example of sonochemical method, metal-based nanoparticles dispersed on graphene sheet using metal precursor and GO. This typical process is carried out *via* reduction and exfoliation of graphene at 20 kHz ultrasound sonicator. Where the shear forces produced by the acoustic cavitations are enough to overwhelm the van der Waals forces between the graphene sheets and prevent their aggregation. During the sonolysis

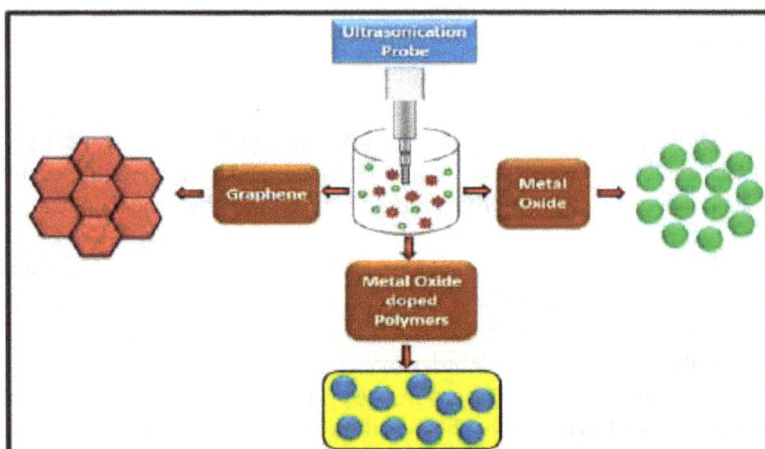

Figure 2. Schematic representation of synthesis of different material via sonochemical method.

process, the cavitations effect produces two types of radicals namely oxidative and reductive radicals which can be tuned to initiate an explicit chemical reaction. In last few years, several research groups have demonstrated surfactant-based metal NPs using different reducing agents which can be synthesized by sonication method at frequencies ranging from 20 to 1,000 kHz. In recent years, ultrasound inspired synthesis of bimetallic NPs with core shell morphological which can be produced by reduction of the metal precursor. The optimum ultrasound frequency for reduction has been determined to be around 200 kHz. Sonication at this frequency provides higher radical concentrations conducive to faster chemical reductions. As a result of the ultrasonic irradiation effect, dissolved particles are homogeneously dispersed in the mixture preventing agglomeration of the nanoparticles (Chaudhary et al. 2021, Potbhare et al. 2019).

Anandan et al., reported sonochemical advances for the simultaneous graphene oxide reduction and stacking of Pt, Sn, and Pt/Sn bimetallic NPs on rGO nanosheets (Anandan et al. 2012). TEM images showed that the adopted method can produce a more uniformly distributed Pt, Sn, and Pt/Sn NPs on the rGO sheets, and the contents of the hybrid materials were confirmed by EDX measurements.

Potbhare et al., discussed sonochemical synthesis of Ag/graphene NCs with the aid of an environmentally friendly bacterial reducing agent, namely *pseudomonas aeruginosa*, was quite successful in terms of controlling and determining experimental time, size, and shape (Potbhare et al. 2020a).

2.3.2 Synthesis of Metal Oxides NMs/NCs

Focussing on green synthesis methods, a diversely applicable, simple, and cost effective sonochemical synthetic route has been executed for metal/metal oxide NPs and NCs synthesis by many researchers. The shape and size of nanoparticles could be achieved by acoustic cavitations which involve nucleation followed by growth of particle from fine molecular distributions and then collapsing micro bubbles pre-existing forms by the virtue of ultrasonic waves in liquid phase (Nagvenkar et al. 2019).

Ohayon and Aharon have reported quick sonosynthesis of well crystalline TiO_2, WO_3, and V_2O_5NPs from metal chlorides with aromatic alcohol in Ar atmosphere at optimum low-down temperature in a non-aqueous system. PXRD, HR-SEM and BET revealed exclusive crystallization patterns of shapes and sizes of these sonosynthesized nanoparticles. TiO_2 particles were found to be '*quasi*' zero dimensional, spherically shaped of three to seven nm size. '*Quasi*' one dimensional ellipsoidal V_2O_5 particles were found to have sizes of 150 to 200 nm in length and 40 to 60 nm in width, whereas square shaped WO_3 particles were found to be '*quasi*' two dimensional sizing 30 to 50 nm. The obtained findings were also validated by ESR and NMR.

To utilize effective biocidal ability of ZnO NPs, their stabilization on dispersion in solution was enhanced by ultra-sonication incorporating polyvinyl alcohol as the stabilizer. This ultrasonic induced one step synthesized zinc oxide PVA nanocomposites were found to be highly stable and efficient biocides against pathogens *Escherichia coli* and *Staphylococcus aureus* compared to zinc oxide exclusively (Nagvenkar et al. 2016).

Hosseni and Khodadadipoor reported ultrasonic calcinating of zinc acetate, nickel acetate, and iron chloride as precursors yielding ZnO, NiO, and α-Fe_2O_3 nanoparticles. Powder XRD and field emission SEM are used to characterise the finished goods. TG and DSC were used to assess the catalysing efficacy of these obtained oxides for thermo degradation of ammonium perchlorate. Insertion of just two percent of these NPs resulted into merging of two exothermic peaks into single peak. Dropping in high thermal decomposition peak of ammonium perchlorate by around 120°C and 105°C is also observed in the presence of ZnO and NiO NPs, respectively (Hosseni and Khodadadipoor 2018).

Further, Liu and co-workers sonochemically synthesized nontoxic Au NCs with near IR emission and Au@AgNCs with good yellow fluorescence. These highly biocompatible nanocomposites proved to be highly water soluble and can be best implemented for selective determination of Cu (II) ions (Liu et al. 2011).

Zahra and Mohammad obtained a novel nanocomposite by silver nanoparticles decorated on reduced graphene oxide (rGO-Ag) through a facile, single step and green method of sonosynthesis. Sonication was carried out on low frequency ultrasonication of 20 kHz under optimum parameters like sonication duration, sonication amplitude and temperature. Here, simultaneous reduction of GO and Ag^+ ions was achieved by polyol without any external capping agents. Rather polyol is playing a dual role of reducing agent as well as solvent phase of reaction. Further, these ultrasonically obtained nanocomposites were compared with those by conventional. When results obtained compared with those by conventional methods, uniform pattern of silver NPs on rGO and superior catalysed action in reducing 4-nitrophenol into 4-aminophenol in the presence of sodium borohydride was demonstrated by sonosynthesized nanocomposites. UV-Vis, FT-IR spectroscopy, XRD, and TEM were used to confirm the morphology of NCs, and it showed that they were 18 nm in size (Mohammadi and Entezari 2018).

Highly stable, reproducible sensors from Fe_3O_4@rGO NCs were developed for analysis of uric acid biomarker for gout and kidney stone. Fe_3O_4@rGO nanocomposites were synthesized by ultrasonication. These electrostatically stabilised Fe_3O_4 NPs were uniformly decorated on rGO and their morphology was substantiated by XRD, high resolution SEM and TEM, elemental mapping, and EDX. Modified Fe_3O_4@ rGO sensor was further implemented to determine traces of uric acid in blood serum and urine samples (Balasubramanian et al. 2019).

2.3.3 Synthesis of Metal Oxides Polymer NMs/NCs

Traditional sol-gel methods employed to prepare polymer and NCs from metal alkoxides involve reactions of hydrolysis and condensations. However, due to the immiscibility of water and metal alkoxides, a common solvent (i.e., methanol, ethanol) was frequently employed. In a different approach, Esquivias and Zarzycki reported a procedure, that they called sonogels, as an alternative method in sol-gel processing (Esquivias and Zarzycki 1988). The authors successfully employed ultrasonic radiation to initiate the hydrolysis of alkoxides in the presence of water and an acid catalyst (i.e., HCl). The resulting sonogels were three times denser than the gels obtained with traditional procedures. Moreover, sonogels presented higher reticulation and more close porosity (Esquivias and Zarzycki 1988, De la Rosa-Fox et al. 1990). Furthermore, sonochemistry can be recognized as environment friendly since it diminishes the number of solvents required for the synthesis of different materials. Regarding the mechanism of ultrasonic wave polymerization, it was recognized early on that the cavitations phenomena generate high pressures due to the collapse of vapor bubbles. In addition, hot spot temperatures increases have been observed. Furthermore, the hydrolysis rate increases as an inherently sophisticated dissolution between alkoxides and water, which increases the reaction within the immiscible phases (Paccola et al. 2015, Vollet et al. 2016).

This sonochemistry procedure or procedures adapted from it, were further employed to obtain sol-gel matrices with thermal and chemical stability, physical rigidity, and highly porous structure. Moreover, these interesting characteristics were further employed to immobilize different biological components with limited or controlled release of the immobilised species (i.e., enzymes, bacteria, yeast, among others). Indeed, upon immobilization the biological species are in a protective microenvironment that allows easy recovery and reuse. For example, the urease enzyme was immobilized in silica NCs prepared by sonication of a mixture of tetraethoxysilane, water and HCl. The resulting homogeneous sol was afterwards mixed with the enzyme in a buffered solution to obtain the NCs (Desimone et al. 2008). This sonochemistry synthesis preserved enzyme stability and activity. However, during the hydrolysis of alkoxides alcohol molecules are released which can generate detrimental effects on living cells. Thus, it was required to eliminate the alcohol generated during the hydrolysis step before the immobilization of living cells. In this sense, Desimone et al., adapted the procedure to eliminate ethanol excess employing N_2 and subsequently immobilize *Escherichia coli*, producer of recombinant proteins (Desimone et al. 2005). This procedure would be further employed to preserve the viability of different bacterial strains for applications in biotechnological or biochemical processes.

The sono gel networks also work as a protective microenvironment. There is evidence that confirms that sono gel immobilisation protects *Saccharomyces cerevisiae* yeast against the toxic effect of ethanol (Desimone et al. 2002). It was suggested that the polyhydroxylated silane network establishes hydrogen bonding interactions with water, generating a highly structured water layer that surrounds and protects the microorganism present in the pores of the sono gel network. Similarly, *Saccharomyces cerevisiae* yeast were protected against the toxic effects of five organic solvents (i.e., ethanol, propanol, butanol, pentanol, and octanol) (Desimone et al. 2003). The survival rate of immobilised *Saccharomyces cerevisiae* has a direct correlation with the hydrophobicity of the solvent, reaching the maximum with the most hydrophobic solvent assayed (e.g., octanol). Other attempts to immobilize mammalian cells in sonogels highlights the necessity to use polyol grafted silane routes, as well as the addition of different organic polymers, to improve the viability of immobilized cells (Desimone et al. 2011, Catalano et al. 2012).

More recently, inorganic nanoparticles and NCs were prepared with the aid of sonochemistry for drug delivery applications. The sonochemical methods succeed to shorten reaction times, diminish nanoparticles size and improve uniformity in nanoparticles size distribution (Liu et al. 2021).

2.4 Mechanism of Sonochemistry

Cavitations are used in sonochemistry to achieve a chemical conversion. Acoustic cavitation is the mechanism that causes sonochemical phenomena in liquids (Bhangu and Ashokkumar 2016). When ultrasound waves travel through a liquid medium, they produce a pattern of rarefactions and compressions that provide energy to the liquid phase. Rarefaction and compression cycles apply positive and negative pressure to the liquid, pulling or pushing the molecules towards or away from one another (Doktycz and Suslick 1990). The acoustic cavitations phenomenon is divided into three stages: (i) nucleation, (ii) bubble growth (expansion), (iii) implosive collapse respectively.

In initial stage, cavitational nuclei are formed by trapping microbubbles in micro crevices of microscopic particles inside the liquid, where cavities are created based on the liquid purity and type (Shut and Mozzharov 2017). In second stage microbubble is formed and a small cavity develops quickly because of inertial effects when the intensity is too strong. When the acoustic intensity is low, the cavity expands by rectified diffusion, and lasts for some more acoustic cycles before expanding again (Suslick 1990). In the last stage, when a cavity becomes too grown to absorb sufficient energy to support itself, the surrounding liquid rushes in, causing a rapid implosion and catastrophic collapse (Jambrak et al. 2014, Mason and Lorimer 2002a). Chemical reactions in sonochemistry proceeds through radical intermediates, even though homogeneous sonication can degrade many organic substances, the decomposition rates may be too slow for practical applications (Ameta et al. 2018).

There are three primary reaction sites in homogeneous sonochemistry: the cavitations bubble, the interface between gas bubbles and the surrounding liquid,

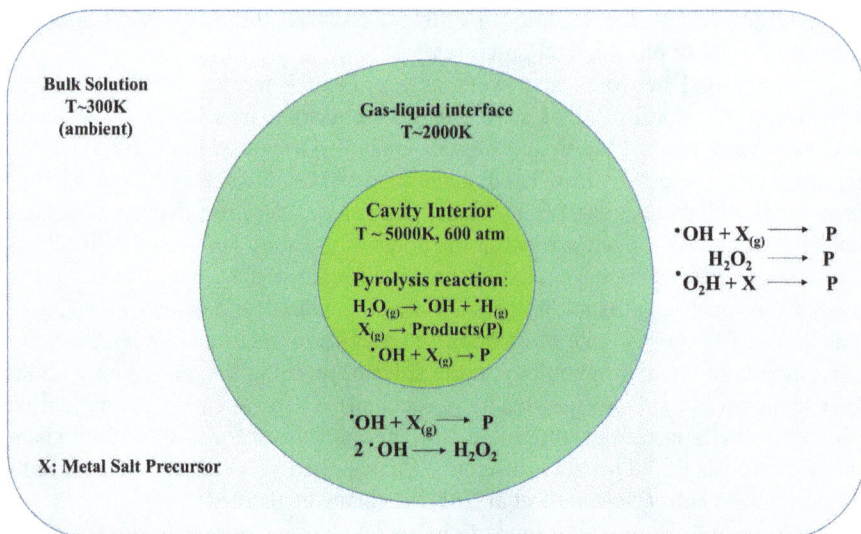

Figure 3. Schematic representation of main reaction sites in homogenous sonochemistry.

and the bulk solution, as shown in Figure 3 (Hekimoglu 2020). Bulk liquid, gas-liquid interface or cavity of highly volatile bubble act as the site for decomposition of organic molecules (Weavers et al. 2005). The quantity of not combined free radicals that have migrated into the bulk solution from collapsing cavities and the interface limits oxidation processes in the bulk solution (Wood et al. 2017). Low volatility hydrophilic chemicals are unlikely to migrate to the bubble or interfacial region, preferring instead to stay in the bulk solution. Highly volatile hydrophobic chemicals quickly diffuse into cavitation bubbles, where pyrolysis and oxidation with the ˙OH radical are the most common reactions. Finally, the gas-liquid interface is the most likely reaction site for water pollutants, and the reactions happen at extremely rapid rates due to the high concentration of ˙OH and significant pressure gradients and temperatures. The rate of ˙OH radical transfer from the bubble interior into the bulk solution, which is a function of pressure intensity, cavities duration, and reactor geometry determines the fate of highly soluble water pollutants in the bulk solution as well as the frequency of the ultrasonic waves used (Lorimer et al. 1991, Mason and Peter 2002b).

2.5 Applications of Sono Mediated Synthesized NMs/NCs

Ultrasound (20 kHz to 10 MHz) is a good tool to modify chemical reactions (synthesis, degradation, catalysis, etc.) and sonochemistry is considered as one of the older techniques to prepare nano sizes materials. There is a huge interest to prepare and modify nanomaterials (nanorods, nanowires, nanorings, nanoplates, nanotubes, nanocones, nanobelts, etc.) and nanocomposites using sonications. Various nanostructured materials (metals, oxides, alloys, carbides, sulfides, catalysts, nano colloids) can be produced by this technique. The Figure 4 represents various

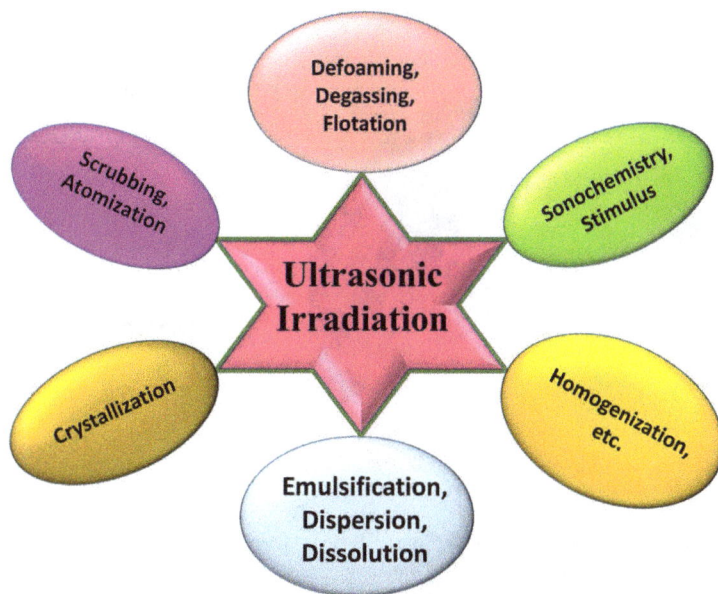

Figure 4. Various applications of ultrasound in chemistry, material science and manufacturing process.

applications of ultrasound in chemistry, material science, and manufacturing processes (Zhanfeng et al. 2021).

Sonochemical synthesis is very useful to form biomaterials like protein microsphere, which has large biomedical applications in the medical field as drug delivery, sonography, MRI, etc. (Suslick and Price 1999). K. Suslick et al., prepared iron particles (Fe(CO)$_5$, three to eight nm) on silica surface using ultrasonic irradiation. In fact, iron particles were prepared using ultrasonic cavitations and then deposited on the silica surface. TEM images also support that Fe(CO)$_5$ was highly dispersed in SiO$_2$ (Suslick and Price 1999, Suslick et al. 1994, 1996). A brief review of sono mediated synthesized nanomaterials was published by A. Gedanken. The novel techniques like sonochemistry, sono electrochemistry, and microwave heating to prepare nanosized materials were discussed (Gedanken 2004). R. F. Elsupikhe et al., studied catalytic activities of Ag-NPs prepared using ultrasound assisted green synthesis route with different amounts (0.1 to 0.3 weight percentage with increment of 0.05 weight percentage) of κ-carrageenan (Elsupikhe et al. 2015). The κ-carrageenan was used as a stabilizer. As prepared Ag/κ-carrageenan were analyzed using FT-IR, UV-Vis, XRD, SEM, and TEM. UV-Vis spectra showed the presence of the surface Plasmon bands ~ 402 to 420 nm, which indicates the formation of Ag-NPs. The XRD patterns of all samples confirmed the formation of Ag nanoparticles (Ag-NPs) and formed Ag/κ-carrageenan NPs possess only pure silver crystalline phase; no impurity was detected. SEM images showed that surface morphology has changed with the κ-carrageenan while TEM analysis and particle size distribution indicated that the amount (number) of Ag-NPs increased with the weight percentage of stabilizer. This review covers the preparation and modification of nanostructured inorganic materials that can be directly used for the catalytic process; it emphasizes

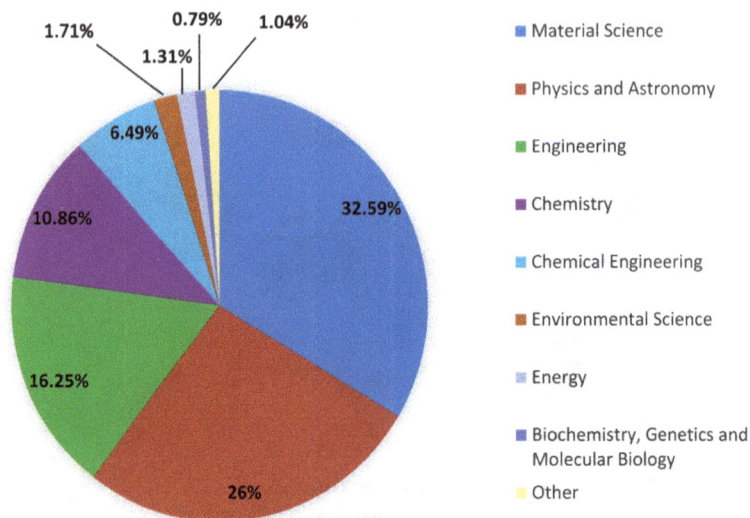

Figure 5. Use of spinels in various fields. Source: Scopus; Topic: "Spinel Ferrite"; Searching time: 15 June 2021. *Reprint with the permission of journal/magazine via Elsevier.*

highly on applications of sonochemically produced materials during catalytic reactions like degradation, oxidations, esterification, and hydrogenation.

Magnetic nanoparticles are widely used in various fields. Spinel ferrites (SFs) have a special interest because of their unique magnetic and dielectric properties. Some advantages of SFs are: good catalytic activity, chemical stability, excellent magnetic properties, modest chemical composition, tunable size and shape, electrical resistivity, low cost, interesting band gap, mechanical hardness, thermal stability, high specific surface area, high Curie temperature. SFs are ferromagnetic and show semiconductor properties. R. Yadav et al., have investigated the effect of ultrasonic power, and sonication time on physical properties of manganese ferrite nanoparticles; the study suggests that prepared nanoparticles can be used to prepare the high frequency device. (Yadav et al. 2020). Other MFe_2O_4 (where 'M' is transition metal ions like Co, Ni, Mn, Fe, Mg, Cu, Zn, etc.) spinel nanoparticles have also attracted huge attention as heterogeneous Fenton catalyst (also in heterogeneous advance oxidation process) because of low cost, large surface area, better catalytic activity, surface oxygen mobility nature, electric conductivity, high chemical, thermal and mechanical stability. The SFs are widely used magnetic materials in various fields (Soufi et al. 2021).

Spinal based nanocomposites such as ferrites and others are used to remove water pollutants (Tatarchuk et al. 2019), as adsorbents to remove heavy metals and dyes (Konicki et al. 2013, Sonkusare et al. 2020, Song et al. 2011, Umekar et al. 2022, Wang et al. 2012, Zhang et al. 2010), as drug delivery and release (Guo et al. 2015, Valente et al. 2017), as MRI contrast agents, in biomedical applications, hyperthermia (Fuentes-Garcia et al. 2020, Gonçalves et al. 2021, Ratkovski et al. 2020, Ravichandran and Velumani 2020) Figure 5.

$La_2Sn_2O_7/g\text{-}C_3N_4$ nanocomposites were prepared in presence of *broccoli* extract using the sonochemical technique for removal of dye (Talebzadeh et al. 2021).

Prepared nanocomposites samples were characterized using UV-Vis spectroscopy, XRD, SEM, EDX, and BET. $La_2Sn_2O_7$/g-C_3N_4 with 30 percent $La_2Sn_2O_7$ has shown better efficiency (99 %) compared to pristine $La_2Sn_2O_7$ nanoparticles (72 %) and g-C_3N_4 (91 %).

Recently, Yousefi et al., have prepared nanocomposites Dy_2NiO_5/Dy_2O_3 and $BaDy_2NiO_5$/NiO with core almond (as a capping agent) using ultrasonic assisted green synthesis method. The structural, optical, electric, magnetic, and photocatalytic properties were investigated. The bandgap of the $BaDy_2NiO_5$ heated at 800°C for two hr was found 2.77 eV. The particle size was decreased with increasing the sonication power from 15 to 50 Watt. The study indicated that present nanocomposites can be used as a photocatalyst to remove dye from water (Yousefi et al. 2021).

Polyaniline (PANI)–nitrogen substituted carbon dot (N@CDs) nanocomposites were prepared using the ultrasonication technique and studied by Maruthapandi et al., for antimicrobial activity as well as photocatalytic degradation of four different dyes namely congo red, methylene blue, Rhodamine-B, and crystal violet (Maruthapandi et al. 2021).

Abazari et al., studied magnetic Fe_3O_4@bio metallic organic framework (magntic@bio-MFO) using *in situ* sonosynthesis technique under ultrasound irradiation for examining cytotoxicity effects on *Leishmania major in* both *vivo* and *vitro* conditions (Abazari et al. 2018).

3. Conclusion and Future Perspective

Ultrasound techniques have been established for the synthesis of nanostructured materials like graphene-based materials, metal/metal oxide NPs and polymer nanocomposites. Sonochemical methods provide ultrasounds irradiation *via* acoustic cavitations. These acoustic cavitations generate bubbles that can effectively accumulate the diffused energy of ultrasound, and upon collapse, an enormous concentration of energy is released to heat the contents of the bubble. In this chapter, we enlightened the different syntheses of materials *via* sonochemical techniques. A high amount of ultrasound is accountable for a physical and chemical effect that is favourable to the preparation or reconstruction of different nanomaterials. The diverse action of mechanisms makes it a tremendous aspect applicable for many futuristic materials, including graphene, metal and metal oxide catalysts, polymers, crystallization, and anisotropic materials. With simple variations of reaction conditions and precursor compositions, a myriad of nanostructured materials with controlled morphologies, structures, and compositions have been successfully prepared by the application of high intensity ultrasound. Nevertheless, sonochemical method is considered as one of the fastest, simplist, cost effective, green, safe, and eco-friendly approaches for production of nanomaterials. There is a lot of scope in the future for industrial scale manufacture of graphene, metal oxide and polymer-based materials using ultrasound method. These materials show very promising, exciting properties and applications in the field of nanoscience and nanotechnology. The applications of ultrasound in materials chemistry are diverse, and there remains much to explore in the sonochemical synthesis of nanostructured materials. The major challenges that face a wider application of sonochemistry, however, includes issues of scale up and energy efficiency.

References

Abazari, R., R.A. Mahjoub, S. Molaie, F. Ghaffarifar, E. Ghasemi, A.M.Z. Slawin et al. 2018. The effect of different parameters under ultrasound irradiation for synthesis of new nanostructured $Fe_3O_4@$ bio-MOF as an efficient anti-leishmanial in vitro and in vivo conditions. Ultrason Sonochem. 43: 248–261. doi.org/10.1016/j.ultsonch.2018.01.022.

Afreen, S., K. Muthoosamy and S. Manickam. 2018. Sono-nano chemistry: a new era of synthesising polyhydroxylated carbon nanomaterials with hydroxyl groups and their industrial aspects. Ultrason. Sonochem. 51: 451–461. doi.org/10.1016/j.ultsonch.2018.07.015.

Ameta, S.C., R. Ameta and G. Ameta. 2018. Sonochemistry: an emerging green technology. CRC Press, 371–374. ISBN: 978-1-315-10274-0.

Anandan, S., A. Manivel and M. Ashokkumar. 2012. One-step sonochemical synthesis of reduced graphene oxide/pt/sn hybrid materials and their electrochemical properties, Fuel Cell 12: 95–962. doi.10.1002/fuce.201200048.

Baladi M., F. Soofivand, M. Valian and M. Salavati-Niasari. 2019. Sonochemical-assisted synthesis of pure Dy_2ZnMnO_6 nanoparticles as a novel double perovskite and study of photocatalytic performance for wastewater treatment. Ultrason. Sonochem. 57: 172–184. doi.org/10.1016/j.ultsonch.2019.05.022.

Balasubramanian, S., M. Govindasamy, S.-F. Wang, R.J. Ramalingam, H. Al-Lohedan and T. Maiyalagan. 2019. Novel sonochemical synthesis of Fe_3O_4 nanospheres decorated on highly active reduced graphene oxide nanosheets for sensitive detection of uric acid in biological samples. Ultrason. Sonochem. 58: 104618. doi.org/10.1016/j.ultsonch.2019.104618.

Bang, J.H. and K.S. Suslick. 2010. Applications of ultrasound to the synthesis of nanostructured materials. Adv. Mater. 22: 1039–1059. doi.org/10.1002/adma.200904093.

Bayramia, A., S. Alioghli, S.R. Pouran, A. Habibi-Yangjeh, A. Khataee and S. Ramesh. 2019. A facile ultrasonic-aided biosynthesis of ZnO nanoparticles using vacciniumarctostaphylos L. leaf extract and its antidiabetic, antibacterial, and oxidative activity evaluation. Ultrason, Sonochem. 55: 57–66. doi.org/10.1016/j.ultsonch.2019.03.010.

Bhangu, S.K. and M. Ashokkumar. 2016. Theory of sonochemistry. pp. 1–28. *In*: Colmenares, J.C. and Chatel G. [eds.]. Sonochemistry: From Basic Principles to Innovative Applications. Topic in Current Chemistry Collections. Springer. International Publishing Switzerland. doi.org/10.1007/978-3-319-54271-3_1.

Blake, J.R. and D.C. Gibson. 1987. Cavitation bubbles near boundaries. Annu. Rev. Fluid Mech. 19: 99–123. doi.org/10.1146/annurev.fl.19.010187.000531.

Catalano, P.N., N.S. Bourguignon, G.S. Alvarez, C. Libertun, L.E. Diaz, M.F. Desimone et al. 2012. Sol-gel immobilized ovarian follicles: collaboration between two different cell types in hormone production and secretion. J. Mater. Chem. 22: 11681–11687. doi:10.1039/c2jm30888f.

Chatel, G. 2016. Sonochemistry: new opportunities for green chemistry. World Scientific Publishing Company France. ISBN: 9781786341273.

Chatel, G. 2018a. How sonochemistry contributes to green chemistry? Ultrason. Sonochem. 40: 117–122. dx.doi.org/10.1016/j.ultsonch.2017.03.029.

Chatel, G. 2018b. Sonochemistry in nanocatalysis: the use of ultrasound from the catalyst synthesis to the catalytic reaction. Curr. Opin. Green Sustain. Chem. 15: 1–6. doi: 10.1016/j.cogsc.2018.07.004.

Chaudhary, R.G., A.K. Potbhare, P.B. Chouke, A.R. Rai, R.K. Mishra, M.F. Desimone et al. 2020. Graphene-based materials and their nanocomposites with metal oxides: biosynthesis, electrochemical, photocatalytic, and antimicrobial applications. Magnetic Oxides Composites-II, 83: 79–116. doi.10.21741/9781644900970-4.

Chaudhary, R.G., A.K. Potbhare, S.T. Aziz, M.S. Umekar, S.S. Bhuyar and A. Mondal, A. 2021. Phytochemically fabricated reduced graphene oxide-ZnO NCs by sesbania biospinosa for photocatalytic performances. Mater. Today: Proc. 36: 756–762. doi. 10.1016/j.matpr.2020.05.821.

Cobley, A.J., T.J. Mason and J. Robinson. 2008. Sonochemical surface modification: a route to lean, green and clean manufacturing?. J. Applied Surface Finishing 3: 190–196. ISSN No. 0360–3164.

Cravotto, G. and P. Cintas. 2006. Power ultrasound in organic synthesis: moving cavitational chemistry from academia to innovative and large-scale applications. Chem. Soc. Rev. 35: 180–196. DOI: 10.1039/b503848k.

Cravotto, G. and P. Cintas. 2007. Forcing and controlling chemical reactions with ultrasound. Angew. Chem. Int. Ed. 46: 5476–5478. doi.org/10.1002/anie.200701567.

De la Rosa-Fox, N., L. Esquivias, A.F. Craievich and J. Zarzycki. 1990. Structural study of silica sonogels. J. Non. Cryst. Solids 121: 211–215. doi.10.1016/0022-3093(90)90134-8.

Desimone, M.F., J. Degrossi, M. D'Aquino and L.E. Diaz. 2002. Ethanol tolerance in free and sol-gel immobilised saccharomyces cerevisiae. Biotechnol. Lett. 24: 1557–1559. doi:10.1023/A:1020375321009.

Desimone, M.F., J. Degrossi, M. D'Aquino and L.E. Diaz. 2003. Sol-gel immobilisation of *Saccharomyces cerevisiae* enhances viability in organic media. Biotechnol. Lett. 25(9): 671–674. doi:10.1023/A:1023481304479.

Desimone, M.F., De M. C. Marzi, G.J. Copello, J., Fernández, M.M., E. L Malchiodi and. L.E. Diaz 2005. Efficient preservation in a silicon oxide matrix of *Escherichia coli*, producer of recombinant proteins. Appl. Microbiol. Biotechnol. 68(6): 745–752. doi:10.1007/s00253-005-1912-7.

Desimone, M.F., S.B. Matiacevich, M.d.P. Buera and L.E. Díaz. 2008. Effects of relative humidity on enzyme activity immobilized in sol-gel-derived silica NCs. Enzyme Microb. Technol. 42(7): 583–588. doi:10.1016/j.enzmictec.2008.03.009.

Desimone, M.F., M.C. De Marzi, G.S. Alvarez, I. Mathov, L.E. Diaz and E.L. Malchiodi. 2011. Production of monoclonal antibodies from hybridoma cells immobilized in 3D sol-gel silica matrices. J. Mater. Chem. 21: 13865–13872. doi:10.1039/c1jm11888a.

Dheyab M.A., A.A. Aziz, M.S. Jameel, P.M. Khaniabadi and B. Mehrdel. 2020. Mechanisms of effective gold shell on Fe_3O_4 core nanoparticles formation using sonochemistry method. Ultrason. Sonochem. 64: 104865–104871. doi.org/10.1016/j.ultsonch.2019.104865.

Dheyab M.A, A.A. Aziz, M.S. Jameel, P.M. Khaniabadi and B. Mehrdel. 2021. Sonochemical-assisted synthesis of highly stable gold nanoparticles catalyst for decoloration of methylene blue dye. Inorg. Chem. Comm. 127: 108551–108558. doi.org/10.1016/j.inoche.2021.108551.

Doktycz, S.J. and K.S. Suslick. 1990. Interparticle collisions driven by ultrasound. Science. 247: 1067–1069. doi.org/10.1126/science.2309118.

Elsupikhe, R.F., K. Shameli, M.B. Ahmad, N.A. Ibrahim and N. Zainudin. 2015. Green sonochemical synthesis of silver nanoparticles at varying concentrations of κ-carrageenan. Nanoscale Res. Lett. 10: 302–309. DOI 10.1186/s11671-015-0916-1.

Ersan, A.C., A.S. Kipcak, M.Y. Ozen and N. Tugrul. 2020. An accelerated and effective synthesis of zinc borate from zinc sulfate using sonochemistry. Main Group Met. Chem. 43: 7–14. doi.org/10.1515/mgmc-2020-0002.

Esquivias, L. and, J. Zarzycki. 1988. Sonogels: An alternative method in sol-gel processing. pp. 255–270. *In:* Mackenzie, J.D. and Ulrich D.R. [eds.]. Ultrastructure Processing of Advanced Ceramics. John Wiley and Sons INC.

Fuentes-Garcia, J.A., A.C. Alavarse, A.C.M. Maldonado, A. Toro-Córdova, M.R. Ibarra and G.F. Goya. 2020. Simple sonochemical method to optimize the heating efficiency of magnetic nanoparticles for magnetic fluid hyperthermia. ACS Omega 5: 26357–26364. doi.org/10.1021/acsomega.0c02212.

Gedanken, A. 2004. Using sonochemistry for the fabrication of nanomaterials. Ultrason. Sonochem. 11: 47–55. doi.org/10.1016/j.ultsonch.2004.01.037.

Gong, C. and D.P. Hart. 1998. Ultrasound induced cavitation and sonochemical yields. J. Acoust. Soc. Am. 104: 2675–2682. doi.org/10.1121/1.423851.

Gonçalves, J. M., D.P. Rocha, M. Silva, P.R. Martins, E. Nossol, L. Agnes et al. 2021. Feasible strategies to promote the sensing performances of spinel MCo_2O_4 (M = Ni, Fe, Mn, Cu and Zn) based electrochemical sensors: a review. J. Mater. Chem. 9: 7852–7887. doi.org/10.1039/D1TC01550H.

Gumeci, C., D.U. Cearnaigh, D. J.Jr. Casadonte and C. Korzeniewsk. 2013. Synthesis of $PtCu_3$ bimetallic nanoparticles as oxygen reduction catalysts via a sonochemical method. J. Mater. Chem. A. 1: 2322–2330. doi.org/10.1039/C2TA00957.

Guo, H., W. Chen, X. Sun, Y.N. Liu, J. Li and J. Wang. 2015. Theranostic magnetoliposomes coated by carboxymethyl dextran with controlled release by low-frequency alternating magnetic field, Carbohydr. Polym. 118: 209–217. doi:10.1016/ j.carbpol.2014.10.076.

Hekimoglu, B.S. 2020. A review on sonochemistry and its environmental applications, acousti. 2: 766–775. dx.doi.org/10.3390/acoustics2040042.

Islam, H.M, M. Paul, O.S. Burheim and B.G. Pollet. 2019. Recent developments in the sonoelectrochemical synthesis of nanomaterials. Ultrason, Sonochem. 59: 104711–104718. doi.org/10.1016/j.ultsonch.2019.104711.

Jambrak, A.R., T.J. Mason, V. Lelas, L. Paniwnyk and Z. Herceg. 2014. Effect of ultrasound treatment on particle size and molecular weight of whey protein. J. Food Eng. 121: 15–23. doi.org/10.1016/j.jfoodeng.2013.08.012.

Katoh, R., Y. Tasaka, E. Sekreta, M. Yumura. F. Ikazaki, Y. Kakudate et al. 1999. Sonochemical production of a carbon nanotube. Ultrason. Sonochem. 6: 185–187. doi.org/10.1016/S1350-4177(99)00016-4.

Hosseini, S.G. and Z. Khodadadipoor. 2018. Sonochemical synthesis of ZnO, NiO and α-Fe_2O_3 nanoparticles and their catalytic activity for thermal decomposition of ammonium perchlorate. Indian J. Chem. Sec. A 57, 449–453. DOI: 10.56042/ijca.v57i3.18852.

Konicki, W., D. Sibera, E. Mijowska, Z. Lendzion-Bieluń and U. Narkiewicz. 2013. Equilibrium and kinetic studies on acid dye acid red 88 adsorption by magnetic $ZnFe_2O_4$ spinel ferrite nanoparticles. J. Colloid Inter. Sci. 398: 152–160, doi:10.1016/j.jcis.2013.02.021.

Liang, J.H., X. Jiang, G. Liu, Z.X. Deng, J. Zhuang and F.L. Li. 2003. Characterization and synthesis of pure ZrO_2 nanopowders via sonochemical method Mater. Res. Bull. 38: 161–168. doi.10.1016/S0025-5408(02)01007-3.

Liu, H., X. Zhang, X. Wu, L. Jiang, C. Clemens Burda and J.-J. Zhu. 2011. Rapid sonochemical synthesis of highly luminescent non-toxic AuNCs and Au@ AgNCs and Cu (II) sensing. Chem. Comm. 47: 4237–4239. DOI: 10.1039/c1cc00103e.

Liu, X., Z. Wu, R. Cavalli and G. Cravotto. 2021. Sonochemical preparation of inorganic nanoparticles and NCs for drug release–a review. Ind. Eng. Chem. Res. 60: 10011–10032. doi:10.1021/acs.iecr.1c01869.

Lorimer, J.P., T.J. Mason and K. Fiddy. 1991. Enhancement of chemical reactivity by power ultrasound: an alternative interpretation of the hot spot. Ultrason. 29: 338–343. doi.org/10.1016/004162 4X(91)90032-4.

Machado, I.V., J.R.N. dos Santos, M.A.P. Januario and A.G. Correa. 2021. Greener organic synthetic methods: sonochemistry and heterogeneous catalysis promoted multicomponent reactions. Ultrason. Sonochem. 78: 105704–105751. doi.org/10.1016/j.ultsonch.2021.105704.

Maruthapandi, M., A. Saravanan, P. Manohar, J.H.T. Luong and A. Gedanken. 2021. Photocatalytic degradation of organic dyes and antimicrobial activities by polyaniline–nitrogen doped carbon dot nanocomposite. Nanomaterials 11(5): 1128–1137. doi.org/10.3390/nano11051128.

Mason, T.J. and J.P. Lorimer. 2002a. Applied sonochemistry: the uses of power ultrasound in chemistry and processing. Weinheim: Wiley. ISBN 3-527-30205-0.

Mason, T.J. and Peter D. 2002b Practical sonochemistry: power ultrasound usage and application. Woodhead Publishing Ltd. ISBN: 978-1-898563-83-7. //dx.doi.org/10.1533/9781782420620.

Mason, T.J. 2003. Sonochemistry and sonoprocessing: the link, the trends and (probably) the future. Ultrason. Sonochem. 10: 175–179. doi.org/10.1016/S1350-4177(03)00086-5.

Mohammadi, Z. and M.H. Entezari. 2018. Sono-synthesis approach in uniform loading of ultrafine Ag nanoparticles on reduced graphene oxide nanosheets: an efficient catalyst for the reduction of 4-Nitrophenol. Ultrason. Sonochem. 44: 1–13. doi.org/10.1016/j.ultsonch.2018.01.020.

Monsef, R., M. Ghiyasiyan-Arani, O. Amiri and M. Salavati-Niasari. 2019. Sonochemical synthesis, characterization and application of $PrVO_4$ nanostructures as an effective photocatalyst for discoloration of organic dye contaminants in wastewater. Ultrason. Sonochem. 61: 104822–104835. doi.org/10.1016/ j.ultsonch.2019.104822.

Nagata, Y., Y. Watananabe, S.-I. Fujita, T. Dohrnaru and S. Taniguchi. 1992. Formation of colloidal silver in water by ultrasonic irradiation. J. Chem. Soc. Chem. Commun. 21: 1620–1622. doi.org/10.1039/C39920001620.

Nagvenkar, A.P., A. Deokar, I. Perelshtein and A. Gedanken. 2016. A one-step sonochemical synthesis of stable ZnO–PVA nanocolloid as a potential biocidal agent. J. Mater. Chem. B. 4: 2124–2132. DOI: 10.1039/C6TB00033A.

Nagvenkar, A.P., I. Perelshtein, Y. Piunno, P. Mantecca and A. Gedanken. 2019. Sonochemical one-step synthesis of polymer-capped metal oxide nanocolloids: antibacterial activity and cytotoxicity. ACS omega 4: 13631–13639. dx.doi.org/10.1021%2Facsomega.9b00181.

Nobre, F.X., I.S. Bastos, R.O. dos Santos Fontenelle, E.A. Araújo Júnior, M.L. Takeno, L. Manzato et al. 2019. Antimicrobial properties of α-Ag$_2$WO$_4$ rod-like microcrystals synthesized by sonochemistry and sonochemistry followed by hydrothermal conventional method. Ultrason, Sonochem. 58: 104620–104629. doi.org/10.1016/j.ultsonch.2019.104620.

Ohayon, E. and A. Gedanken. 2010. The application of ultrasound radiation to the synthesis of nanocrystalline metal oxide in a non-aqueous solvent. Ultrason. Sonochem. 17: 173–178. DOI:10.1016/j.ultsonch.2009.05.015.

Ohl, C.D., T. Kurz, R. Geisler, O. Lindau and W. Lauterborn. 1999. Philosophical transactions of the royal society. mathematical, physical and engineering sciences 357: 269–294. //doi.org/10.1098/rsta.1999.0327.

Paccola, C.E.T., C.M. Awano, F.S. de Vicente, D.A. Donatti, M. Yoshida and D.R. Vollet. 2015. Kinetics of oxalic acid catalyzed and ultrasound-assisted hydrolysis of 3-glycidoxypropyltrimethoxysilane and structural characteristics of the resulting aged sonogels. J. Phys. Chem. C 119: 19162–19170. doi:10.1021/acs.jpcc.5b04990.

Palomino, R.L., A.M. Bolarín Miró, F.N. Tenorio, F.S. De Jesús, C.A. Cortés Escobedo and S. Ammar. 2016. Sonochemical assisted synthesis of SrFe$_{12}$O$_{19}$ nanoparticles. Ultrason. Sonochem. 29: 470–475. dx.doi.org/10.1016/j.ultsonch.2015.10.023.

Park, A-H., W. Shi, J.-U. Jung and Y.U. Kwon. 2021. Mechanism study of single-step synthesis of Fe(core)@Pt(shell) nanoparticles by sonochemistry. Ultrason. Sonochem. 77: 105679–105688. doi.org/10.1016/j.ultsonch.2021.105679.

Potbhare, A.K., R.G. Chaudhary, P.B. Chouke, S. Yerpude, A. Mondal, V. Sonkusare et al. 2019. Phytosynthesis of nearly monodisperse CuO nanospheres using phyllanthus reticulatus/conyza bonariensis and its antioxidant/antibacterial assays. Mater. Sci. Eng. C 99: 783–793. doi. 10.1016/j.msec.2019.02.010.

Potbhare, A.K., M.S. Umekar, P.B. Chouke, M.B. Bagade, S.T. Aziz, A.A. Abdala et al. 2020a. Bioinspired graphene-based silver nanoparticles: fabrication, characterization and antibacterial activity. Materials Today: Proceedings 29: 720–725. doi.10.1016/j.matpr.2020.04.212.

Potbhare, A.K., P.B. Chouke, A. Mondal, R.U. Thakare, S. Mondal, R.G. Chaudhary et al. 2020b. Rhizoctonia solani assisted biosynthesis of silver nanoparticles for antibacterial assay. Mater. Today: Proc. 29: 939–945. doi.10.1016/j.matpr.2020.05.419.

Prozorov, T., R. Prozorov and K.S. Suslick. 2004. High velocity interparticle collisions driven by ultrasound. J. Am. Chem. Soc. 126: 13890. doi.org/10.1021/ja049493o.

Ravichandran, M. and S. Velumani. 2020. Manganese ferrite nanocubes as an MRI contrast agent. Mater. Res. Express 7: 016107–016115. doi.org/10.1088/2053-1591/ab66a4.

Qian, D., J.Z. Jiang and P.L. Hansen. 2003. Preparation of ZnO nanocrystals via ultrasonic irradiation. Chem. Comm. 9: 1078–1079. doi.org/10.1039/B301504A.

Qiu, X.-F., J.-J. Zhu and H-Y Chen. 2003. Controllable synthesis of nanocrystalline gold assembled whiskery structures via sonochemical route. J. Cryst. Growth 257: 378–383. doi:10.1016/S0022-0248(03)01467-2.

Ratkovski, G.P., K. Nascimento, G.C. Pedro, D.R. Ratkovski, F. Gorza, R.J. da Silva et al. 2020. Spinel cobalt ferrite nanoparticles for sensing phosphate ions in aqueous media and biological samples. Langmuir 36: 2920–2929. doi: 10.1021/acs.langmuir.9b02901.

Shchukin, D.G., D. Radziuk and H. Mohwald. 2010. Ultrasonic fabrication of metallic nanaomaterials and nanoalloys. Annu. Rev. Mater. Res. 40: 345–362. doi.org/10.1146/annurev-matsci-070909-104540.

Shut, V.N. and S.E. Mozzharov. 2017. Properties of ultrafine copper-containing powders prepared by a sonoelectrochemical method. Inorg. Mater. 53: 883–889. Doi: 10.1134/S0020168517080155.

Song, B.Y., Y. Eom and T.G. Lee. 2011. Removal and recovery of mercury from aqueous solution using magnetic silica nanocomposites. Appl. Surf. Sci. 257: 4754–4759, doi:10.1016/j.apsusc.2010.12.156.

Soufi, A., H. Hajjaoui, R. Elmoubarki, M. Abdennouri, S. Qourzal and N. Barka. 2021. Spinel ferrites nanoparticles: synthesis methods and application in heterogeneous fenton oxidation of organic pollutants–a review. Appl. Surf. Sci. 6: 100145–100163. doi.org/10.1016/j.apsadv.2021.100145.

Sonkusare, V.N., R.G. Chaudhary, G.S. Bhusari, A. Mondal, A.K. Potbhare, R.K. Mishra et al. 2020. Mesoporous octahedron-shaped tricobalt tetroxide nanoparticles for photocatalytic degradation of toxic dyes. ACS Omega 5: 7823–7835. doi.org/10.1021/acsomega.9b03998.

Suslick, K.S., D.A. Hammerton and R.E. Cline. 1986. Sonochemical hot spot. J. Am. Chem. Soc. 108: 5641–5642. doi:10.1021/ja00278a055.

Suslick, K.S. 1990. Sonochemistry. Science 247: 1439–1445. Doi: 10.1126/science.247.4949.1439.

Suslick, K.S., M.W. Grinstaff, A.A. Cichowlas and S.-B. Choe. 1992. Effect of cavitation conditions on amorphous metal synthesis. Ultrasonics 30: 168–172. doi.org/10.1016/0041-624X(92) 90068-W.

Suslick, K.S, M. Fang, T. Hyeon and A.A. Cichowlas. 1994. Molecularly designed nanostructured materials, MRS symp. Proc. 351: 443–48. *In*: Gonsalves, K.E., G.M. Chow, T.O. Xiao and R.C. Cammarata [eds.]. Pittsburgh: Mater. Res. Soc. DOI: 10.1007/978-94-015-9215-4_24.

Suslick, K.S., T. Hyeon, M. Fang and A.A. Cichowlas. 1996. pp. 197–211. In Advanced Catalysts and Nanostructured Materials. New York: Academic.

Suslick, K.S. and G.P. Price. 1999. Applications of ultrasound to materials chemistry. Annu. Rev. Mater. Sci. 29: 295–326. doi:10.1146/annurev.matsci.29.1.295.

Talebzadeh, Z., M. Masjedi-Arani, O. Amiri and M. Salavati-Niasari. 2021. $La_2Sn_2O_7$/g-C_3N_4 nanocomposites: rapid and green sonochemical fabrication and photo-degradation performance for removal of dye contaminations. Ultrason. Sonochem. 77: 105678–105691. //doi.org/10.1016/j. ultsonch.2021.105678.

Tatarchuk, T., B. Al-Najar, M. Bououdina and M.A. Aal. 2019. Catalytic and photocatalytic properties of oxide spinels, Handbook of Ecomaterials 3: 1701–1750. doi.org/10.1007/978-3-319-48281-1_158-1.

Umekar, M.S., G.S. Bhusari, A.K. Potbhare, A. Mondal, B.P. Kapgate, M.F. Desimone et al. (2021a). Bioinspired reduced graphene oxide based nanohybrids for photocatalysis and antibacterial applications. Current Pharmaceutical Biotechnology, 22 (13): 1759–1781.doi. 10.2174/13892010 22666201231115826.

Umekar, M., R. Chaudhary, G. Bhusari and A. Potbhare. 2021b. Fabrication of zinc oxide-decorated phytoreduced graphene oxide via clerodendrum infortunatum. Emerging Materials Research, 10(1): 75–84.doi.10.1680/jemmr.19.00175.

Umekar, M.S., G.S. Bhusari, T. Bhoyar, V. Devthade, B.P. Kapgate, A.K. Potbhare et al. 2022. Graphitic carbon nitride-based photocatalyst for environmental remediation of organic pollutants, current nanoscience. 12(19): 3494-.doi: 10.2174/1573413718666220127123935.

Valente, F., L. Astolfi, E. Simoni, S. Danti, V. Franceschini, M. Chicca et al. 2017. Nanoparticle drug delivery systemsfor inner eartherapy: an overview. Drug Deliv. Sci. Technol. 39: 28–35. doi:10.1016/j.jddst.2017.03.003.

Vollet, D.R., L.A. Barreiro, C.E.T. Paccola, C.M. Awano, F.S. de Vicente, M. Yoshida et al. 2016. A kinetic modeling for the ultrasound-assisted and oxalic acid-catalyzed hydrolysis of 3-glycidoxypropyltrimethoxysilane. J. Sol-Gel Sci. Technol. 80: 873–880. doi:10.1007/s10971-016-4157-2.

Wang, L., J. Li, Y. Wang, L. Zhao and Q. Jiang. 2012. Adsorption capability for congo red on nanocrystalline MFe_2O_4 (M = Mn, Fe, Co, Ni) spinelferrites. J. Chem. Eng. 181–182: 72–79, doi:10.1016/j. cej.2011.10.088.

Wang, G.Z., B.Y. Geng. X.M. Huang, Y.W. Wang, G.H. Li and L.D. Zhang. 2003. A convenient ultrasonic irradiation technique for in situ synthesis of zinc sulphide nanocrystallites at room temperature Appl. Phys. A–Mater. Sci. Process. 77: 933–936. DOI: 10.1007/s00339-002-2033-0.

Weavers, L.K., G.Y. Pee, A.J. Frim, L. Yang and J.F. Rathman. 2005. Ultrasonic destruction of surfactants: Application to Industrial Wastewater. Water Environ. Res. 77: 259–265. Doi: https://www.jstor.org/ stable/25045868.

Wood, R.J., J. Lee and M.J. Bussemaker. 2017. A parametric review of sonochemistry: control and augmentation of sonochemical activity on aqueous solution. Ultrason. Sonochem. 38: 352–370. dx.doi.org/10.1016/j.ultsonch.2017.03.030.

Wu, S-H. and D-H. Chen. 2003. Synthesis and characterization of nickel nanoparticles by hydrazine reduction in ethylene glycol. J. Colloid Interface Sci. 259: 282–286. doi:10.1016/S0021-9797(02)00135-2.

Xu, H., B.W. Zeiger and K.S. Suslick. 2013. Sonochemical synthesis of nanomaterials. Chem. Soc. Rev. 42: 2555–2567. DOI: 10.1039/c2cs35282f.

Yadav, R.S., I. Kuřitka, J. Vilcakova, T. Jamatia, M. Machovsky, D. Skoda et al. 2020. Impact of sonochemical synthesis condition on the structural and physical properties of $MnFe_2O_4$ spinel ferrite nanoparticles. Ultrason. Sonochem. 61: 104839–104853. doi.org/10.1016/j.ultsonch.2019.104839.

Yeung, S.A., R. Hobson, S. Biggs and F. Grieser. 1993. Formation of gold sols using ultrasound. J. Chem. Soc. Chem. Commun. 4: 378–379. doi.org/10.1039/C39930000378.

Yousefi, S.R., A. Sobhani, H. Abbas and A.M. Salavati-Niasari. 2021. Green sonochemical synthesis of $BaDy_2NiO_5/Dy_2O_3$ and $BaDy_2NiO=/NiO$ nanocomposites in the presence of core almond as a capping agent and their application as photocatalysts for the removal of organic dyes in water, RSCAdv. 11: 11500–11512. doi.org/10.1039/D0RA10288A.

Yu, Y., Q.Y. Zhang and X.G. Li. 2003. Reduction process of transition metal ions by zinc powder to prepare transition metal nanopowder. Acta. Phys. Chim. Sin. 19: 436–440. doi.org/10.3866/PKU.WHXB20030512

Zhang, S., H. Niu, Y. Cai, X. Zhao and Y. Shi. 2010. Arsenite and arsenate adsorption on coprecipitated bimetal oxide magnetic nanomaterials: $MnFe_2O_4$ and $CoFe_2O_4$. J. Chem. Eng. 158: 599–607. doi:10.1016/j.cej.2010.02.013.

Zhanfeng, Li, J. Dong, H. Zhang, Y. Zhang, H. Wang, X. Cui et al. 2021. Sonochemical catalysis as a unique strategy for the fabrication of nano-/micro-structured inorganics, Nanoscale Adv. 3: 41–72, DOI: 10.1039/ D0NA00753F.

CHAPTER 7

Sonochemical Based Processes for Treatment of Water and Wastewater
Future Challenges

Kirill Fedorov,[1] Manoj P. Rayaroth,[2,3] Xun Sun,[4] Reza Darvishi Cheshmeh Soltani,[5] Shirish Sonawane,[2,6,] Noor Samad Shah,[7] Varsha Srivastava,[8] Zhaohui Wang[9,10,11] and Grzegorz Boczkaj[1,12,*]*

1. Introduction

As a result of rapid socio economic advancement, industrialization, and population growth, a substantial amount of organic contaminants are continuously being discharged into wastewater triggering severe environmental concerns (Hube and Wu 2021,

[1] Gdańsk University of Technology, Faculty of Civil and Environmental Engineering, Department of Sanitary Engineering, Gdańsk, Poland.

[2] Gdańsk University of Technology, Faculty of Chemistry, Department of Process Engineering and Chemical Technology, Gdańsk, Poland.

[3] Bigelow Laboratory for Ocean Sciences, 60 Bigelow Dr, East Boothbay, ME, 04544, USA.

[4] Key Laboratory of High Efficiency and Clean Mechanical Manufacture, Ministry of Education, National Demonstration Center for Experimental Mechanical Engineering Education at Shandong University, School of Mechanical Engineering, Shandong University, 17923, Jingshi Road, Jinan, Shandong Province, 250061, People's Republic of China.

[5] Department of Environmental Health Engineering, School of Health, Arak University of Medical Sciences, Arak, Iran.

[6] Department of Chemical Engineering, National Institute of Technology Warangal, Telangana State 506004, India.

[7] Department of Environmental Sciences, COMSATS University Islamabad, Vehari Campus 61100, Pakistan.

[8] Research Unit of Sustainable Chemistry, Faculty of Technology, University of Oulu, FI-90014 Oulu, Finland.

[9] Shanghai Key Lab for Urban Ecological Processes and Eco-Restoration, School of Ecological and Environmental Sciences, East China Normal University, Shanghai 200241, China.

[10] Shanghai Engineering Research Center of Biotransformation of Organic Solid Waste, Shanghai 200241, China.

[11] Technology Innovation Center for Land Spatial Eco-restoration in Metropolitan Area, Ministry of Natural Resources, 3663 N. Zhongshan Road, Shanghai 200062, China.

[12] EkoTech Center, Gdansk University of Technology, G. Narutowicza St. 11/12, 80–233 Gdansk, Poland

* Corresponding authors: shirish@nitw.ac.in; grzegorz.boczkaj@pg.edu.pl

Oliveira et al. 2022, Ribeiro et al. 2022, Zhang et al. 2022). The occurrence of various organic contaminants such as phenols, pesticides, dyes, pharmaceuticals, microplastics, plasticizers, and other industrial chemicals have been reported in water resources. Such pollutants may cause stress on the aquatic ecosystem as well as are a risk to human health and elevate a plethora of environmental issues (Agarkoti et al. 2021, Bernabeu et al. 2011, Hammouda et al. 2017, Montoya-Rodríguez et al. 2020, Serna-Galvis et al. 2022, Wiest et al. 2021).

Due to their persistent propensity, these pollutants can remain in the environment for a long time, causing acute and chronic damage to the organisms exposed, as well as posing significant issues for water supply treatment (Singh and Mishra 2022, Tian et al. 2021, Yu et al. 2022). Additionally, discharged pollutants can interact with organic contaminants already present in wastewater to form transformation metabolites, which possess higher toxicity than the parent compounds. One of the most significant impediments to global socio economic progress is lack of fresh water, which has spurred the reclamation of water from disparate sources in order to meet water supply demands and alleviate the escalating issue of water scarcity across several regions of the world (Dong et al. 2022, Singh and Mishra 2022, Titchou et al. 2021). Furthermore, new stringent water discharge compliances burgeon concerns about the treatment of refractory organic pollutants in water and wastewater (Rayaroth et al. 2022).

Apart from environmental consequences, the economic constraints are also associated with the establishment of an effluent treatment plant for the disposal of hazardous industrial effluents, especially for small and medium size enterprises. Traditional wastewater treatment plants (WWTPs) are not designed to remove the trace concentration of emerging contaminants (ECs) and have been found ineffective in their removal (Bernabeu et al. 2011, Wiest et al. 2021). Some emerging contaminants could be degraded or eliminated in traditional WWTPs, but the preponderance of residues likely pass through the wastewater treatment plant effluents (Eniola et al. 2022, Mashile et al. 2022). Moreover, conventional treatment methods cannot achieve a complete mineralization of most of the emerging pollutants (Wiest et al. 2021). In the aquatic environment, WWTPs are by far the most significant sources of ECs and their transformation byproducts (Eniola et al. 2022, Ribeiro et al. 2022). Henceforth, WWTPs effluents create a menace to the environment and adversely affect flora and fauna (Ribeiro et al. 2022, Serna-Galvis et al. 2019).

There has been massive growth in a broad array of treatment approaches to address this issue and many endeavors have been undertaken to enhance the degrading efficiency and mineralization of organic pollutants. To remove organic contaminants from wastewaters, various physical, chemical, and biological treatment strategies such as sedimentation, adsorption, membrane separation, precipitation, coagulation, and activated sludge based biological treatment have been investigated (Pang et al. 2011, Rodriguez et al. 2011). Effective chemical and physicochemical treatment approaches have emerged in recent decades to address the limitations of traditional treatment technologies (Agarkoti et al. 2021, Eniola et al. 2022).

Advanced oxidation processes (AOPs) proved to be promising in the treatment of organic contaminants laden wastewater (Babu et al. 2019, Ma et al. 2021, Singh and Mishra 2022). Due to *in situ* generation of reactive oxygen species (ROS) such as hydroxyl radicals (˙OH), sulfate radicals ($SO_4^{\cdot-}$), superoxide anion radicals ($O_2^{\cdot-}$), singlet oxygen ($^1O_2$), AOPs are capable to convert recalcitrant organic compounds into low molecular biodegradable compounds or CO_2, H_2O (Bernabeu et al. 2011, Bhuta 2014, Ma et al. 2021, Yu et al. 2022).

Traditional AOPs based on ˙OH radicals are widely accepted due to their superior oxidation potential (E^0 (˙OH/H_2O) = 2.8 V) (Babu et al. 2019, Liu et al. 2021). AOPs are associated with various advantages such as high mineralization efficiency, strong oxidation capacity, and less secondary pollution. AOPs, including Fenton's process, photo Fenton, ozonation, photocatalytic oxidation, electrochemical oxidation reactions, UV/H_2O_2 and $SO_4^{\cdot-}$ radical based AOPs (SR-AOPs) are effective in the degradation of organic pollutants through generation of highly active radicals (Hammouda et al. 2017, Minh et al. 2019, Titchou et al. 2021, Wang et al. 2020, Yu et al. 2022). Although AOP based technologies can be utilized to reduce aqueous pollutant concentrations up to tolerable levels, the poor stability of oxidants, narrow operating hydrogen potential and low process efficiencies are still major limitations prohibiting widespread application (Dong et al. 2022). Additionally, the generation of undesirable toxic transformation products is also a matter of concern. Therefore, an approach for resolving the current problem must be devised.

Cavitation has emerged as a competent oxidative method for industrial wastewater treatment in the last few decades (Agarkoti et al. 2021). Cavitation involves the production, evolution, and rapid collapse of bubbles at several sites in the reactor in nanoseconds, releasing a massive amount of energy (Agarkoti et al. 2021, Hartmann et al. 2008, Ku et al. 2005). The efficient collapse of bubbles, which leads to an increase in the release of ˙OH radicals, causes the formation of hotspots. The shock waves created by cavity collapse can break complicated bulk organic molecules into smaller intermediates, making them more vulnerable to ˙OH radical attack (Agarkoti et al. 2021, Sutkar and Gogate 2009). The sonochemical treatment approach includes the cavitation process which is induced by high frequency ultrasound (US) waves (Montoya-Rodríguez et al. 2020), whereas organic pollutants are degraded through thermal decomposition or radical reactions (Hartmann et al. 2008, Ku et al. 2005).

Due to limited ˙OH radical generation, industrial wastewater treatment utilizing cavitation alone has been proven ineffective. However, when cavitation is used in combination with other AOPs, the production of ˙OH radicals is augmented, which boosts the degradation of pollutants and hence the treatment efficacy. In recent years, sonochemical based AOPs have attracted great concern in the treatment of organic contaminants (Hartmann et al. 2008, Ku et al. 2005, Yano et al. 2005, Yasuda 2021).

A wide spectrum of organic contaminants in wastewater is degraded by sonochemical treatment (Harada 2001, Yano et al. 2005). In the sonochemical treatment, US irradiation generates ROS without addition of chemicals and secondary pollution at normal temperatures and pressures (Yasuda 2021). However, several factors such as water matrix, pollutant concentration, and nature of pollutants may have a significant impact on the efficiency of sonochemical systems, which must be taken into consideration (Serna-Galvis et al. 2022). Many researchers have investigated the combination of UV and US irradiation with a variety of oxidizing agents and catalysts to enhance the degradation efficiency and minimize treatment costs (Grčić et al. 2013, Kyzas et al. 2022, Lin et al. 2008, Moghni et al. 2022, Philip et al. 2022, Uma et al. 2020). Sonochemical based treatment approaches have been visualized as an option to deal with wastewater and decrease aquatic pollution (Serna-Galvis and Torres-Palma 2021).

The sonochemical degradation of organic contaminants employing US/SR-AOPs, US/H_2O_2, and US/O_3 is addressed in this book chapter. The role of some advanced sonocatalysts such as layered double hydroxides (LDHs), MXene, metal organic frameworks (MOFs), and carbonaceous catalysts are incorporated in this chapter. Additionally, a discussion regarding the formation of undesirable byproducts in the sonochemical based treatment approaches is also provided. Microplastic degradation via sonochemical treatment is outlined in this chapter, as microplastic pollution is a major concern, and sonochemical treatments have emerged as a viable option for microplastic degradation in recent years. This book chapter presents an overall abridgment of the organic pollutant degradation by using sonochemical based water treatment approaches.

2. Sonochemical Based Processes for Treatment of Water and Wastewater Opportunities and Challenges: A Future Perspective

2.1 Sonochemical Degradation of Organic Pollutants and Activation of Oxidants

The use of US in water treatment has been recognized as a competitive technology due to its unique properties, which allow US based technologies to effectively degrade various organic pollutants. US based technologies provide numerous advantages, such as facilitated mass transfer, accelerated rate of chemical reactions, lower consumption of reagents and generation of reactive oxygen species (ROS). The latter allows qualifying US as a part of advanced oxidation processes (AOPs) and is responsible for the oxidative degradation capacity of US towards organic pollutants. The generation of ROS in aqueous media has resulted due to acoustic cavitation induced by the input of US. Under the US field with sufficient intensity, the molecules in aqueous media can no longer remain the intramolecular forces and cavitation bubbles are formed. Subsequent growth of cavitation bubbles leads to the adiabatic collapse of cavities, which is accompanied with an emission of a large magnitude of energy. The implosion of bubbles occurs in a short time interval creating so-called "hot spots", regions with extreme temperature (> 5,000 K) and pressure (> 1,000 atm) inside the collapsing bubble (Adewuyi 2001, Suslick 1990). These conditions are sufficient to cause direct pyrolytic degradation of organic pollutants within the bubble interior or induce thermal disassociation of water to yield ROS and hydrogen peroxide (Equations 1–6) (Adewuyi 2001, Chowdhury and Viraraghavan 2009, Makino et al. 1982):

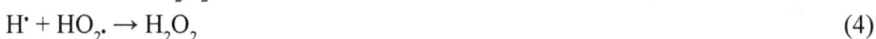

$$H_2O + \text{cavitation} \rightarrow H^{\bullet} + {}^{\bullet}OH \tag{1}$$

$$H^{\bullet} + O_2 \rightarrow HO_2^{\bullet} \tag{2}$$

$${}^{\bullet}OH + {}^{\bullet}OH \rightarrow H_2O_2 \tag{3}$$

$$H^{\bullet} + HO_2{}^{\bullet} \rightarrow H_2O_2 \tag{4}$$

$$HO_2^{\bullet} + HO_2^{\bullet} \rightarrow H_2O_2 + O_2 \tag{5}$$

$$H_2O_2 + cavitation \rightarrow {}^{\bullet}OH + {}^{\bullet}OH \tag{6}$$

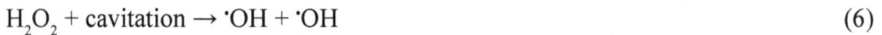

Except for primary radicals, the collapse of cavitation bubbles produces a variety of molecules, such as NO_2, H_2O_2 and $O^{\bullet}$ radicals. Formed NO_2 can further react with water resulting in nitric acid, which is reflected in hydrogen potential drop when treating air saturated water (Chatel and Colmenares 2017). Although H_2O_2 can also contribute to the oxidative degradation of organic pollutants, the lower energy requirement (180 kJ/mol) of O-O bond makes it as a potential precursor of ${}^{\bullet}OH$ radicals, as the cleavage of H-O bond in water requires 460 kJ/mol (Stefan 2017). Due to the high oxidation capacity and non selective chemistry, ${}^{\bullet}OH$ radicals are the major reactive species responsible for the degradation of organic pollutants under US. Particularly, the oxidation potential of 2.8 V allows ${}^{\bullet}OH$ radicals to convert hazardous organic pollutants into less toxic intermediates or to mineralize them producing CO_2 and H_2O. With an aim of the practical implementation in water treatment processes, the effectiveness of US has been evaluated towards the degradation of organic pollutants (Kotronarou et al. 1991), including dyes, pharmaceuticals (Gao et al. 2022), volatile organic compounds (VOCs), and pesticides (Matouq et al. 2008). However, requirements in high energy input to reach a satisfactory degradation efficiency of pollutants makes single use of US economically disadvantageous. Numerous external additives were tested in combination with US to increase the degradation extent and shorten the treatment time. In most cases, a synergistic effect between the additive and US was observed in the developed hybrid systems. This was due to the altered combination of paths that occur in the individual processes of the additive and US. As a result, the degradation efficiency of the hybrid process is superior to the cumulative effect of individual processes. The synergistic effect (ξ) can be evaluated using the rate constant (k) of processes as shown in Equation 7 (Fedorov et al. 2022):

$$\xi = \frac{{}^k additive + US}{{}^k additive + {}^k US} \tag{7}$$

The coefficient ξ is considered as synergistic, when $\xi > 1$, while it is cumulative or antagonistic, if $\xi =$ one or $\xi <$ one, respectively.

2.1.1 Combination of US with Ozone

Ozone (O_3) is a well known oxidant commonly used in water treatment systems. Despite the relatively high oxidation potential (2.08 V), the reaction of O_3 with organic pollutants is slow and selective towards specific compounds (e.g., saturated carboxylic acids and inactivated aromatics) (Gągol et al. 2018a). The effective application of O_3 in practice is also hindered due to its poor solubility in water. Ozone at basic conditions (hydrogen potential $\geq$ nine) reacts with HO^- anions to yield ${}^{\bullet}OH$ radicals and, thus, classified as AOP (Equation 8). In such case, the effectiveness of ozonation is significantly higher compared with the ozonation at acidic pH, where O_3 react with H^+ as described in Equation 9.

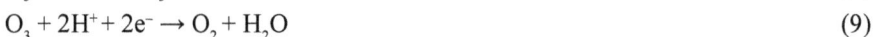

$$O_3 + HO^- \rightarrow O_2^{\bullet-} + {}^{\bullet}OH \tag{8}$$

$$O_3 + 2H^+ + 2e^- \rightarrow O_2 + H_2O \tag{9}$$

Owing to the physical effect of US and acoustic cavitation induced by US, the narrow operating hydrogen potential and low solubility of O_3 in water can be solved. In particular, the effect of US promotes the mass transfer in US/O_3 system, where the diffusion of O_3 from a carrier gas to the bulk liquid is facilitated by the local turbulence, microstreaming, and the shock waves. This increases the concentration of O_3 in the system, where diffusion of O_3 into cavitation bubble interior results in pyrolytic decomposition of O_3 to form ˙OH radicals (Equations 10,11) (Weavers and Hoffmann 1998):

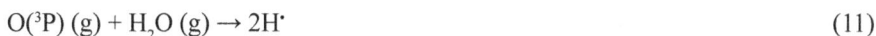

$$O_3 \text{ (g)} + \text{cavitation} \rightarrow O_2 \text{ (g)} + O(^3P)(g) \tag{10}$$

$$O(^3P) \text{ (g)} + H_2O \text{ (g)} \rightarrow 2H˙ \tag{11}$$

The hybrid process of US/O_3 with various experimental parameters and configurations has been studied for the degradation of water pollutants. Thus, the degradation of diclofenac using simultaneous action of US and O_3 was greater than the effect of sole US and O_3 (Naddeo et al. 2009). The rate constant of diclofenac TOC removal was determined as 0.211 min^{-1} for US/O_3, while it was 0.106 min^{-1} and 0.190 min^{-1} for the individual processes of US and O_3, respectively. However, the hybrid US/O_3 showed a negligible synergy between the joint techniques resulting in $\xi = 0.296$. Higher ξ value of 2.8 was observed for the degradation of 10 mmol/L p-aminophenol using 0.3 W/mL US and 5.3 g/L O_3 at hydrogen potential = 11 (He et al. 2007). The enhanced degradation of p-aminophenol in combined process was attributed to the additional production of reactive radicals due to sonochemical and thermal decomposition of O_3 in collapsing bubbles. The impact of additional reactive species formed in combined US/O_3 system has been highlighted for the degradation of aryl-azo-naphthol dyes (Gültekin and Ince 2006). Besides ˙OH radicals, the generation of the secondary oxidative species (such as $O_2^{˙-}$ and $HO_2˙$) radicals was proposed. Despite the lower oxidation potential than ˙OH radicals, $O_2^{˙-}$ and $HO_2˙$ radicals contributed to the synergy in US/O_3 providing an alternative pathway of oxidation. These radicals are formed due to the reaction of O_3 and H_2O_2 with ˙OH radicals as follows, Equations 12–15:

$$O_3 + HO˙ \rightarrow HO_2˙ + O_2 \tag{12}$$

$$2HO˙ \rightarrow H_2O_2 \tag{13}$$

$$H_2O_2 + ˙OH \rightarrow H_2O + HO_2˙ \tag{14}$$

$$HO_2˙ \leftrightarrow H^+ + O_2^{˙-} \tag{15}$$

In terms of practical implementation of sonochemical ozonation in large-scale water treatment, US/O_3 system suffers from challenges associated with energy consumption along with the purchase of ultrasonic devices (e.g., sonoprobes, transducers). Moreover, to achieve the amount of dissolved O_3 sufficient for satisfactory degradation efficiency, it requires high flow rates of O_3 due to the limited adsorption of O_3 in water. This demands the development of new technologies addressed to reduce the amount of O_3 utilization and waste.

2.1.2 Combination of US with Hydrogen Peroxide

Low cost, high oxidation potential (1.78 V) and solubility in water made hydrogen peroxide (H_2O_2) one of the most studied oxidants in degradation of water pollutants. The sole use of H_2O_2 does not lead to the generation of ROS, thereby the degradation of refractory and recalcitrant water pollutants cannot be attained. Hybrid H_2O_2 based processes demonstrate an oxidation capacity sufficient to degrade a broad range of organic pollutants with a peroxy (-O-O-) bond cleavage to yield ˙OH radicals. A combination of various technologies (e.g., UV, O_3, Fe^{2+}) with H_2O_2 have been developed and evaluated towards organic pollutants by generating ˙OH radicals. In the presence of US, the extreme conditions of pressure and temperature induced by collapsing microbubbles cause the conversion of H_2O_2 to various ROS as described in Equations 16–18:

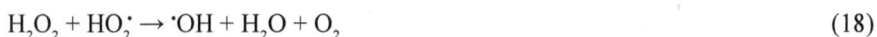

$$H_2O_2 + \text{cavitation} \rightarrow 2\text{˙OH} \tag{16}$$

$$H_2O_2 + \text{˙OH} \rightarrow HO_2^{\bullet} + H_2O \tag{17}$$

$$H_2O_2 + HO_2^{\bullet} \rightarrow \text{˙OH} + H_2O + O_2 \tag{18}$$

Formed ROS initiate the chain of radical reactions with water pollutants, where ˙OH radicals account for the predominant role. Therefore, the sonochemical effect of the hybrid US/H_2O_2 system is mainly contributed by ˙OH radicals, where the increased generation of ˙OH radicals increase the degradation extend of pollutants. Bagal and Gogate demonstrated the effectiveness of US/H_2O_2 towards the oxidative degradation of 20 mg L^{-1} alachlor solution using ultrasonic horn (20 kHz, 100 W) at hydrogen potential = three. It was found that the increase of H_2O_2 concentration from 0.025 to 0.2 g/L led to the increase of alachlor degradation. However, the significant increase of alachlor degradation was observed until the optimal H_2O_2 loading of 0.07 g/L, beyond which the increase was negligible. A similar trend was observed for the degradation of 2,4-dinitrophenol, p-nitrophenol, humic acid, bisphenol-A, acetamiprid and 3-methylpyridine. Such effect of H_2O_2 addition was assigned to the radical recombination reactions, whereas the excess of H_2O_2 and ROS serve as a scavenger of reactive species (Equations 19–21).

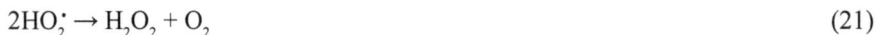

$$2\text{˙OH} \rightarrow H_2O_2 \tag{19}$$

$$HO_2^{\bullet} + \text{˙OH} \rightarrow H_2O + O_2 \tag{20}$$

$$2HO_2^{\bullet} \rightarrow H_2O_2 + O_2 \tag{21}$$

The optimal concentration of H_2O_2 in the combined US/H_2O_2 process depends on the type of the treated pollutant and process parameters. Thus, the optimal concentration of H_2O_2 for the degradation of 3-methylpyridine using US (25 kHz, 100 W) was found as 10 mg/L, while the maximum extend of methyl parathion degradation was attained at 200 mg/L of H_2O_2 under US (20 kHz, 270 W). These results indicate the importance of determination of the optimal load of H_2O_2 in US/H_2O_2 systems, where the activation of H_2O_2 dominates over the depletion of reactive species. Additionally, the effective utilization of H_2O_2 can be affected by undesired decomposition during the diffusion to gas-liquid interface of the cavitating bubble.

In such scenario, the load of H_2O_2 will be consumed without formation of reactive species as shown in Equation 22.

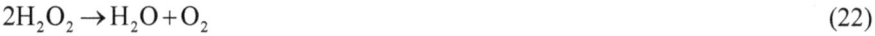

$$2H_2O_2 \rightarrow H_2O + O_2 \tag{22}$$

In terms of safety and transportation, the self decomposition of H_2O_2 presents a serious concern regarding the wide spread use of H_2O_2 in water treatment facilities. In order to prevent the decomposition of H_2O_2 and the acidification of treated waters, sodium percarbonate ($Na_2CO_3 \times 1.5\ H_2O_2$) was recently proposed as a safe alternative to H_2O_2 in US assisted processes employing H_2O_2 as an oxidant (Eslami et al. 2020). Besides, multi coupled systems involving transition metal ions, zero valent metals, metal oxides and carbonaceous materials have been reported for effective activation of H_2O_2.

2.1.3 Combination of US with SR-AOPs

Recently, the growing interest of the scientific community is focused on sulfate ($SO_4{}^{\cdot-}$) radicals as an alternative to short living and nonselective $\cdot OH$ radicals. These SR-AOPs commonly rely on the activation of persulfate (PS) and peroxymonosulfate (PMS) anions. $SO_4{}^{\cdot-}$ radicals along with a relatively high oxidation potential (2.6 V) demonstrate a lifetime of 30 to 10 μs, while $\cdot OH$ radicals exist 20 ns in water (Fedorov et al. 2021, Ghanbari and Moradi 2017). Unlike $\cdot OH$ radicals, $SO_4{}^{\cdot-}$ radicals possess selectivity towards certain organic compounds with electron rich functional groups. The mechanism of sonochemical degradation of pollutants by $SO_4{}^{\cdot-}$ radicals includes hydrogen abstraction, single electron transfer and double bond addition. Energy input (e.g., heat, UV and radiolytic irradiation), carbonaceous materials and transition metals are used to accomplish the activation of PS and PMS. US assisted activation of both PS and PMS is accompanied with the cleavage of peroxy bond to yield radicals as described in Equations 23, 24. In US/PS and US/PMS processes the radicals can also be formed due to the interaction of precursor anions with primary radicals (Equations 25–29).

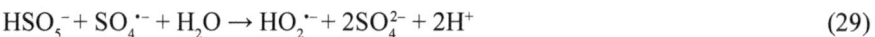

$$S_2O_8{}^{2-} + \text{cavitation} \rightarrow 2SO_4{}^{\cdot-} \tag{23}$$

$$HSO_5{}^- + \text{cavitation} \rightarrow SO_4{}^{\cdot-} + \cdot OH \tag{24}$$

$$S_2O_8{}^{2-} + \cdot OH \rightarrow SO_4{}^{\cdot-} + HSO_4{}^- + \frac{1}{2}O_2 \tag{25}$$

$$S_2O_8{}^{2-} + H^\cdot \rightarrow SO_4{}^{\cdot-} + HSO_4{}^- \tag{26}$$

$$HSO_5{}^- + \cdot OH \rightarrow SO_5{}^{\cdot-} + H_2O \tag{27}$$

$$HSO_5{}^- + SO_4{}^{\cdot-} \rightarrow SO_5{}^{\cdot-} + HSO_4{}^- \tag{28}$$

$$HSO_5{}^- + SO_4{}^{\cdot-} + H_2O \rightarrow HO_2{}^{\cdot-} + 2SO_4{}^{2-} + 2H^+ \tag{29}$$

In the case of US/PMS, the secondary radicals (e.g., $SO_5{}^{\cdot-}$ and $HO_2{}^{\cdot-}$) are known as less capable species which suppress the degradation of pollutants. The unfavorable conversion of $\cdot OH$ and $SO_4{}^{\cdot-}$ radicals into less reactive species occurs because of excessive loading of PMS. The overload of PS and PMS also leads to the exceeded number of radicals in the system, resulting in self recombination. To avoid the predominance of such interactions, it is necessary to determine the optimal loading of the oxidants in US/PS and US/PMS processes. For instance, the degradation of furfural varying the concentration of PS from 0.05 mM to 0.5 mM under US (130 kHz) showed the increase of furfural degradation within the range of 0.05 to 0.3 mM (Kermani et al. 2020). Addition of 0.5 mM PS decreased the degradation efficiency, similarly, the increase PMS concentration from 0.05 to 0.3 mM, the degradation efficiency of furfural was increased from 8.1 to 44.9 percent, respectively. When 0.5 mM of PMS was added, the degradation efficiency merely increased to 45.1 percent as the increased concentration of PMS led to the consumption of $SO_4^{\cdot-}$ by unreacted PMS (Equation 28). On the other hand, higher concentration of PS and PMS under US can be preferable. To undergo activation under US, hydrophilic molecules of PS and PMS need to reach the liquid sheath surrounding the cavitation bubbles from the bulk solution (Xu et al. 2020). Higher concentration of PS and PMS promotes the diffusion of the oxidants between phases, once more indicating the importance of determination of the optimal concentration. The improved performance of US/PS and PMS processes has been reported with the involvement of metal ions, solid catalysts based on metal oxides, carbonaceous materials (Fedorov et al. 2020), oxidants and when the additional energy input was applied.

2.2 Recent Developments on Sonocatalysts

The application of US has gained significant attention for the degradation of various organic pollutants in the aquatic medium (De Andrade et al. 2021, Lu et al. 2021). However, utilization of US alone requires a considerable amount of electrical power due to the immense loss of energy in thermal dissipation to attain a rapid decomposition rate, thereby inhibiting the wide utilization of US alone for the practical water and wastewater treatment. To solve this problem, development of heterogeneous sonocatalysts is proposed as one of the most practical and promising strategies (Qiu et al. 2018). Heterogeneous sonocatalytic processes have exhibited great potential in water and wastewater treatment. In the past decade, these operations using heterogeneous sonocatalysts have been widely utilized for the removal of refractory organic pollutants (Yousef Tizhoosh et al. 2020). A variety of heterogeneous sonocatalysts have been applied for treatment techniques and their development and application have been increasing year by year. Titanium dioxide (TiO_2) and zinc oxide (ZnO) are used conventionally as heterogeneous sonocatalyst for the sonocatalysis. Nanostructured TiO_2 and ZnO are used as sonocatalyst and the formation of ROS is demonstrated under ultrasonic irradiation (Wang et al. 2011). In the presence of such sonocatalysts, generation of ROS such as ˙OH radicals is remarkably increased as a result of the intensified formation of nucleation sites for cavitation phenomenon (formation, enlargement and violent collapse of cavitation bubbles) in the solution under ultrasonication (Abdullah and Ling 2010, Soltani et al. 2019a). To inhibit washout of nanostructures from the reactor, ZnO and TiO_2 nanoparticles have been immobilized on the surface of clay-like substances (Darvishi Cheshmeh Soltani et al. 2016, Khataee et al. 2015, Soltani et al. 2019b) and carbonaceous compounds (Gholami et al. 2019a, Nuengmatcha et al. 2016, Soltani et al. 2019a). Additionally, enhanced sonocatalytic degradation, of target pollutants, is also reported in the presence of suitable supports. This enhancement is associated with (1) improved transfer rate of the pollutant to the reactive sites of the nanocatalyst, (2) enhanced formation of nuclei to create cavitation bubble during "hot spot" phenomenon generating extra hydroxyl free radicals and (3) limited aggregation of nanostructures in the aquatic phase (Soltani et al. 2019b). Magnesium oxide (MgO) in nano size is also used as heterogeneous sonocatalyst of the sonocatalysis for the environmental remediation by means of degradation and removal of persistent organic pollutants (Darvishi Cheshmeh Soltani et al. 2018). Other semiconductors including $LuFeO_3$ (Zhou et al. 2015), AgBr (Wu et al. 2013), Bi_2O_3 (Chen et al. 2016) and BiOI (Song et al. 2012) are also recommended and employed as effective sonocatalysts. In the following, recently developed sonocatalysts are mentioned.

2.2.1 *Layered Double Hydroxides (LDHs)*

Layered double hydroxides (LDHs) containing metal ions are efficient and promising sonocatalysts because of high interlayer ion exchange capacity, abundant reactive sites, convenient synthetic procedure, high structural stability, immense surface area and remarkable recyclability (Karim et al. 2022, Khataee et al. 2019, Naderi and Darvishi Cheshmeh Soltani 2021). Particularly, LDHs have high ability to preserve various biomolecules in the interlayers with excellent biocompatibility and low toxicity (Nas 2021). LDHs are considered as catalyst for water and wastewater reclamation because of their appropriate structure interacting with organic pollutants in the aquatic phase. Figure 1 shows various morphologies of LDHs.

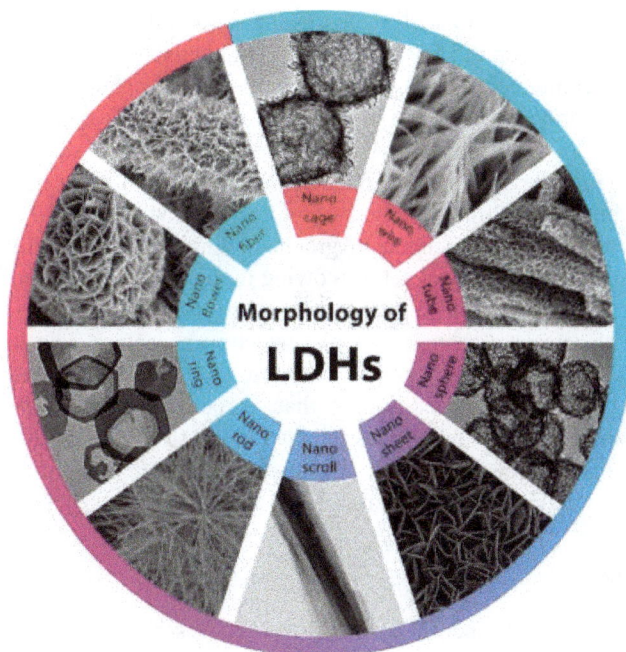

Figure 1. Scanning electron microscopy (SEM) and transition electron microscopy (TEM) images of LDHs (Karim et al. 2022).

LDHs are applied for not only catalytic processes but also for polymerization, energy storage, medication and adsorption process for the sequestration of both organic and inorganic compounds (Naderi and Darvishi Cheshmeh Soltani 2021). LDHs have the formula of $[M^{II}_{1-x} M^{III}_{x} (OH)_2]^{x+} [An^-]_{x/y} \cdot yH_2O$. In this formula, M^{II} represents divalent cations such as Mg^{2+}, Zn^{2+}, Cu^{2+}, Ni^{2+} and M^{III} represents trivalent cations such as Al^{3+}, Cr^{3+} and Fe^{3+}. A^{n-} symbolizes interlayer anions such as NO_3^-, SO_4^{2-}, CO_3^{2-} and Cl^-. In addition, y represents the number of water molecules in the structure of a layered compound (Khataee et al. 2019).

LDHs with layered structure are extensively used as excellent sonocatalyst for the catalytic decomposition and conversion of organic pollutants through destructive reactions. LDHs typically have low quantum yield under ultraviolet or visible light irradiation due to weak electron hole transfer, rapid recombination rate of electron hole pairs and poor mobility of charge carriers (Sadeghi Rad et al. 2022a). Thus, their performance is enhanced using doping approaches, integration with other semiconductors and even immobilization onto suitable media. In this regard, LDHs are doped using vanadium (Keyikoglu et al. 2022) and europium (Saska Romero et al. 2022) elements in order to hinder recombination rate of electron hole pairs. The sonoluminescence phenomenon is one of the main mechanisms during the sonocatalytic processes due to the generation of light with a short wavelength in the micro bubbles collapse by the ultrasonic irradiation. This accelerates the production of active free radicals by charge separation and formation of electron hole pairs (Abdi et al. 2022). Using doping agents in the structure of LDHs hinders or inhibits the recombination of as generated electron hole pairs. LDHs are successfully combined with other semiconductors to achieve composite sonocatalysts with higher catalytic activity. The combination of LDHs with other semiconductors such as WO_3 leads to the restricted recombination of as generated electron hole pairs during the catalytic process. This improves charge carrier mobility of the main layered semiconductor (Khataee et al. 2020). Moreover, LDHs are immobilized onto suitable beds such as graphene (Khataee et al. 2019, Sadeghi Rad et al. 2022a), carbon nanotube (Sadeghi Rad et al. 2022b), graphitic carbon nitride (Zou et al. 2017) and biochar (Gholami et al. 2020). Application of graphene and carbon nanotube as support results in the enhanced catalytic activity owing to their unique electron conducting characteristics and high specific surface area (Nas 2021). In a biochar gallery, the Si and Al with vacant d-orbitals can play electron acceptors' role in enhancing electron transfer and hindering recombination rate of as generated electron hole pairs (Gholami et al. 2019a). In addition to that, carbon based materials have high mechanical and chemical stability, excellent porous structure and high electrical conductivity making them effective candidates for the immobilization and also improvement of LDHs for the sonocatalysis (Sadeghi Rad et al. 2022a).

2.2.2 *Metal Organic Frameworks (MOFs)*

MOFs have remarkable properties such as large specific surface area, excellent porous structure, designable pore size, synergistic effect between functional groups of the ligand and metal ions, and high capacity to easily incorporate particular functionalities per species without altering the framework configuration (Jun et al. 2020b, Ryu et al.

2021). MOFs have different environmental applications, for example, in adsorption process (Jun et al. 2020a), catalytic processes (Abdolalian et al. 2021) and membrane technologies (Xiao et al. 2021). There are different modifications of MOFs which are shown in Figure 2. Sonocatalytic processes have also benefited from MOFs. It is used as a sonocatalyst in sonocatalytic degradation of pharmaceutical compounds (Jun et al. 2019a). High porosity and consequently high specific surface area are notable properties of MOFs increasing the reactive sites on surface of the catalyst. In this manner, ultrasonic irradiation accelerates the rate of mass transfer into the porous structure of MOFs with less restriction (Abdi et al. 2022). The main aspect of MOFs in the sonocatalytic systems is associated with their semi conductivity properties. In this regard, MOFs as sonocatalyst can be excited by energy higher than or equal to the band gap energy (energy gap between conduction band (CB) and valence band (VB)). Consequently, holes (h^+) and electrons (e^-) are in the VB and CB, respectively. Superoxide ($O_2^{\cdot-}$) and $\cdot$OH radicals can be formed when oxidant and dissolved oxygen react with e^- in the CB. In fact, holes (h^+) in the VB react with water molecules and hydroxyl ions oxidizing them to generate $\cdot$OH radicals (Abdi et al. 2022). The layers of MOFs efficiently inhibit close π–π pile providing a porous structure for oxygen/water per target organic pollutant interaction. The effective metal to ligand charge transfer facilitates water dissociation creating ROS such as $\cdot$OH radicals under US irradiation. Finally, as generated radicals react with the target pollutant (Zhu et al. 2022).

Moreover, MOFs have been modified to enhance their sonocatalytic characteristics. For example, MIL-101, as a common type of MOFs, is incorporated with graphene oxide and magnetic $ZnFe_2O_4$ for the sonocatalytic degradation of organic dye as model organic pollutant. Under ultrasonication and sonoluminescence phenomenon, a charge separation and generation of e^-/h^+ pairs are expected. Moving

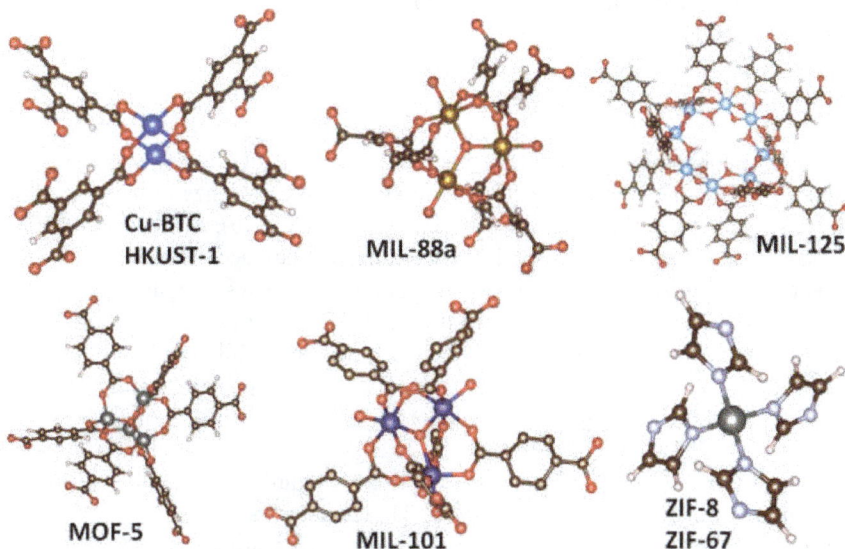

Figure 2. Molecular structures of various MOFs (Abdi et al. 2022).

e^-/h^+ pairs between CB/VB of MIL-101 and $ZnFe_2O_4$ inhibits recombination of e^-/h^+ pairs. Furthermore, the presence of graphene oxide improves the formation of active radicals because of its high surface conductivity and considerable electron acceptance properties (Nirumand et al. 2018).

2.2.3 MXenes

For the first time in 2014 , MXene was proposed as a heterogeneous semiconductor by combining conductivity of transition metal carbides with a hydrophilic nature owing to its hydroxyl or oxygen terminated surface (Jun et al. 2020c, Persson and Rosen 2019). It is classified as a two dimensional inorganic substance. MXene has remarkable performance when employed for the decomposition of hazardous organic pollutants through different AOPs such as heterogeneous sonocatalysis (Saravanakumar et al. 2021). This material is composed of transition metal carbides, nitrides and carbonitrides thick layers. MXene based semiconductors exhibit a series of excellent characteristics, including adjustable band gap structure, high electrical conductivity, thermal stability, dispersibility, large specific surface area and hydrophobicity (Carey and Barsoum 2021, Vasseghian et al. 2022). Generally, MXene synthesis is based on an etching method of –A- multilayers on a "MAX" part in which M indicates transition metals, A is Si or Al elements (A-group) and X represents N or C elements (Jun et al. 2019b, Persson and Rosen 2019, Vasseghian et al. 2022). The combined forms of MXene, including $TiO_2@C$ (Huang et al. 2019), $Ti_3C_2@Bi_2WO_6$ (Cao et al. 2018) and $BiOBr@Ti_3C_2$ (Li et al. 2020), are prepared to extend its applicability for environmental implementation as well as energy storage capacity. It is theoretically demonstrated that Ti_3C_2 MXene effectively hinders the recombination of e^-/h^+ pairs. In addition, integration of MXene with other semiconductors like TiO_2 leads to enhanced sonocatalytic effectiveness (Ding et al. 2021).

2.2.4 Novel Carbonaceous Sonocatalysts

Typically, graphene based carbonaceous materials, including bare graphene (GR), graphene oxide (GO) and reduced graphene oxide (rGO), are utilized as carbonaceous catalysts for water treatment processes (Gholami et al. 2019b). Among them, GO is efficiently used as sonocatalyst for the degradation of pharmaceuticals under US. The presence of GO intensifies formation of cavitation bubbles resulting in the enhanced generation of ˙OH radicals (Al-Hamadani et al. 2018). Other carbonaceous materials such as graphitic carbon nitride ($g-C_3N_4$) and buckminsterfullerene (C60) are used for the sonocatalysis. However, these carbonaceous compounds with particular characteristics are only employed as support of other sonocatalysts, including TiO_2 (Meng and Oh 2011) and CdS (Meng et al. 2012), rather than the main sonocatalyst. Recently, carbon dots (CDs) such as graphene quantum dots (GQDs) are used as innovative carbonaceous sonocatalysts. These are fluorescent nanostructures with quasi spherical shapes smaller than 10 nm which are proposed as efficient semiconductor not only for photocatalytic systems but also for the sonocatalysis (Selim et al. 2020). Totally, CDs have multiple applications in sensors, drug delivery, photo catalysis, optoelectronic instruments and bioimaging because of their unique characteristics and superior biocompatibility (Ren et al. 2021, Sajjadi et al. 2019). These compounds are comprised of graphitized sp^2 carbon cores surrounded by an outer flake covered by surface functional groups such as $-NH_2$, C–O–C, –COOH and –OH. Among them, GQDs are comprised of mono or few sheet graphene as a new group of Zero dimensional nanostructures. Wide absorption spectrum, chemical stability and nontoxicity are the most remarkable characteristics making GQDs a potential semiconductor with significant sonocatalytic activity (Sajjadi et al. 2017). During sonocatalytic processes using US irradiation, the external shell of CDs is exposed to shock of as generated cavitation bubbles, breaking apart the functional groups on the CDs surface. In fact, extreme conditions due to the cavitation phenomenon are responsible for the production of free oxidizing radicals (Ren et al. 2021). Additionally, CDs such as GQDs are combined with other catalysts for the sonocatalysis. As a result, the combination of GQDs with CdSe led to the enhanced sonocatalytic degradation of the target pollutants. Improved catalytic activity of the CdSe is reported in the presence of GODs (Sajjadi et al. 2017).

2.3 Unexpected Byproduct Formation

The sonochemical degradation of organic pollutants occurs mainly through the attack of reactive species such as ˙OH and H˙ radicals. As given in the previous sections, the pollutants can undergo the temperature effect and mechanical shearing depending on the physicochemical feature of the pollutants (Cao et al. 2020, Gogate and Prajapat 2015, Mohod and Gogate 2011). In the radical mediated degradation, the pollutants undergo hydrogen abstraction, one electron oxidation, and electrophilic OH addition (Boczkaj et al. 2014, Boczkaj et al. 2017, Fernandes et al. 2020, Fu et al. 2021, Landge et al. 2021, Wang and Xu 2012). The radical reaction depends on the electron density and the reaction occurred only on the specific sites (Dostanić et al. 2020). On the other hand, the latter reaction occurs randomly along the chain of the compound. All the pathways resulted in the formation of intermediate products and the continuous process caused the complete mineralization as given in Figure 3. As it can be seen, the mechanism was proposed by the identified intermediate products by certain analytical techniques like mass spectrometry (MS). The common intermediate products are hydroxylated derivatives, products of ring cleavage, aliphatic acids, etcetera. The hydroxylated (or phenolic products) are formed by the electrophilic attack of ˙OH radicals to the electron rich center (Nejumal et al. 2014, Sasi et al. 2015). It also attacks the ipso position to release the attached functional groups. The unremitted process of reactive species results in the breaking of the aromatic ring to the formation of aliphatic acids such as maleic acid, malic, pyruvic acid, glyoxylic acid, and oxalic acid (Dirany et al. 2012, Oturan et al. 2018). The inorganic ions such as NO_3^-/NO_2^-, NH_4^+, SO_4^{2-}, Cl^-, F^- were also released as the final product depending on the functional group of the parent compound.

In addition to the above reactive species, O_3 is formed by the sonochemical dissociation of dissolved oxygen as given in the following equations (Wang and Xu 2012). A product corresponding to the ozone mediated degradation was reported

Pollutants   Toxic / less toxic

↓ ROS

Hydroxylated products   Elimination of functional groups   Side chain oxidation

Toxic / less toxic

↓ Further hydroxylation

Aromatic products
                                    Toxic / less toxic

↓ Ring opening

Low molecular weight organic acids   Less toxic

$CO_2 + H_2O$ + inorganic ions   Complete mineralisation

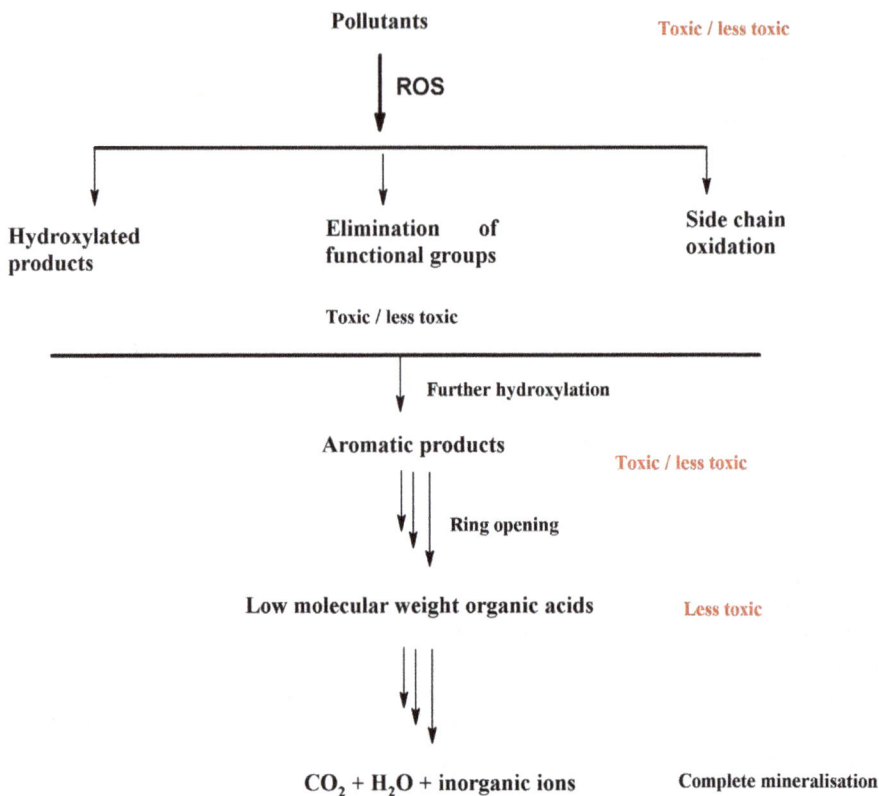

Figure 3. Degradation pathway for the removal of organic contaminants by the ROS (Rayaroth et al. 2016).

during the sonochemical degradation of ranitidine, which is not formed in other AOPs such as UV photolysis of H_2O_2 or $S_2O_8^{2-}$ (Elias et al. 2019).

$$O_2 \rightarrow 2O^{\bullet} \tag{30}$$

$$O^{\bullet} + H_2O_2 \rightarrow {}^{\bullet}OH + HO_2^{\bullet} \tag{31}$$

$$O^{\bullet} + O_2 \rightarrow O_3 \tag{32}$$

The formation of nitro product is considered an emerging issue in the sonochemical degradation of pollutants, especially the nitrogen bearing compounds (Rayaroth et al. 2022). This nitro product formation was reported for pollutants such as *para*-aminosalicylic acid (PAS) and diphenylamine (DPhA). These nitro products are formed by the *in situ* generated nitrite per nitrate ions (Rayaroth et al. 2018, Yao et al. 2021).

The evolution of the nitrogen species was reported in the previous studies by the acoustic nitrogen fixation processes (Mišík and Riesz 1996, Supeno and Kruus 2002, Virtanen and Ellfolk 1950, Yao et al. 2018)2-(4-carboxyphenyl. This process converts the atmospheric nitrogen into nitrate per nitrite ions. It involved several pathways as given in Figure 4 (Rayaroth et al. 2022). The atomic oxygen is initially formed by the thermal dissociation of O_2 cavitating bubbles. The atomic oxygen

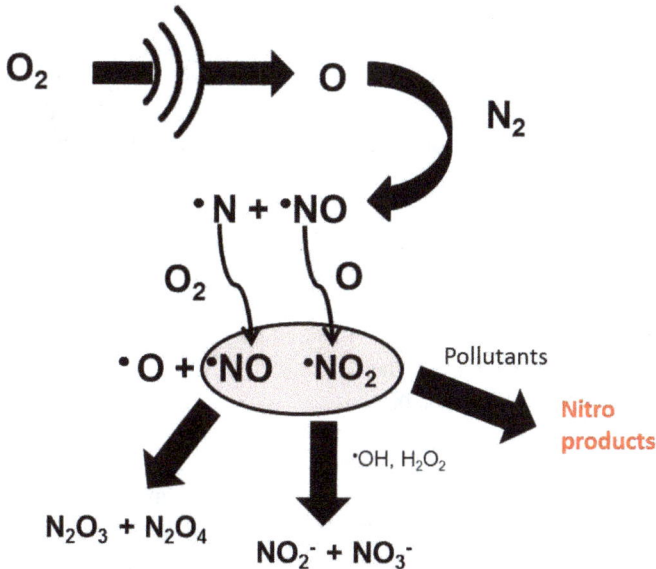

Figure 4. Acoustic nitrogen fixation and the subsequent formation of reactive nitrogen species.

then reacts with the dissolved nitrogen to form the reactive N species (RNs) such as NO˙, and NO$_2$˙. These species then hydrolyzed to form nitrite and nitrate ions and discharged into the liquid region (Glick et al. 1957, Wakeford et al. 1999). These ions can scavenge the reactive oxygen species to form the reactive species in the liquid medium.

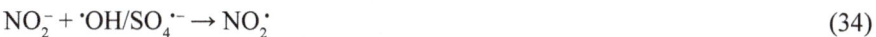

$$NO_3^- + \cdot OH/SO_4^{\cdot-} \rightarrow NO_3^{\cdot} \tag{33}$$

$$NO_2^- + \cdot OH/SO_4^{\cdot-} \rightarrow NO_2^{\cdot} \tag{34}$$

The RNs possessed oxidizing properties and were able to transform the organic pollutants (Wu et al. 2020). Among the RNs, NO$_3$˙ has a strong redox potential of 2.3 to 2.5 V. On the other hand, the redox potential value of NO$_2$˙, and NO˙ are 1.03 and 0.39 V, respectively, with nitrating and nitrosating behavior (Neta et al. 1988). These species react with the pollutants bearing electron donating functional groups to form the nitro products. The nitro products are mutagenic and highly toxic. The nitroso derivatives of DPhA and the nitration byproducts of PAS, such as nitroaniline, and 3-amino-4-nitrophenol are more toxic than the parent compound. Therefore, the formation of such kinds of nitro derivatives should be minimized by the proper hybrid processes.

2.4 Cavitation Process for Microplastics

Microplastics (MPs) are the recent class of pollutants characterized by less than smaller than five mm in diameter and large surface area (Lebreton et al. 2017, Richardson and Kimura 2017). The major sources of microplastics are the chemical products that are used for daily purposes (e.g., cleansing agents, toothpaste) and the fragmentation of large plastics (Andrady 2011, Fendall and Sewell 2009) with research efforts focusing on both the macroplastic (>5 mm. They are poorly biodegraded and accumulate in the environmental matrices. Plastic materials such as polypropylene (PP), polyethylene (PE), polystyrene (PS), polyvinyl chloride (PVC), polycarbonate (PC), polyamides (PA), polyester (PES), and polyethylene terephthalate (PET) microplastics are toxic to the living organisms. These MPs intensify the bioaccumulation of pollutants by the strong interaction with them. The widespread of MPs in the various matrices caused the presence of these pollutants in the human bodies and living organism. Like other pollutants physical, chemical and biological treatments have been employed for the removal of MPs as well (e.g., UV/H$_2$O$_2$, UV per chlorine photolysis, ozonolysis, photocatalysis, heat activated PS) (Ricardo et al. 2021, Talvitie et al. 2017). In these AOPs, the degradation mostly changes the surface properties of the MPs (Kim et al. 2022). In some cases, the deformed MPs would adsorb the organic chemicals more than the untreated ones. However, the cavitation based AOPs were not tested vigorously for MPs.

The physical and chemical effects of cavitation are suitable for various chemical applications. The upgrading heavy oils is one of the important applications in these aspects. The US was effectively applied for the depolymerization reaction (Gogate and Prajapat 2015, Jellinek and White 1951, Mohod and Gogate 2011) especially considering the fact that the reduction in the molecular weight (also the intrinsic viscosity. The depolymerization by cavitation involves (i) the disaggregation of polymer clusters to form more flexible structure by the shear force generated by cavitation, (ii) the fragmentation of the longer chain into smaller chain polymers by the mechanical stress, and (iii) the radical mediated oxidation of polymers as reported in other AOPs. Various factors influencing the polymer degradation by ultrasonic degradation are ultrasonic intensity, frequency, molecular weight, initial concentration of polymer solution, type of solvent, pH, operating temperature, functional group, and dissolved gases. The selection of frequency in the degradation process depends on the solubility. Typically, low frequency is suitable for the less soluble polymers. This condition is enough to drag the polymer molecule in to the cavitating bubble reaction. On the other hand, the high frequency is likely to speed up the process in the case of water soluble polymers. The MPs are polymers and therefore the same degradation mechanism can be applied for their removal as well. In addition, the synergistic effect of cavitation with other processes such as UV, O$_3$, PS and photocatalysis speed up the degradation of MPs by facilitating the mechanical effect and the subsequent oxidation, Figure 5 (Yan et al. 2015).

2.5 Scaling Up Aspects

The development of a cost effective technology for wastewater treatment is very important for achieving a successful operation at industrial scale operation (Gogate 2002). The operating cost of ultrasonic treatment processes is high due to the poor efficiency of the conversion of electrical energy into cavitation energy. The scaled up versions of these reactors require a huge amount of energy, which in turn results in a high capital cost of equipment and higher operating and maintenance costs. On the other hand, hydrodynamic cavitation reactors are more energy efficient as compared to ultrasonic reactors, and the scale up of these reactors is

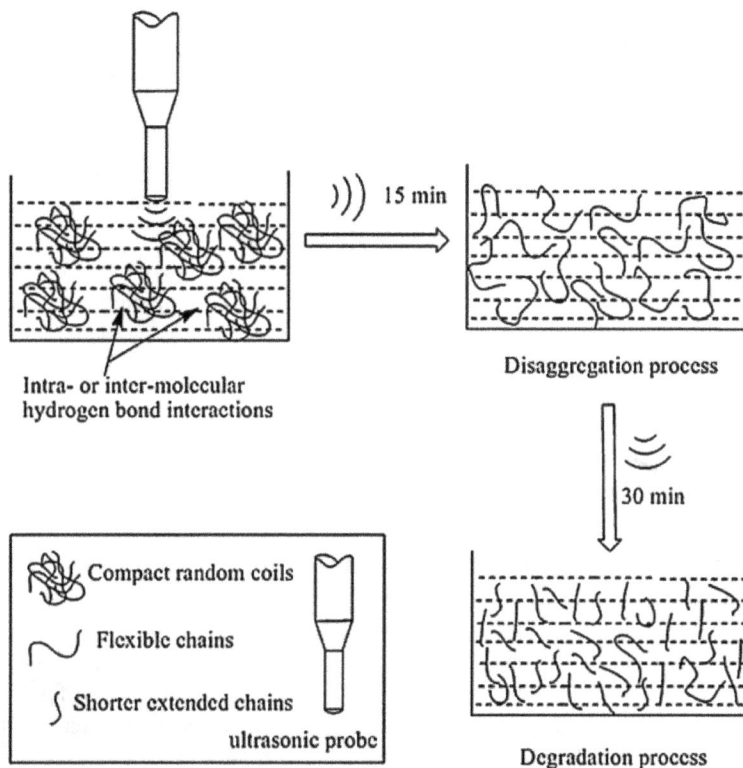

Figure 5. Various steps involved in the degradation of polymers by sono cavitation process.

also easy. The efficient design of reactors helps in reducing the treatment cost. In the case of combined operation, the optimization of operating parameters such as flow rates, inlet pressures, and loadings of oxidants may help in reducing the overall treatment cost as it reduces the electrical cost and the cost of chemicals. To break chemical bonds, cavitation reactors must be able to convert mechanical and electrical energy into the cavitation phenomenon. A higher degradation rate is caused by the efficient generation of many cavities during the process. The more effective energy conversion, the higher cost efficiency. The kinetic and thermal effects during the collapse of cavities enhances the possibility of chemical bond cleavage and degradation of pollutants. The overall cost of the process includes installation costs, energy requirement costs, overall maintenance, and operation costs, and many more. All these must be considered during the exact implementation of the processes. The energy required for laboratory scale equipment is often exceptionally low. The application of batch ultrasonic reactors equipped with ultrasonic transducers is sufficient at a lower volume, but in the case of large scale applications, additional solutions are required (Gągol et al. 2018b). Mostly flow cell systems are used with recirculation of treated liquid in the system.

The required dimensions of the reactor could be estimated by using the quantity of waste to be treated and required flowrate to be handled during the sonochemical approach. The flow rate of the waste needs to be treated indirectly concerned with the energy input to the reactor. The reactor accessories requirement also induces the additional capital in the process. The energy consumption associated with large capacity reactors to treat the larger amount of waste has very limited exposure. The energy dissipation per unit volume of the wastewater (W/L) needs to optimize during actual implementations. The overall cost of the treatment will include all the effective additions such as chemical requirements, manpower, land per space requirements, and energy associated with processes. The disposal of waste needs also to be considered while evaluating the cost of treatment. The sonochemical approach to wastewater treatment is a robust method that generates the utmost equivalent effect during each cycle on cost of required energy inputs. The appropriate optimization techniques are needed to optimize the processes before their actual application on the large scale of industrial waste.

3. Conclusion and Recommendations

Different types of hybrid processes based on the combination of US and AOPs, including sonocatalysts were reviewed. Recent advances and limitations in the usage of US based hybrid processes were discussed in terms of the degradation of organic pollutants in water. Studies on the combined processes of US/O_3, US/SR-AOPs, US/ H_2O_2 revealed their high effectiveness towards the degradation of organic pollutants. The performance of such processes is generally ascertained by the optimal loading of the oxidant. However, the dose of the oxidant to attain a sufficient extent of pollutant's degradation is still high, which is associated with lower economic feasibility. To improve the utilization rate of oxidants to produce ROS, several strategies, such as the employment of solid catalysts and the assessment of various US operation parameters can be considered. The latter requires a profound investigation on the effect of US parameters (e.g., frequency, power density) on the evolution of cavitational events, which are responsible for the cleavage of peroxy bonds to yield ROS.

Application of sonocatalysts is an interesting route of US based AOPs for the degradation of organic pollutants. The mechanism relying behind the sonocatalytic activity proceeds in a similar manner to widely known photocatalysis and the role of sonocatalysts under acoustic cavitation is boiled down to the formation of nucleation sites. The impact of acoustic cavitation along with other parameters (e.g., hydrogen potential, catalyst surface properties) on the activity of sonocatalysts, sonocatalyst pollutant interaction and generation of ROS is still unclear. Hence, the discussion on the usage of sonocatalyst for the degradation of water pollutants opens a promising prospective scope of research.

To control the secondary pollution, it is important to conduct the analysis of degradation products. As forementioned above, the degradation products may appear to be more harmful than the parent compound, especially for nitrogen containing organic compounds. This fact leads to the concern regarding the toxicity of transformation products yielded throughout the degradation of other class of pollutants. Recently, the hybrid US/AOPs processes are tested for the degradation of diverse types of MPs. Although, the degradation of MPs occurs on the surface and the damaged part can act as a carrier of contaminants in the environment, the overall oxidation mechanism of microplastics and their transformation products are poorly studied. Therefore, more effort must put on the determination of by products and their toxicity. The proposed mechanism pathway of MP degradation can encourage the scientific community to develop new hybrid processes and optimize the process parameters to assure the highest mineralization rate of MPs.

Acknowledgements

The authors gratefully acknowledge financial support from the National Science Centre for project UMO-2017/25/B/ST8/01364. prof. Shirish H. Sonawane would like to acknowledge financial support from Polish National Agency for Academic Exchange (NAWA) under the Ulam programme (grant number: PPN/ULM/2020/1/00037/U/00001).

References

Abdi, J., A.J. Sisi, M. Hadipoor and A. Khataee. 2022. State of the art on the ultrasonic-assisted removal of environmental pollutants using metal-organic frameworks. J Hazard Mater. 424: 127558.

Abdolalian, P., A. Morsali and S.K. Tizhoush. 2021. Sono-synthesis of basic metal-organic framework for reusable catalysis of organic reactions in the eco-friendly conditions. J Solid State Chem. 303: 122525.

Abdullah, A.Z. and P.Y. Ling. 2010. Heat treatment effects on the characteristics and sonocatalytic performance of TiO_2 in the degradation of organic dyes in aqueous solution. J Hazard Mater. 173(1–3): 159–167.

Adewuyi, Y.G. 2001. Sonochemistry: environmental science and engineering applications. Ind. Eng. Chem. Res. 40(22): 4681–4715.

Agarkoti, C., P.D. Thanekar and P.R. Gogate. 2021. Cavitation based treatment of industrial wastewater: a critical review focusing on mechanisms, design aspects, operating conditions and application to real effluents. J. Environ Manage. 300: 113786–113816.

Al-Hamadani, Y.A.J., G. Lee, S. Kim, C.M. Park, M. Jang, N. Her et al. 2018. Sonocatalytic degradation of carbamazepine and diclofenac in the presence of graphene oxides in aqueous solution. Chemosphere. 205: 719–727.

Andrady, A.L. 2011. Microplastics in the marine environment. Mar Pollut Bull. 62(8): 1596–1605.

Babu, D.S., V. Srivastava, P. V Nidheesh and M.S. Kumar. 2019. Detoxification of water and wastewater by advanced oxidation processes. Sci Total Environ. 696: 133961–133970.

Bernabeu, A., R.F. Vercher, L. Santos-Juanes, P.J. Simón, C. Lardín, M.A. Martínez et al. 2011. Solar photocatalysis as a tertiary treatment to remove emerging pollutants from wastewater treatment plant effluents. Catal Today. 161(1): 235–240.

Bhuta, H. 2014. Chapter 4 - Advanced treatment technology and strategy for water and wastewater management. pp. 193–213. In: Ranade, V V, Bhandari Recycling and Reuse VMBT-IWT, [eds.]. Oxford: Butterworth-Heinemann;

Boczkaj, G., A. Przyjazny and M. Kamiński. 2014. New procedures for control of industrial effluents treatment processes. Ind Eng Chem Res. 53(4): 1503–1514.

Boczkaj, G., P. Makoś, A. Fernandes and A. Przyjazny. 2017. New procedure for the examination of the degradation of volatile organonitrogen compounds during the treatment of industrial effluents. J Sep Sci. 40(6): 1301–1309.

Cao, H., W. Zhang, C. Wang and Y. Liang. 2020. Sonochemical degradation of poly- and perfluoroalkyl substances – a review. Ultrason Sonochem. 69: 105245–105255.

Cao, S., B. Shen, T. Tong, J. Fu and J. Yu. 2018. 2D/2D Heterojunction of ultrathin MXene/Bi2WO6 nanosheets for improved photocatalytic CO_2 reduction. Adv Funct Mater. 28(21): 1800136–1800146.

Carey, M. and M.W. Barsoum. 2021. MXene polymer nanocomposites: a review. Mater Today Adv. 9: 100120–100133.

Chatel, G. and J.C. Colmenares. 2017. Sonochemistry: from basic principles to innovative applications. Top Curr Chem. 375(1): 8–11.

Chen, X., J. Dai, G. Shi, L. Li, G. Wang and H. Yang. 2016. Sonocatalytic degradation of rhodamine B catalyzed by β-Bi$_2$O$_3$ particles under ultrasonic irradiation. Ultrason Sonochem. 29: 172–177.

Chowdhury, P. and T. Viraraghavan. 2009. Sonochemical degradation of chlorinated organic compounds, phenolic compounds and organic dyes – a review. Sci Total Environ. 407(8): 2474–2492.

Darvishi Cheshmeh Soltani, R., A.R. Khataee and M. Mashayekhi. 2016. Photocatalytic degradation of a textile dye in aqueous phase over ZnO nanoparticles embedded in biosilica nanobiostructure. Desalin Water Treat. 57(29): 13494–13504.

Darvishi Cheshmeh Soltani, R., M. Mashayekhi, A. Khataee, M.J. Ghanadzadeh and M. Sillanpää. 2018. Hybrid sonocatalysis/electrolysis process for intensified decomposition of amoxicillin in aqueous solution in the presence of magnesium oxide nanocatalyst. J Ind Eng Chem. 64: 373–382.

De Andrade, F.V., R. Augusti and G.M. de Lima. 2021. Ultrasound for the remediation of contaminated waters with persistent organic pollutants: a short review. Ultrason Sonochem. 78: 105719–105732.

Ding, Z., M. Sun, W. Liu, W. Sun, X. Meng and Y. Zheng. 2021. Ultrasonically synthesized N-TiO$_2$/Ti$_3$C$_2$ composites: enhancing sonophotocatalytic activity for pollutant degradation and nitrogen fixation. Sep Purif Technol. 276: 119287–119298.

Dirany, A., I. Sirés, N. Oturan, A. Özcan and M.A. Oturan. 2012. Electrochemical treatment of the antibiotic sulfachloropyridazine: kinetics, reaction pathways and toxicity evolution. Environ Sci Technol. 46(7): 4074–4082.

Dong, G., B. Chen, B. Liu, L.J. Hounjet, Y. Cao, S.R. Stoyanov et al. 2022. Advanced oxidation processes in microreactors for water and wastewater treatment: development, challenges, and opportunities. Water Res. 211: 118047–118067.

Dostanić, J., M. Huš and D. Lončarević. 2020. Effect of substituents in hydroxyl radical-mediated degradation of azo pyridone dyes: theoretical approaches on the reaction mechanism. J Environ Sci. 98: 14–21.

Elias, M.T., J. Chandran, U.K. Aravind and C.T. Aravindakumar. 2019. Oxidative degradation of ranitidine by UV and ultrasound: identification of transformation products using LC-Q-ToF-MS. Environ Chem. 16(1): 41–54.

Eniola, J.O., R. Kumar, M.A. Barakat and J. Rashid. 2022. A review on conventional and advanced hybrid technologies for pharmaceutical wastewater treatment. J Clean Prod.: 131826–131850.

Eslami, A., F. Mehdipour, K.Y.A. Lin, H. Sharifi Maleksari, F. Mirzaei and F. Ghanbari. 2020. Sono-photo activation of percarbonate for the degradation of organic dye: the effect of water matrix and identification of by-products. J Water Process Eng. 33: 100998–101006.

Fedorov, K., M. Plata-Gryl, J.A. Khan and G. Boczkaj. 2020. Ultrasound-assisted heterogeneous activation of persulfate and peroxymonosulfate by asphaltenes for the degradation of BTEX in water. J Hazard Mater. 397: 122804–122811.

Fedorov, K., X. Sun and G. Boczkaj. 2021. Combination of hydrodynamic cavitation and SR-AOPs for simultaneous degradation of BTEX in water. Chem Eng J. 417: 128081–128096.

Fedorov, K., K. Dinesh, X. Sun, R. Darvishi Cheshmeh Soltani, Z. Wang, S. Sonawane et al. 2022. Synergistic effects of hybrid advanced oxidation processes (AOPs) based on hydrodynamic cavitation phenomenon – A review. Chem Eng J. 432: 134191–134223.

Fendall, L.S. and M.A. Sewell. 2009. Contributing to marine pollution by washing your face: microplastics in facial cleansers. Mar Pollut Bull. 58(8): 1225–1228.

Fernandes, A., P. Makoś, Z. Wang and G. Boczkaj. 2020. Synergistic effect of TiO$_2$ photocatalytic advanced oxidation processes in the treatment of refinery effluents. Chem Eng J. 391: 123488–123501.

Fu, Y., L. Wang, W. Peng, Q. Fan, Q. Li, Y. Dong et al. 2021. Enabling simultaneous redox transformation of toxic chromium(VI) and arsenic(III) in aqueous media—a review. J Hazard Mater. 417: 126041–126057.

Gągol, M., A. Przyjazny and G. Boczkaj. 2018a. Effective method of treatment of industrial effluents under basic pH conditions using acoustic cavitation – a comprehensive comparison with hydrodynamic cavitation processes. Chem Eng Process - Process Intensif. 128: 103–113.

Gągol, M., A. Przyjazny and G. Boczkaj. 2018b. Highly effective degradation of selected groups of organic compounds by cavitation based AOPs under basic pH conditions. Ultrason Sonochem. 45: 257–266.

Gao, Y., J. Zhou, Y. Rao, H. Ning, J. Zhang, J. Shi et al. 2022. Comparative study of degradation of ketoprofen and paracetamol by ultrasonic irradiation: mechanism, toxicity and DBP formation. Ultrason Sonochem. 82: 105906–105916.

Ghanbari, F. and M. Moradi. 2017. Application of peroxymonosulfate and its activation methods for degradation of environmental organic pollutants: review. Chem Eng J. 310: 41–62.

Gholami, P., L. Dinpazhoh, A. Khataee and Y. Orooji. 2019a. Sonocatalytic activity of biochar-supported ZnO nanorods in degradation of gemifloxacin: synergy study, effect of parameters and phytotoxicity evaluation. Ultrason Sonochem. 55: 44–56.

Gholami, P., A. Khataee, R.D.C. Soltani and A. Bhatnagar. 2019b. A review on carbon-based materials for heterogeneous sonocatalysis: fundamentals, properties and applications. Ultrason Sonochem. 58: 104681–104698.

Gholami, P., L. Dinpazhoh, A. Khataee, A. Hassani and A. Bhatnagar. 2020. Facile hydrothermal synthesis of novel Fe-Cu layered double hydroxide/biochar nanocomposite with enhanced sonocatalytic activity for degradation of cefazolin sodium. J Hazard Mater. 381: 120742–120757.

Glick, H.S., J.J. Klein and W. Squire. 1957. Single-pulse shock tube studies of the kinetics of the reaction $N2+O2 \rightleftarrows 2NO$ between 2000–3000°K. J Chem Phys. 27(4): 850–857.

Gogate, P.R. and A.L. Prajapat. 2015. Depolymerization using sonochemical reactors: a critical review. Ultrason Sonochem. 27: 480–494.

Gogate, P.R. 2002. Cavitation: an auxiliary technique in wastewater treatment schemes. Adv Environ Res. 6(3): 335–358.

Grčić, I., S. Papić and N. Koprivanac. 2013. Sonochemical effectiveness factor (eUS) in the reactors for wastewater treatment by sono-Fenton oxidation: novel considerations. Ultrason Sonochem. 20(4): 1037–1045.

Gültekin, I. and N.H. Ince. 2006. Degradation of aryl-azo-naphthol dyes by ultrasound, ozone and their combination: effect of α-substituents. Ultrason Sonochem. 13(3): 208–214.

Hammouda, S. Ben, F. Zhao, Z. Safaei, V. Srivastava, D. Lakshmi Ramasamy et al. 2017. Degradation and mineralization of phenol in aqueous medium by heterogeneous monopersulfate activation on nanostructured cobalt based-perovskite catalysts ACoO3 (A = La, Ba, Sr and Ce): Characterization, kinetics and mechanism study. Appl Catal B Environ. 215: 60–73.

Harada, H. 2001. Sonophotocatalytic decomposition of water using TiO_2 photocatalyst. Ultrason Sonochem. 8(1): 55–58.

Hartmann, J., P. Bartels, U. Mau, M. Witter, W. v. Tümpling, J. Hofmann et al. 2008. Degradation of the drug diclofenac in water by sonolysis in presence of catalysts. Chemosphere. 70(3): 453–461.

He, Z., S. Song, H. Ying, L. Xu and J. Chen. 2007. p-Aminophenol degradation by ozonation combined with sonolysis: operating conditions influence and mechanism. Ultrason Sonochem. 14(5): 568–574.

Huang, H., Y. Song, N. Li, D. Chen, Q. Xu, H. Li et al. 2019. One-step in-situ preparation of N-doped $TiO_2@C$ derived from Ti_3C_2 MXene for enhanced visible-light driven photodegradation. Appl Catal B Environ. 251: 154–161.

Hube, S. and B. Wu. 2021. Mitigation of emerging pollutants and pathogens in decentralized wastewater treatment processes: a review. Sci Total Environ. 779: 146545–146562.

Jellinek, H.H.G. and G. White. 1951. The degradation of long-chain molecules by ultrasonic waves. I. Theoretical. J Polym Sci. 6(6): 745–756.

Jun, B.M., S. Kim, J. Heo, N. Her, M. Jang, C.M. Park et al. 2019a. Enhanced sonocatalytic degradation of carbamazepine and salicylic acid using a metal-organic framework. Ultrason Sonochem. 56: 174–182.

Jun, B.-M., S. Kim, J. Heo, C.M. Park, N. Her, M. Jang et al. 2019b. Review of MXenes as new nanomaterials for energy storage/delivery and selected environmental applications. Nano Res. 12(3): 471–487.

Jun, B.M., J. Heo, N. Taheri-Qazvini, C.M. Park and Y. Yoon. 2020a. Adsorption of selected dyes on Ti3C2Tx MXene and Al-based metal-organic framework. Ceram Int. 46(3): 2960–2968.

Jun, B.M., Y.A.J. Jun, Al-Hamadani, A. Son, C.M. Park, M. Jang et al. 2020b. Applications of metal-organic framework based membranes in water purification: a review. Sep Purif Technol. 247: 116947–116966.

Jun, B.M., J. Han, C.M. Park and Y. Yoon. 2020c. Ultrasonic degradation of selected dyes using Ti_3C_2Tx MXene as a sonocatalyst. Ultrason Sonochem. 64: 104993–105002.

Karim, A.V., A. Hassani, P. Eghbali and P. V. Nidheesh. 2022. Nanostructured modified layered double hydroxides (LDHs)-based catalysts: a review on synthesis, characterization, and applications in water remediation by advanced oxidation processes. Curr Opin Solid State Mater Sci. 26(1): 100965–100990.

Kermani, M., M. Farzadkia, M. Morovati, M. Taghavi, S. Fallahizadeh, R. Khaksefidi et al. 2020. Degradation of furfural in aqueous solution using activated persulfate and peroxymonosulfate by ultrasound irradiation. J Environ Manage. 266: 110616–110626.

Keyikoglu, R., A. Khataee, H. Lin and Y. Orooji. 2022. Vanadium (V)-doped ZnFe layered double hydroxide for enhanced sonocatalytic degradation of pymetrozine. Chem Eng J. 434: 134730–134739.

Khataee, A., M. Sheydaei, A. Hassani, M. Taseidifar and S. Karaca. 2015. Sonocatalytic removal of an organic dye using TiO2/Montmorillonite nanocomposite. Ultrason Sonochem. 22: 404–411.

Khataee, A., T. Sadeghi Rad, S. Nikzat, A. Hassani, M.H. Aslan, M. Kobya et al. 2019. Fabrication of NiFe layered double hydroxide/reduced graphene oxide (NiFe-LDH/rGO) nanocomposite with enhanced sonophotocatalytic activity for the degradation of moxifloxacin. Chem Eng J. 375: 122102–122115.

Khataee, A., A. Fazli, F. Zakeri and S.W. Joo. 2020. Synthesis of a high-performance Z-scheme 2D/2D WO3@CoFe-LDH nanocomposite for the synchronic degradation of the mixture azo dyes by sonocatalytic ozonation process. J Ind Eng Chem. 89: 301–315.

Kim, S., A. Sin, H. Nam, Y. Park, H. Lee and C. Han. 2022. Advanced oxidation processes for microplastics degradation: a recent trend. Chem Eng J Adv. 9: 100213–100221.

Kotronarou, A., G. Mills and M.R. Hoffmann. 1991. Ultrasonic irradiation of p-nitrophenol in aqueous solution. J Phys Chem. 95(9): 3630–3638.

Ku, Y., Y.-H. Tu and C.-M. Ma. 2005. Effect of hydrogen peroxide on the decomposition of monochlorophenols by sonolysis in aqueous solution. Water Res. 39(6): 1093–1098.

Kyzas, G.Z., N. Mengelizadeh, M. khodadadi Saloot, S. Mohebi and D. Balarak. 2022. Sonochemical degradation of ciprofloxacin by hydrogen peroxide and persulfate activated by ultrasound and ferrous ions. Colloids Surfaces A Physicochem Eng Asp. 642: 12862–-128735.

Landge, V.K., S.H. Sonawane, M. Sivakumar, S.S. Sonawane, G. Uday Bhaskar Babu and G. Boczkaj. 2021. S-scheme heterojunction Bi$_2$O$_3$-ZnO/Bentonite clay composite with enhanced photocatalytic performance. Sustain Energy Technol Assessments. 45: 101194–101202.

Lebreton, L.C.M., J. van der Zwet, J.-W. Damsteeg, B. Slat, A. Andrady and J. Reisser. 2017. River plastic emissions to the world's oceans. Nat Commun. 8(1): 15611–15620.

Li, Z., H. Zhang, L. Wang, X. Meng, J. Shi, C. Qi et al. 2020. 2D/2D BiOBr/Ti$_3$C$_2$ heterojunction with dual applications in both water detoxification and water splitting. J Photochem Photobiol A Chem. 386: 112099.

Lin, J., X. Zhao, D. Liu, Z. Yu, Y. Zhang and H. Xu. 2008. The decoloration and mineralization of azo dye C.I. acid red 14 by sonochemical process: rate improvement via fenton's reactions. J Hazard Mater. 157(2): 541–546.

Liu, L., Z. Chen, J. Zhang, D. Shan, Y. Wu, L. Bai et al. 2021. Treatment of industrial dye wastewater and pharmaceutical residue wastewater by advanced oxidation processes and its combination with nanocatalysts: a review. J Water Process Eng. 42: 102122–102140.

Lu, X., W. Qiu, J. Peng, H. Xu, D. Wang, Y. Cao et al. 2021. A review on additives-assisted ultrasound for organic pollutants degradation. J Hazard Mater. 403: 123915–123926.

Ma, D., H. Yi, C. Lai, X. Liu, X. Huo, Z. An et al. 2021. Critical review of advanced oxidation processes in organic wastewater treatment. Chemosphere. 275: 130104–130130.

Mailler, R., J. Gasperi, Y. Coquet, A. Buleté, E. Vulliet, S. Deshayes et al. 2016. Removal of a wide range of emerging pollutants from wastewater treatment plant discharges by micro-grain activated carbon in fluidized bed as tertiary treatment at large pilot scale. Sci Total Environ. 542: 983–996.

Makino, K., M.M. Mossoba and P. Riesz. 1982. Chemical effects of ultrasound on aqueous solutions. Evidence for hydroxyl and hydrogen free radicals (.cntdot.OH and .cntdot.H) by spin trapping. J Am Chem Soc. 104(12): 3537–3539.

Mashile, G.P., S.K. Selahle, A. Mpupa, A. Nqombolo and P.N. Nomngongo. 2022. Chapter 8 - Remediation of emerging pollutants through various wastewater treatment processes. pp. 137–150. *In*: Dalu T, Tavengwa NTBT-EFP, [eds]. Elsevier.

Matouq, M.A., Z.A. Al-Anber, T. Tagawa, S. Aljbour and M. Al-Shannag. 2008. Degradation of dissolved diazinon pesticide in water using the high frequency of ultrasound wave. Ultrason Sonochem. 15(5): 869–874.

Meng, Z. Da and W.C. Oh. 2011. Sonocatalytic degradation and catalytic activities for MB solution of Fe treated fullerene/TiO2 composite with different ultrasonic intensity. Ultrason Sonochem. 18(3): 757–764.

Meng, Z. Da, L. Zhu, J.G. Choi, C.Y. Park and W.C. Oh. 2012. Sonocatalytic degradation of rhodamine B in the presence of C60 and CdS coupled TiO2 particles. Ultrason Sonochem. 19(1): 143–150.

Minh, T. Do, M.C. Ncibi, V. Srivastava, S.K. Thangaraj, J. Jänis et al. 2019. Gingerbread ingredient-derived carbons-assembled CNT foam for the efficient peroxymonosulfate-mediated degradation of emerging pharmaceutical contaminants. Appl Catal B Environ. 244(October 2018): 367–384.

Mišík, V. and P. Riesz. 1996. Nitric oxide formation by ultrasound in aqueous solutions. J Phys Chem. 100(45): 17986–17994.

Moghni, N., H. Boutoumi, H. Khalaf, N. Makaoui and G. Colón. 2022. Enhanced photocatalytic activity of TiO_2/WO_3 nanocomposite from sonochemical-microwave assisted synthesis for the photodegradation of ciprofloxacin and oxytetracycline antibiotics under UV and sunlight. J Photochem Photobiol A Chem. 428: 113848–113859.

Mohod, A.V. and P.R. Gogate. 2011. Ultrasonic degradation of polymers: effect of operating parameters and intensification using additives for carboxymethyl cellulose (CMC) and polyvinyl alcohol (PVA). Ultrason Sonochem. 18(3): 727–734.

Montoya-Rodríguez, D.M., E.A. Serna-Galvis, F. Ferraro and R.A. Torres-Palma. 2020. Degradation of the emerging concern pollutant ampicillin in aqueous media by sonochemical advanced oxidation processes - parameters effect, removal of antimicrobial activity and pollutant treatment in hydrolyzed urine. J Environ Manage. 261: 110224–110230.

Naddeo, V., V. Belgiorno, D. Ricco and D. Kassinos. 2009. Degradation of diclofenac during sonolysis, ozonation and their simultaneous application. Ultrason Sonochem. 16(6): 790–794.

Naderi, M. and R. Darvishi Cheshmeh Soltani. 2021. Hybrid of ZnFe layered double hydroxide/nano-scale carbon for activation of peroxymonosulfate to decompose ibuprofen: thermodynamic and reaction pathways investigation. Environ Technol Innov. 24: 101951–101964.

Nas, M.S. 2021. AgFe$_2$O$_4$/MWCNT nanoparticles as novel catalyst combined adsorption-sonocatalytic for the degradation of methylene blue under ultrasonic irradiation. J Environ Chem Eng. 9(3): 105207–105220.

Nejumal, K.K., P.R. Manoj, U.K. Aravind and C.T. Aravindakumar. 2014. Sonochemical degradation of a pharmaceutical waste, atenolol, in aqueous medium. Environ Sci Pollut Res. 21(6): 4297–4308.

Neta, P., R.E. Huie and A.B. Ross. 1988. Rate constants for reactions of inorganic radicals in aqueous solution. J Phys Chem Ref Data. 17(3): 1027–1284.

Nirumand, L., S. Farhadi, A. Zabardasti and A. Khataee. 2018. Synthesis and sonocatalytic performance of a ternary magnetic MIL-101(Cr)/RGO/ZnFe$_2$O$_4$ nanocomposite for degradation of dye pollutants. Ultrason Sonochem. 42: 647–658.

Nuengmatcha, P., S. Chanthai, R. Mahachai and W.C. Oh. 2016. Sonocatalytic performance of ZnO/graphene/TiO2 nanocomposite for degradation of dye pollutants (methylene blue, texbrite BAC-L, texbrite BBU-L and texbrite NFW-L) under ultrasonic irradiation. Dye Pigment. 134: 487–497.

Oliveira, T.M.B.F., F.W.P. Ribeiro, S. Morais, P. de Lima-Neto and A.N. Correia. 2022. Removal and sensing of emerging pollutants released from (micro)plastic degradation: strategies based on boron-doped diamond electrodes. Curr Opin Electrochem. 31: 100866–100874.

Oturan, N., C.T. Aravindakumar, H. Olvera-Vargas, M.M. Sunil Paul and M.A. Oturan. 2018. Electro-fenton oxidation of para-aminosalicylic acid: degradation kinetics and mineralization pathway using Pt/carbon-felt and BDD/carbon-felt cells. Environ Sci Pollut Res. 25(21): 20363–20373.

Pang, Y.L., A.Z. Abdullah and S. Bhatia. 2011. Review on sonochemical methods in the presence of catalysts and chemical additives for treatment of organic pollutants in wastewater. Desalination. 277(1): 1–14.

Persson, P.O.Á. and J. Rosen. 2019. Current state of the art on tailoring the MXene composition, structure, and surface chemistry. Curr Opin Solid State Mater Sci. 23(6): 100774–1000799.

Philip, J.M., C.M. Koshy, U.K. Aravind and C.T. Aravindakumar. 2022. Sonochemical degradation of DEET in aqueous medium: complex by-products from synergistic effect of sono-fenton – new insights from a HRMS study. J Environ Chem Eng. 10(3): 107509–107522.

Qiu, P., B. Park, J. Choi, B. Thokchom, A.B. Pandit and J. Khim. 2018. A review on heterogeneous sonocatalyst for treatment of organic pollutants in aqueous phase based on catalytic mechanism. Ultrason Sonochem. 45: 29–49.

Rayaroth, M.P., U.K. Aravind and C.T. Aravindakumar. 2016. Degradation of pharmaceuticals by ultrasound-based advanced oxidation process. Environ Chem Lett. 14(3): 259–290.

Rayaroth, M.P., U.K. Aravind and C.T. Aravindakumar. 2018. Role of in-situ nitrite ion formation on the sonochemical transformation of para-aminosalicylic acid. Ultrason Sonochem. 40: 213–220.

Rayaroth, M.P., C.T. Aravindakumar, N.S. Shah and G. Boczkaj. 2022. Advanced oxidation processes (AOPs) based wastewater treatment - unexpected nitration side reactions - a serious environmental issue: a review. Chem Eng J. 430: 133002–133016.

Ren, W., H. Wang, Q. Chang, N. Li, J. Yang and S. Hu. 2021. Origin of sonocatalytic activity of fluorescent carbon dots. Carbon N Y. 184: 102–108.

Ribeiro, A.R.L., D. Hermosilla, M.A. Mueses, R. Xiao and D. Mantzavinos. 2022. Advanced oxidation technologies for water/wastewater treatment: advances, gaps and challenges - editorial. Chem Eng J Adv. 10: 100272.

Ricardo, I.A., E.A. Alberto, A.H. Silva Júnior, D.L.P. Macuvele, N. Padoin, C. Soares et al. 2021. A critical review on microplastics, interaction with organic and inorganic pollutants, impacts and effectiveness of advanced oxidation processes applied for their removal from aqueous matrices. Chem Eng J. 424: 130282–130296.

Richardson, S.D. and S.Y. Kimura. 2017. Emerging environmental contaminants: challenges facing our next generation and potential engineering solutions. Environ Technol Innov. 8: 40–56.

Rodriguez, S., A. Santos and A. Romero. 2011. Effectiveness of AOP's on abatement of emerging pollutants and their oxidation intermediates: nicotine removal with fenton's reagent. Desalination. 280(1): 108–113.

Ryu, U.J., S. Jee, P.C. Rao, J. Shin, C. Ko, M. Yoon et al. 2021. Recent advances in process engineering and upcoming applications of metal–organic frameworks. Coord Chem Rev. 426: 213544–213617.

Sadeghi Rad, T., A. Khataee, S. Arefi-Oskoui, S. Sadeghi Rad, Y. Orooji, E. Gengec et al. 2022a. Graphene-based ZnCr layered double hydroxide nanocomposites as bactericidal agents with high sonophotocatalytic performances for degradation of rifampicin. Chemosphere. 286: 131740–131752.

Sadeghi Rad, T., A. Khataee, S. Sadeghi Rad, S. Arefi-Oskoui, E. Gengec, M. Kobya et al. 2022b. Zinc-chromium layered double hydroxides anchored on carbon nanotube and biochar for ultrasound-assisted photocatalysis of rifampicin. Ultrason Sonochem. 82: 105875–10885.

Sajjadi, S., A. Khataee and M. Kamali. 2017. Sonocatalytic degradation of methylene blue by a novel graphene quantum dots anchored CdSe nanocatalyst. Ultrason Sonochem. 39: 676–685.

Sajjadi, S., A. Khataee, R. Darvishi Cheshmeh Soltani and A. Hasanzadeh. 2019. N, S co-doped graphene quantum dot–decorated Fe_3O_4 nanostructures: preparation, characterization and catalytic activity. J Phys Chem Solids. 127: 140–150.

Saravanakumar, K., A. Fayyaz, S. Park, Y. Yoon, Y.M. Kim and C.M. Park. 2021. Hierarchical $CoTiO_3$ microrods on Ti_3C_2Tx MXene heterostructure as an efficient sonocatalyst for bisphenol a degradation. J Mol Liq. 344: 117740–117750.

Sasi, S., M.P. Rayaroth, D. Devadasan, U.K. Aravind and C.T. Aravindakumar. 2015. Influence of inorganic ions and selected emerging contaminants on the degradation of methylparaben: a sonochemical approach. J Hazard Mater. 300: 202–209.

Saska Romero, J.H., G.P. Saito, F. Cagnin, M.A. Cebim and M.R. Davolos. 2022. Europium-doped Zn-Al-LDH intercalated with 4-biphenylcarboxylate anion and undoped Zn-Al-LDH intercalated with its anionic Eu(III) complex: structural and UV or X-ray excited luminescence properties. Opt Mater (Amst). 124: 111703-111712.

Selim, A., S. Kaur, A.H. Dar, S. Sartaliya and G. Jayamurugan. 2020. Synergistic effects of carbon dots and palladium nanoparticles enhance the sonocatalytic performance for rhodamine b degradation in the absence of light. ACS Omega. 5(35): 22603–22613.

Serna-Galvis, E.A., A.M. Botero-Coy, D. Martínez-Pachón, A. Moncayo-Lasso, M. Ibáñez, F. Hernández et al. 2019. Degradation of seventeen contaminants of emerging concern in municipal wastewater effluents by sonochemical advanced oxidation processes. Water Res. 154: 349–360.

Serna-Galvis, E.A. and R.A. Torres-Palma. 2021. Chapter 10 - Recent developments in sonochemical treatments of contaminated wastewaters. p. 299–315. *In*: Sharma SKBT-GC and WRR and A [eds.]. Adv Green Sustain Chem.Elsevier.

Serna-Galvis, E.A., J. Porras and R.A. Torres-Palma. 2022. A critical review on the sonochemical degradation of organic pollutants in urine, seawater, and mineral water. Ultrason Sonochem. 82: 105861–105874.

Singh, A. and B.K. Mishra. 2022. Chapter 2 - Treatment aspect of an emerging pollutant from pharmaceutical industries using advanced oxidation process: past, current, and future trends. pp. 23–44. *In*: Shah, M., Rodriguez-Couto S, Biswas JBT-D in WTR and P. [eds.]. Elsevier.

Soltani, R.D.C., M. Mashayekhi, M. Naderi, G. Boczkaj, S. Jorfi and M. Safari. 2019a. Sonocatalytic degradation of tetracycline antibiotic using zinc oxide nanostructures loaded on nano-cellulose from waste straw as nanosonocatalyst. Ultrason Sonochem. 55: 117–124.

Soltani, R.D.C., Z. Miraftabi, M. Mahmoudi, S. Jorfi, G. Boczkaj and A. Khataee. 2019b. Stone cutting industry waste-supported zinc oxide nanostructures for ultrasonic assisted decomposition of an anti-inflammatory non-steroidal pharmaceutical compound. Ultrason Sonochem. 58: 104669.

Song, L., S. Zhang and Q. Wei. 2012. Porous BiOI sonocatalysts: hydrothermal synthesis, characterization, sonocatalytic, and kinetic properties. Ind Eng Chem Res. 51(3): 1193–1197.

Stefan, M.I. 2017. Advanced oxidation processes for water treatment: fundamentals and applications. IWA publishing.

Supeno and P. Kruus. 2002. Fixation of nitrogen with cavitation. Ultrason Sonochem. 9(1): 53–59.

Suslick, K.S. 1990. Sonochemistry. Science 247(4949): 1439–1445.

Sutkar, V.S. and P.R. Gogate. 2009. Design aspects of sonochemical reactors: Techniques for understanding cavitational activity distribution and effect of operating parameters. Chem Eng J. 155(1): 26–36.

Talvitie, J., A. Mikola, A. Koistinen and O. Setälä. 2017. Solutions to microplastic pollution – removal of microplastics from wastewater effluent with advanced wastewater treatment technologies. Water Res. 123: 401–407.

Tian, K., L. Hu, L. Li, Q. Zheng, Y. Xin and G. Zhang. 2021. Recent advances in persulfate-based advanced oxidation processes for organic wastewater treatment. Chinese Chem Lett. 33(10): 4461–4477.

Titchou, F.E., H. Zazou, H. Afanga, J. El Gaayda, R. Ait Akbour, P.V. Nidheesh et al. 2021. Removal of organic pollutants from wastewater by advanced oxidation processes and its combination with membrane processes. Chem Eng Process - Process Intensif. 169: 108631–108652.

Uma, K., B. KrishnaKumar, G.-T. Pan, T.C.-K. Yang and J.-H. Lin. 2020. Enriched silver plasmon resonance activity on the sonochemical synthesis of ZnO flowers with α-Fe$_2$O$_3$ as an efficient catalyst for photo-Fenton reaction and photo-oxidation of ethanol. J Water Process Eng. 34: 101089–101098.

Vasseghian, Y., E.N. Dragoi, F. Almomani and V.T. Le. 2022. A comprehensive review on MXenes as new nanomaterials for degradation of hazardous pollutants: deployment as heterogeneous sonocatalysis. Chemosphere. 287: 132387–132396.

Virtanen, A.I. and N. Ellfolk. 1950. Nitrogen fixation in an ultrasonic field. J Am Chem Soc. 72(2): 1046–1047.

Wakeford, C.A., R. Blackburn and P.D. Lickiss. 1999. Effect of ionic strength on the acoustic generation of nitrite, nitrate and hydrogen peroxide. Ultrason Sonochem. 6(3): 141–148.

Wang, J., Y. Guo, B. Liu, X. Jin, L. Liu, R. Xu et al. 2011. Detection and analysis of reactive oxygen species (ROS) generated by nano-sized TiO2 powder under ultrasonic irradiation and application in sonocatalytic degradation of organic dyes. Ultrason Sonochem. 18(1): 177–183.

Wang, J.L. and L.J. Xu. 2012. Advanced oxidation processes for wastewater treatment: formation of hydroxyl radical and application. Crit Rev Environ Sci Technol. 42(3): 251–325.

Wang, Z., V. Srivastava, S. Wang, H. Sun, S.K. Thangaraj, J. Jänis et al. 2020. UVC-assisted photocatalytic degradation of carbamazepine by Nd-doped Sb2O3/TiO2 photocatalyst. J Colloid Interface Sci. 562: 461–469.

Weavers, L.K. and M.R. Hoffmann. 1998. Sonolytic decomposition of ozone in aqueous solution: mass transfer effects. Environ Sci Technol. 32(24): 3941–3947.

Wiest, L., A. Gosset, A. Fildier, C. Libert, M. Hervé, E. Sibeud et al. 2021. Occurrence and removal of emerging pollutants in urban sewage treatment plants using LC-QToF-MS suspect screening and quantification. Sci Total Environ. 774: 145779–145791.

Wu, Y., L. Song, Shujuan Zhang, X. Wu, Shuna Zhang, H. Tian et al. 2013. Sonocatalytic performance of AgBr in the degradation of organic dyes in aqueous solution. Catal Commun. 37: 14–18.

Wu, Y., L. Bu, X. Duan, S. Zhu, M. Kong, N. Zhu et al. 2020. Mini review on the roles of nitrate/nitrite in advanced oxidation processes: radicals transformation and products formation. J Clean Prod. 273: 123065–123074.

Xiao, Y., W. Zhang, Y. Jiao, Y. Xu and H. Lin. 2021. Metal-phenolic network as precursor for fabrication of metal-organic framework (MOF) nanofiltration membrane for efficient desalination. J Memb Sci. 624: 119101–119110.

Xu, L., X. Wang, Y. Sun, H. Gong, M. Guo, X. Zhang et al. 2020. Mechanistic study on the combination of ultrasound and peroxymonosulfate for the decomposition of endocrine disrupting compounds. Ultrason Sonochem. 60: 104749–104756.

Yan, J.K., J.J. Pei, H. Le Ma and Z. Bin Wang. 2015. Effects of ultrasound on molecular properties, structure, chain conformation and degradation kinetics of carboxylic curdlan. Carbohydr Polym. 121: 64–70.

Yano, J., J. Matsuura, H. Ohura and S. Yamasaki. 2005. Complete mineralization of propyzamide in aqueous solution containing TiO_2 particles and H2O2 by the simultaneous irradiation of light and ultrasonic waves. Ultrason Sonochem. 12(3): 197–203.

Yao, J., L. Chen, X. Chen, L. Zhou, W. Liu and Z. Zhang. 2018. Formation of inorganic nitrogenous byproducts in aqueous solution under ultrasound irradiation. Ultrason Sonochem. 42: 42–47.

Yao, J., H. Zhang, L. Chen, W. Liu, N. Gao, S. Liu et al. 2021. The roles of sono-induced nitrosation and nitration in the sono-degradation of diphenylamine in water: mechanisms, kinetics and impact factors. J Hazard Mater. 402: 123720.

Yasuda, K. 2021. Sonochemical green technology using active bubbles: degradation of organic substances in water. Curr Opin Green Sustain Chem. 27: 100411–100416.

Yousef Tizhoosh, N., A. Khataee, R. Hassandoost, R. Darvishi Cheshmeh Soltani and E. Doustkhah. 2020. Ultrasound-engineered synthesis of WS_2@CeO_2 heterostructure for sonocatalytic degradation of tylosin. Ultrason Sonochem. 67: 105114–105125.

Yu, C., Z. Xiong, H. Zhou, P. Zhou, H. Zhang, R. Huang et al. 2022. Marriage of membrane filtration and sulfate radical-advanced oxidation processes (SR-AOPs) for water purification: current developments, challenges and prospects. Chem Eng J. 433: 133802–133822.

Zhang, Y., Y.-G. Zhao, F. Maqbool and Y. Hu. 2022. Removal of antibiotics pollutants in wastewater by UV-based advanced oxidation processes: influence of water matrix components, processes optimization and application: a review. J Water Process Eng. 45: 102496–102519.

Zhou, M., H. Yang, T. Xian, R.S. Li, H.M. Zhang and X.X. Wang. 2015. Sonocatalytic degradation of RhB over LuFeO3 particles under ultrasonic irradiation. J Hazard Mater. 289: 149–157.

Zhu, Z.-H., Y. Liu, C. Song, Y. Hu, G. Feng and B.Z. Tang. 2022. Porphyrin-based two-dimensional layered metal–organic framework with sono-/photocatalytic activity for water decontamination. ACS Nano. 16(1): 1346–1357.

Zou, Y., P. Wang, W. Yao, Xiangxue Wang, Y. Liu, D. Yang et al. 2017. Synergistic immobilization of UO22+ by novel graphitic carbon nitride @ layered double hydroxide nanocomposites from wastewater. Chem Eng J. 330: 573–584.

CHAPTER 8

Sonochemistry in Foods of Animal Origin

Iván Adrián Garcia-Galicia,[1] *Luis Manuel Carrillo-Lopez,*[1,2]
Mariana Huerta-Jimenez,[1,2] *Monserrath Felix-Portillo*[1]
and *Alma Delia Alarcon-Rojo*[1,*]

1. General Introduction

In the last two decades, high intensity ultrasound (HIU) has received a large interest as a potential technology to modify food matrices to enhance their physicochemical, structural, functional, and sensorial properties. Ultrasound is considered a mechanical energy generated by the vibration of acoustic waves, above the audible threshold in humans (Alarcon-Rojo et al. 2019, Berlan and Mason 1992, Mason et al. 2010). The spectrum of acoustic waves has been listed by J.D.N. Cheeke in the following frequency ranges: < 20 Hz as Infrasound; from 20 Hz to < 20 kHz, as audible sound; from 20 kHz < to < one GHz, as ultrasound; and > one GHz as hyper sound (Cheeke 2000).

Ultrasound waves have been historically used in the food industry. Initially, as a non-destructive, non-invasive technology (intensity < one W/cm² and frequency > one MHz), to determine the acoustic properties of foods and to identify deterioration or modifications in their physical structure. Later, ultrasound has been continuously studied and utilized as an interventionist technology (Intensity = five to 1,000 W/cm² and Frequency = 20 to 100 kHz) to modify physicochemical attributes of foods (Chemat et al. 2011, Mason et al. 2010).

In high intensity ultrasonication, the acoustic intensity (I) of the sonication plays an important parameter, since it can cause the phenomena known as cavitation and shock waves (Suslick and Flannigan 2008). Acoustic cavitation is described as the formation of a cloud of bubbles in a liquid media, by effect of an acoustic field. Those tiny liquid bubbles grow over several compression and rarefaction cycles

[1] Facultad de Zootecnia y Ecología, Universidad Autónoma de Chihuahua, Perif. Francisco R. Almada km 1, Chihuahua, Chih., 31453, México.
 Emails: igarciag@uach.mx; monserrath.felix@uach.mx; aalarcon@uach.mx
[2] Consejo Nacional de Ciencia y Tecnología. Av. Insurgentes Sur 1582, Col. Credito Constructor, Del. Benito Juárez, Ciudad de México, C.P. 03940. México.
 Emails: mhuertaj@uach.mx; lmcarrillo@uach.mx
* Corresponding author: aalarcon@uach.mx

of the acoustic field, until they cannot absorb more energy and produce implosion, releasing a big amount of energy in a very small physical space. The violent collapse of bubbles produces other forms of physical forces such as: micro streaming, agitation, micro jetting, shocks and sonoluminescence (Ashokkumar 2011). On these principles, high intensity ultrasonication applied on foods can induce physical and chemical effects, by producing micro and macro structural changes. HIU has been evaluated on a wide range of foods, from liquids to solids, from vegetables to processed meats. The major beneficial effects of HIU application on foods are enzyme activation or inhibition, tenderization, higher homogenization, better emulsification and crystallization, improvement of organoleptic properties such as colour and flavour, reduction of microbial counts, and improvement of mass transfer (Alarcon-Rojo et al. 2019, Almanza-Rubio et al. 2016, Arvanitoyannis et al. 2017, Bilek and Turantas 2013, Carrillo López et al. 2021a).

2. Sonochemistry in Foods of Animal Origin

2.1 Ultrasonication in Meat and Meat Products

HIU application in the meat industry has shown clear benefits in processes such as microbial inhibition, tenderization, freezing, storing, thawing, curing, tumbling, and cooking. Relevant effects of HIU upon foods have been extensively described in several review articles during the last years (Alarcón-Rojo et al. 2015, 2019, Al-Hilphy et al. 2020, Boateng and Nasiru 2019, Cichoski et al. 2019, Gómez-Salazar et al. 2021, Jayasooriya et al. 2004, Nowak et al. 2017). The latest research has focused on three main subjects: (1) the elucidation of the microscopic or molecular changes produced in the tissue by the cavitation or other physical forces derived from ultrasonication, (2) the identification of intrinsic factors of the muscle per meat interfering on the cavitation effect, and (3) the standardization of HIU optimal conditions to achieve positive effects on the meat quality components.

This section summarizes results of ultrasonication on quality properties of meat and meat products during the last five years of research worldwide. Firstly, we examine the scientific evidence of the physicochemical, functional, and structural effects of sonochemistry (specifically HIU) on fresh meat, and the consequent results on meat quality. Practical uses of sonochemistry on industrial processes such as freezing thawing and curing of fresh meat will also be described. We then review the applications of ultrasonication in the meat transformation processes to enhance technological characteristics of meat products (ham, bacon, burgers, etc.,) or to improve their quality per sensorial features.

2.2 Ultrasonication of Fresh Meat

The most recent findings about the HIU effects on physicochemical and functional traits in fresh meat were presented in the section of Effects on Physicochemical and Functional Characteristics of Fresh Meat. HIU has been applied on meat of different animal species, including beef (Dang et al. 2022, Fallavena et al. 2020, Garcia-Galicia et al. 2020a, 2020b), pork (García-Galicia et al. 2020a, Visy et al. 2021), rabbit (Carrillo-Lopez 2021b, Reyes-Villagrana et al. 2020) and chicken (Li Y et al. 2020, Zhang et al. 2020).

The parameters for meat ultrasonication treatment in the referenced studies range in frequency from 20 to 40 KHz, in intensity from 5.09 to 90 W/cm^2, and in time from five to 120 min.

2.2.1 Effects on Physicochemical and Functional Characteristics of Fresh Meat

The main advantage of ultrasonicating fresh meat is an increase in tenderness. To explain the physicochemical mechanisms to achieve such meat tenderization, several studies have been conducted. Very recently, D.S. Dang demonstrated that HIU triggers a greater activity of enzymes involved in meat tenderization, such as calpain-1 and caspase-3, through an increase of cytosolic calcium and mitochondrial dysfunction (Dang et al. 2022). They also showed that HIU applied to beef for 40 min promotes the degradation of structural proteins, such as titin, desmin and troponin, which resulted in an improved tenderization of the 13-d aged meat. The recent findings on physicochemical, structural, and functional effects of HIU application in fresh meat from different species are presented in below lines.

1. Bovine *longissimus thoracis et lumborum* was the type of meat or tissue reported, with the experimental parameters; 40 kHz, and 12 W/cm^2. 40 min at four °C; with the objective to elucidate the underlying mechanism triggering calpain-1 and caspase-3 activation upon HIU treatment and relevant findings reported were HIU increases degradation of calpastatin, autolysis of calpain-1, and activity of caspase-3. HIU also increases titin, desmin, and troponin-T degradation. Ultrasonicated beef has higher cytosolic calcium, ROS, and lower mitochondrial oxygen consumption rate. Such effects resulted in a more tender beef with no negative effects on colour, hydrogen potential or cooking losses (Dang et al. 2022).

2. Subprimals of pork legs was the type of meat or tissue reported, with the experimental parameters; I = 22 W/cm^2, F = 37 kHz, t = zero, 10 or 30 min; with the objective to evaluate the effect of HIU on technological and consumer acceptance of pork and relevant findings reported where HIU increased water retention and hydrogen potential, but reduced the shear force in meat after injection, tumbling, ultrasonication, freezing, and thawing. However, it decreases consumer preference due to salty perception (Garcia-Galicia et al. 2022).

3. Rabbit carcasses was the type of meat or tissue reported, with the experimental parameters; F = 40 kHz, I = 9.6 W/cm^2, t = 20 or 40 min; with the objective to evaluate the effect of HIU pre and post freezing (120 hr/−20°C) on yield and physicochemical quality of rabbit meat and relevant findings reported were post freezing. Ultrasonication increases water loss and shear force, further, it reduces redness. Pre freezing ultrasonication increased weight loss and toughening (Carrillo-Lopez et al. 2021b).

4. Vacuum packed beef *Longissimus dorsi* was the type of meat or tissue reported, with the experimental parameters; water bath: F = 37 kHz, 150 W of efficient

working power, t = 30, 60 or 90 min; with the objective to compare HIU or vacuum impregnation pre-treatment on improvement of tenderness of immersion marinated beef and relevant findings reported that HIU did not alter lightness but increased tenderness in beef. Additionally, HIU reduced tyramine levels in uncooked and marinated beef, and increased marinade uptake (Demir et al. 2021).

5. Pork loin circles (diameter of 15 mm and height 80 mm) in brine conditions (200 g /L brine) were the type of meat or tissue reported, with the experimental parameters; F = 20 kHz Max., I = 5.09 W/cm^2, Power: 100 W, microbubbles produced with a gas–liquid mixing pump (Pressure: 0.5 to 0.6 MPa, low rate of air inlet: 100 L/min); with the objective to evaluate the combined effect of HIU cavitation and microbubbles on NaCl content and diffusion, mass balance, water holding capacity, protein denaturation and microstructure of pork meat and relevant findings reported that HIU increases salt diffusion and produced micro pores on myofibers; additionally, HIU reduces water holding capacity with changes on protein (actin and myosin) denaturation by temperature effect (Visy et al. 2021).

6. Chicken breast was the type of meat or tissue reported, with the experimental parameters; HIU assisted immersion freezing (125, 165, 205 and 245 W); with the objective to evaluate different power levels of HIU on freezing rate, microstructure, and water distribution in chicken breast and relevant findings reported that HIU (165 W) reduced freezing times and cutting force; additionally, HIU also reduced lightness, cooking and thawing water losses, due to smaller ice crystals formation, in comparison with immersion freezing, air freezing and control (Zhang et al. 2020).

7. Pork *Longissimus dorsi* was the type of meat or tissue reported, with the experimental parameters; Ultrasonic: 22 KHz, Power consumption 180 W for five min. Flow cavitation: Amplitude = 0.20 MPa I = 20 KHz. Productivity five L/min; with the objective to elucidate the indirect effect of cavitation on colour stability in chilled normal and abnormally autolysed pork with PSE and DFD defects and relevant findings reported were cavitation in three percent brine optimised colour of DFD and PSE pork, by disintegration of liquid sodium chloride-based salting media (Krasulya et al. 2020).

8. Marinated chicken m. *pectoralis major* was the type of meat or tissue reported, with the experimental parameters; HIU assisted tumbling (F: 40 kHz, Power: 140 W); with the objective to apply a practical redesigned tumbler (ultrasound assisted and adjustable pressure) to marinating chicken derived products and relevant findings reported were HIU tumbling improved marinating absorptivity, tenderness, taste, water holding capacity, and micro structure of chicken, with rupture of myofibers, myofibrils, and lysosomes. Besides, HIU accelerated the degradation of myosin light chain. HIU promoted the alpha helix turned into beta bold conformation (Li Y et al. 2020).

9. Fresh bovine *Longissimus lumborum, Infraspinatus and Cleidooccipitalis* was the type of meat or tissue reported, with the experimental parameters; F: 40 kHz,

I = 11 W/cm², Time: zero, 40, 60, and 80 min; with the objective to determine the effect of HIU and aging on quality parameters and microstructure of beef and relevant findings reported that HIU reduced total collagen in muscle, as well toughness depending on muscle (*Infraspinatus* and *L. lumborum* < *Cleidoccipitalis*). HIU accelerates ageing process and together with seven d of aging resulted in a remarkable improved tenderness. Changes in colour, pH and water holding capacity were not negative (Gonzalez-Gonzalez et al. 2020).

10. Lamb meat from hip was the type of meat or tissue reported, with the experimental parameters; Bath: F = 35 kHz, I = one W/cm². HIU emitter: F = 26 kHz, I = W/cm², t = five hr; with the objective to study the effect of ultrasound assisted salting on the grain of lamb and relevant findings reported increase of transverse fissures, microcracks, and cavities in the muscle fibre. Which may allow an easier penetration of brine in the tissue (Krasnikiva et al. 2020).

11. Beef *Longissimus lumborum* was the type of meat or tissue reported, with the experimental parameters; I = 90 W/cm², F = 37 kHz t = 40 min; with the objective to evaluate if HIU application in vacuum packed beef generated physicochemical changes comparable to those occurring in wet ageing and relevant findings reported that HIU and five d ageing increased redness and chroma but reduced water holding capacity during retail display in comparison with 10 d ageing. HIU did not show advantages in tenderness compared to 10 d ageing (Garcia-Galicia et al. 2020a).

12. Beef *Biceps femoris* was the type of meat or tissue reported, with the experimental parameters; I = 26 to 84 W/cm² or 22 to 84 W/cm², T = 18 to 28 or 10 fixed °C; with the objective to evaluate the effect of HIU on beef physicochemical parameters (tenderness, water holding capacity, colour, hydrogen potential, and lipid oxidation) and relevant findings reported that increase of HIU intensity decreased shear force, and increased lipid oxidation, but it does not affect hydrogen potential and water holding capacity (Fallavena et al. 2020).

13. Half rabbit carcasses were the type of meat or tissue reported, with the experimental parameters; F: 24 kHz. I: 12 W/cm², t = 15 min; with the objective to evaluate the effect of HIU on quality traits of whole rabbit carcases and relevant findings reported that HIU reduced shear force, and water holding capacity, but increased lightness and yellowness in the loin. However, these changes are not equal in all muscles. Apparently, HIU effects are not only defined by HIU parameters, but by the type of muscle (Reyes-Villagrana et al. 2020).

14. Beef *Biceps femoris* was the type of meat or tissue reported, with the experimental parameters; F = 20 KHz, Axial radial transmitter. 600 W of power. t = 30, 60 and 120 min; with the objective to evaluate the effect of crossbreeding of Nellore, Angus, and Wagyu cattle and the use of ultrasound on beef wet brining and relevant findings reported that ultrasonication increased the rate and accelerate the mass transfer of NaCl in the muscle (Sanches et al. 2020).

15. Beef *longissimus lumborum* and Semitendinosus steaks was the type of meat or tissue reported, with the experimental parameters; I = zero, 16 or 28 W/cm²;

with the objective to evaluate the combined effect of HIU applied in packed meat and two different types of packaging on quality traits of two beef muscles and relevant findings reported that HIU increased redness, and chroma on vacuum packed beef. HIU increased shear force in *longissimus lumborum* but not in Semitendinosus muscle (Garcia-Galicia et al. 2020b).

Other authors have reported a reduction of toughness in secondary pork cuts, after ultrasonication combined with other processes such as freezing and thawing (Garcia-Galicia et al. 2022), HIU causes changes in the microstructure of pork tissue, described by the formation of micropores in the muscle fibres and an increase of diameter in muscle filaments, concluding that these physical effects may improve the chewability of pork meat (Visy et al. 2021). Tenderization because of HIU treatment in other species like chicken, may be related to the structural damage of myofibrils and myofibers, derived from degradation of lysosomes, which, in turn, degrade myosin light chains into the sarcomeres (Li Y et al. 2020).

Water holding capacity (WHC) in meat is a major concern to consumers and processers. Losses of water in meat, are a matter of economic concern in the meat production industry. Consumers are very susceptible to reject meat with undesirable sensorial traits caused by inadequate water content (i.e., dryness, excessive water in packaging). Processers reject inadequate water holding meat since it increases the risk of economic losses during further processing (Huff-Lonergan and Lonergan 2005). WHC is described as the amount of water the fresh meat may lose by effect of dehydration or further processing, either by evaporation or dripping (Den Hertog-Meischke et al. 1997). Contradictory results have been described about the effect of HIU on WHC. Some authors have shown increase in water holding capacity after ultrasonication, while others have shown increase in water loss (Carrillo-Lopez et al. 2021b, Reyes-Villagrana et al. 2020 Visy et al. 2021, Garcia-Galicia et al. 2020a).

The reported water losses are thought to depend not only on the HIU application, but also on other factors such as the moment of HIU application, the duration of HIU, the post processing of the meat, among others. For instance, it has been shown that if HIU is applied during immersion on salts, there will be a higher mass transfer of salts or curing products into pork and lamb (Krasnikova et al. 2020, Sanches et al. 2020, 2021), increase in water accumulation and retention, and reduction of posterior losses in processes such as freezing and thawing in chicken and pork (Li Y et al. 2020, Zhang et al. 2020, Garcia-Galicia et al. 2022). However, if HIU is applied with no curing products, the microstructural changes in the tissue by cavitation may promote the release of water during the ultrasonication or during the retail display of carcasses of rabbits (Carrillo-Lopez et al. 2021b, Reyes-Villagrana et al. 2020), pork loin (Visy et al. 2021), and beef (Garcia-Galicia et al. 2020a).

Meat colour is one of the most important characteristics for consumers and it depends on the redox state of the natural pigment protein called myoglobin (Suman and Joseph 2013). Hence, any processing methodology applied to meat must be carefully evaluated for possible undesirable colours. HIU may have negative effects on meat colour when the food is directly ultrasonicated and exposed to cavitation forces for more than 20 min, which provokes lixiviation of the superficial liquid and pigments. Nevertheless, this change of colour may not be perceived by consumers when the meat is cooked. On the other hand, when fresh meat is ultrasonicated inside a protective layer (plastic bag, vacuum film, or other packaging), the colour may not be affected (Dang et al. 2022, Demir et al. 2021, Gonzalez-Gonzalez et al. 2020), and even an increase of redness could occur (Garcia-Galicia et al. 2020a). Also, with short times of ultrasonication, the changes in colour may not be significant (Alarcón-Rojo et al. 2015, 2019, Carrillo-Lopez et al. 2017).

An increase of redness in ultrasonicated meat has been attributed to a higher O_2 exposition and association to myoglobin in the tissue surface, as a physical effect of the acoustic waves if lixiviation is avoided. This effect may result in the conversion from myoglobin to oxymyoglobin, which provides the cherry red colour in meat, and increases the redness values. Very recently HIU has been reported to improve the colour of PSE and DFD pork, by increasing the penetration of salts into the tissue (Krasulya et al. 2020).

2.2.2 Ultrasonication During Freezing and Thawing

Lately, some studies have been designed to apply ultrasonication to meat, not only through the immersion of the product in a liquid or brine tank, but into special devices such as tumblers or freezers. Li et al., designed an equipment consisting of an ultrasonic probe (40 kHz, 140 W) attached to the inside of a vacuum tumbler with adjustable pressure (Li Y et al. 2020). They ultrasound tumbled chicken breast and reported a higher marinating absorptivity, tenderness, and more intense flavour in the meat processed with this equipment. These effects were the results of higher degradation of myosin, release of free peptides and amino acids and rupture of microstructures in the tissue (see the recent findings on physicochemical, structural, and functional effects of HIU application in fresh meat from different species in the Ultrasonication of Fresh Meat section).

Ultrasonication during the freezing process has also been lately studied in chicken (Zhang et al. 2020), and pork (Zhang et al. 2019). HIU (165 W) reducing the freezing time and producing smaller ice crystals, which resulted in less water losses when thawing and cooking the meat. In addition, HIU reduced undesirable structure rupture and lightness (L*), which is related to superficial water release. Most importantly, HIU assisted freezing also produced more tender poultry in comparison to air freezing and commercial freezing (Zhang et al. 2020). In pork, Zhang et al. demonstrated that long periods of freezing developed bigger ice crystals with meat quality disadvantages (Zhang et al. 2019). Hence, the HIU (180 W) shortening effect on freezing time is beneficial for characteristics such as water losses during cooking and thawing. HIU reduced the water migration during freezing, promoted a lower lipid oxidation, and a higher redness. However, if the pork is kept for 60 to 180 d storage, the HIU may promote an increase in the cutting force.

2.2.3 Ultrasonication of Fresh Meat and Bacterial Reduction

HIU application in meat as an antimicrobial treatment has been previously described (Ananta et al. 2005, Caraveo et al. 2015, Morild et al. 2011, Piyasena et al. 2003). HIU is considered a non-thermal technology with potential use for microbial reduction in foods (Turantas et al. 2015). HIU reduces coliforms, mesophiles, and *Psychrophiles* bacteria. HIU has the capacity to damage the bacterial membrane by increasing the pressure gradient and temperature in the media (Alarcón-Rojo 2015). Specifically, HIU has been shown to reduce pathogens such as *Escherichia coli* (Inguglia et al. 2018) *Salmonella* ssp. (Sienkiewicz et al. 2017), *Clostridium perfringens* (Evelyn and Silva 2015), *Yersinia enterolitica, Listeria* ssp. (Mikš-Krajnik et al. 2017) and some *Mycobacterium* ssp. (Al Bsoul et al. 2010).

The latest studies evaluating the effect of HIU upon bacteria in meat, meat emulsions or water tank for poultry processing are described below.

1. Type of meat or tissue was bacterial culture in water used for poultry chilling process; the experimental parameters were F = 37 kHz. Nominal power: 330 W, Amplitude: 40 and 100 percent, t = one to nine min., T = 25°C; to evaluate HIU alone or combined with ClO_2 on *Salmonella typhimurium* and *Escherichia coli* inactivation in processing poultry tank water and relevant findings are that HIU (37 kHz, 330 W, and one min) combined with ClO_2 inactivated *Salmonella typhimurium* at 25°C and *Escherichia coli* at chilling conditions. HIU bacterial inactivation was higher, compared to manual stirring in the presence of ClO_2 (Rossi et al. 2021).

2. Type of meat or tissue was bovine *Semitendinosus*; the experimental parameters were F = 40 kHz, I = 11 W/cm²; to explore the impact of HIU on physicochemical characteristics and shelf life of beef (four °C) and relevant findings are that HIU reduced mesophilic, psychrophilic, *Staphylococcus* spp., and coliform bacteria on the surface of the tissue (Valenzuela et al. 2021).

3. Type of meat or tissue was pork meat from supermarket; the experimental parameters were T = 50–70°C, F = 20 – 60 kHz, t = 10–20 min. Sonotrode of a three mm tip diameter. Amplitude 100 percent. Pulse style (acoustic power discharged 10 s, no power discharged three s; to optimize HIU parameters (temperature, pre-treatment time, and frequency) for inactivation of *Bacillus cereus* and achieve the best colour, texture, moisture, and fat losses and relevant findings are optimal parameters for *Bacillus cereus* inactivation (0.47 CFU/g) and physicochemical (moisture loss, 22.42 percent; texture, 134.38 N; and colour, 45.48) pork properties were 20 kHz at 70°C and 13.56 min (Owusu-Ansah et al. 2020).

4. Type of meat or tissue was beef *Longissimus dorsi*; the experimental parameters were I = zero, six, 28 and 90 W/cm²; to study the effect of HIU intensity, time, and storage time on physicochemical, microbiological, and structural properties of bovine and relevant findings are HIU (90 W/cm²) effectively controlled mesophilic and psychrophilic bacteria during storage, whereas decontamination of coliform bacteria was efficient independently of HIU intensity with longer times than 40 min (Carrillo-Lopez et al. 2019).

5. Type of meat or tissue was raw meat emulsion; the experimental parameters were Power: 200, 250, 300, 350, and 400 W. Pulses = zero, 10, 20 and 30 s., t = one,

2.5, five, 7.5, and 10 min; to evaluate the influence of high power ultrasound on natural microflora, *Lactobacillus monocytogenes* and *Lactobacillus delbrueckii* in a raw meat emulsion and relevant findings are HIU effect without wave pulses for each microbe, and a quadratic interaction with time of application and power. Inactivation of 60 percent of microbial population with 10 min a 7.56 s wave pulse and 400 W (Aguilar et al. 2021).

Last year studies have confirmed that in beef (Carrillo-Lopez et al. 2019a, Valenzuela et al. 2021) and pork (Owusu-Ansah et al. 2020). HIU significantly reduces the populations of mesophilic, *Psychrophilic, Staphylococcus, Bacillus cereus* and coliforms, independently of the intensity, but dependently on application time. On the other side, Rossi et al. combined chlorine dioxide with HIU to reduce the counts of *Salmonella tyhimurium* and *Escherichia coli* in poultry chilling water (Rossi et al. 2021). A superior effect of HIU over the traditional stirring was found, concluding that synergism of HIU with the chlorine dioxide (ClO_2) reduced 100 percent of the bacterial population at carcass pre and chilling conditions. Aguilar et al., demonstrated that HIU reduces up to 60 percent the microbial population on a raw meat emulsion, and that the application time has a major effect on the reduction of microorganisms (Aguilar et al. 2021). The optimal conditions used by these authors were 10 min and 400 W of power for the best microbial reduction in the emulsion.

2.3 Ultrasonication Effects on Meat Products

Over time, an enormous variety of processed or semi processed meat products with different sensorial predilection characteristics have been developed worldwide. In some regions there are hundreds of different meat products, with different names and flavours. Despite the diversity of shapes and flavours, many of these products use similar manufacturing technologies. Thus, the processed meat products can be classified as follows: raw (sausages, hamburgers, merguez, longaniza), cured (serrano ham or Parma ham), raw cooked (mortadella, Frankfort, Vienna sausage, meatballs, or meat pies), precooked cooked (liver pâtés, black pudding), raw fermented sausages (sausage and salami type sausages), dried (jerky, biltong) (Heinz and Hautzinger 2007).

Processed meat products are mainly made with meat from one or several species of animals. Fats and/ or condiments are added to this meat during the manufacturing process. In addition, these products go through a treatment of drying, cooking, salting, curing or some other transformation process. The use of power ultrasound during the meat processing phases has been considered in the meat industry, since it contributes to improve the technological properties of the product, as reviewed in the previous section. The main effects of ultrasound in raw meat are tenderization, efficiency of water retention capacity and reduced microbial growth (as seen in sections Effects on Physicochemical and Functional Characteristics of Fresh Meat and Ultrasonication of Fresh Meat and Bacterial Reduction).

This section describes the studies that in recent years have focused on evaluating the effects of ultrasound on the physical, technological, and sensory properties of processed meat products and some of their implications during such processing, as an example are presented below from some studies on meat products.

1. Chicken breast: the experimental parameters were treatments HIU (40 kHz, 300W), t = 10, 20, 30 and 40 min, and salt contents (one, 1.5 and two percent, respectively) and the relevant findings are improved gel properties (Li et al. 2015).

2. Cooked ham: the experimental parameters were treatments T100: 1,5 percent NaCl; T75: 1,12 percent NaCl; T50: 0,7 percent NaCl and T50US: 0,75 percent NaCl and ultrasound 10 min (F = 20 kHz, I = 600 W/cm^2), cooking conditions: 72°C at its thermal center and the use of ultrasound decreased the total fluid release and increased the hardness, increased redness. Caused micro fissures on the myofibrils. The sensory acceptance of restructured cooked ham with 0.75 percent of salt was improved with ultrasound applied. The ultrasound showed good potential for use in the production of healthier meat products (Barretto et al. 2018).

3. Chicken breast: the experimental parameters were 20 mL of chicken actomyosin (CAM) solutions, F: 20 kHz, I = 100, 150 and 200 W and ultrasonic pulsed mode of on time two s and off time four s was given for total 24 min and the decrease in α helix fraction and the improvement of fluorescence intensity, as well as the result of improved surface hydrophobicity and reactive SH groups, exhibited the conformation changes induced by ultrasound effect (Zou et al. 2018).

4. Italian Salami: the experimental parameters were treatments US bath normal mode, F = 25 KHz, I = 500 W cm^{-2}, t = zero, three, six, and nine min; processing storage: 28 and 120 d, respectively and the relevant findings as US treatment does not affect the hydrogen potential and the fermentation process.

   Sonication increased total heme pigments, especially during storage, when the product is exposed to consumer. The lipid and protein oxidative reactions were accelerated by US treatment (de Lima Alves et al. 2017).

5. Meat emulsion: after elaboration meat emulsion were sonicated for zero, nine and 18 min. Ultrasonic bath (normal mode, 60 percent amplitude, 25 KHz frequency, 230 W acoustic power, and 33 W/L volumetric power) and the relevant findings are that 18 min of US in meat emulsion was effective to compensate for defects in cooking yield and emulsion stability caused by the phosphate reduction; 8 min US application made the products more cohesive and did not increase the lipid oxidation (Basso Pinton et al. 2019).

6. Meat emulsion: the experimental parameters were US: Ultrasonic bath (25 kHz, 200 W) for 10.53 min at 74°C and without ultrasound (WUS) and with conventional pasteurization (CT, water bath at 82°C, 16 min to reach 73°C inside of sausages) and the relevant findings are that US treatment inhibited the growth of psychrotrophic and lactic bacteria, reduced the lipid oxidation and improving the psychotropic (Cichoski et al. 2015).

7. Hot dog sausage: the experimental parameters were treatments: Ultrasonic bath (25 kHz, 200 W) for 10.53 min at 74°C and without ultrasound (WUS) and with conventional pasteurization (CT, water bath at 82°C, 16 min to reach 73°C in inside sausages), and the relevant findings was that US treatment inhibited the growth of psychrotrophic and lactic bacteria, reduced the lipid oxidation and improving the pasteurization (Li et al. 2015).

8. Mortadella: experimental parameters four cooking conditions control, TUS100 and TUS50 were cooked with US (25 kHz) and 50 reduction percent of the cooking time of Control, using an amplitude percent of 100 (462 W) and 50 (301 W), respectively; and TWUS: cooking without the application of US and 50 percent reduction of the cooking time of Control and the relevant findings as US bath at (25 kHz and 461 W) reduced the cooking time with no changes in the sensory quality of the processed product (Cichoski et al. 2021).

During the processing of meat products, the application of techniques that guarantee sensory quality and health safety is of great importance. As an effective non thermal food processing, ultrasonic treatment is widely used in the food industry to change the structure of proteins and regulate their aggregation by inducing interactions between ultrasound and media. Initially, its effect on protein aggregation can be observed under shear force, high temperature and high pressure released by cracking of cavitation bubbles in a compression cycle. Then, the protein solution mass transfer can be accelerated by microjets and microbubbles due to bursting of cavitation bubbles. Finally, the intermolecular forces of proteins, such as hydrophobic forces and hydrogen bonds, can be broken by peroxides, but also cross linking might be induced by the active free radicals generated from water molecules during ultrasonication (Ding et al. 2021).

Meat is an important component of the diet. Myofibrillar protein (MP) content accounts for approximately 50 to 60 percent of total muscle protein and contains higher lysine concentration as well as higher digestibility in humans (Chen et al. 2020).

Among the main effects of HIU regarding the development of low salt meat products, there is a higher efficiency of salt diffusion, and a positive effect on the water holding capacity and microbial safety (Kim et al. 2021). The same significant effect is not observed in processed meat products. However, positive effects have been observed in other properties such as gelation (Li et al. 2015), better sensory acceptance of reduced salt cooked ham (Barretto et al. 2018), improvements in the emulsifying properties of proteins (Amiri et al. 2018b, Cheng et al. 2019, Li Z et al. 2020, Zou et al. 2018), enhancement of the technological quality by phosphate reduction, and no increase of the lipid oxidation in meat emulsions (Basso Pinton et al. 2019).

In the food industry, myofibrillar protein is a potentially useful emulsifier for the development of new foods with biological activity (Alarcón-Rojo 2019). Acoustic cavitation of food emulsions is widely applied as the main processing method to improve the quality of a finished product and its organoleptic characteristics, as well as to increase production yield. It has been observed that below 20 kHz, 30 min and 100 W (Amiri 2018a) the emulsifying properties of myofibrillar protein stabilize after ultrasonic treatment. Also, ultrasound treatment at 20 kHz, 450 W, for six min, increases the structural flexibility of the protein and promotes myofibrillar protein adsorption, contributing to emulsifying performance (Li Z et al. 2020). After investigating the physicochemical effects on actomyosin from chicken meat under different ultrasonic treatments (20 kHz, 100 to 200 W, 20 min), Zou et al., concluded that the emulsifying activity index and the stability index of the emulsion gradually decrease with increasing ultrasound power (Zou et al. 2018). Thus, the lack of uniformity of high intensity ultrasonic processing is the main problem in the application of this technology.

In a study reported by Cheng et al., they concluded that double frequency ultrasound had a significant positive effect on the structure of whey protein, that is, better physicochemical properties of whey protein (Cheng et al. 2019). After this observation, Cheng et al., then applied frequency modes (20 and 28 kHz) on chicken meat myofibrillar protein (MP) with different protein concentrations (20, 40 and 60 mg/mL), and reported the structural changes of the MP based on the active sulfhydryl content (ASC), surface hydrophobicity (SUH), particle size and turbidity; as well as emulsifying properties determined by emulsifying activity index (EAI), emulsion stability index (ESI) and creaming stability (Cheng et al. 2019). The results indicate that treatments under different ultrasonic frequencies, especially the superposition of two frequencies, can be an efficient method to improve the emulsifying properties of chicken MP and obtain beneficial results for the development of processed meat products.

In processes such as the maturation of conventionally fermented foods, a large space is usually required for long term storage. Ultrasound has a great potential to improve the efficiency of the maturation stage to produce high quality fermented foods. At high power ($\geq$ 100 W), ultrasound promotes the Maillard reaction, oxidation, esterification, and proteolysis, and consequently the maturation of fermented foods. Ultrasound also elicits improvements in the texture, colour, taste, and flavour of fermented foods (Yu et al. 2021). Such benefits of applying ultrasound under appropriate conditions are thought to be due to the induced stimulation of microorganisms, improving their metabolism (Dai et al. 2017, Nguyen et al. 2009, 2012, Shokri et al. 2020).

The relationship between ultrasound and chemical reactions has been widely studied in sonochemistry (Suslick 1990), but only a few studies have evaluated its impact on the oxidation of lipids and proteins in meat and meat products. In this regard, there are contradictory results, as Stadnik and Dolatowski (Stadnik and Dolatowsky 2011) and McDonnell et al., found no oxidative changes in ultrasound treated beef and pork (McDonnell et al. 2014), respectively, while Chang and Wong reported increased lipid oxidation in the cobia fish (*Rachycentron canadum*) (Chang and Wong 2012), and Cichoski et al., found a reduction in lipid oxidation in sausages treated with ultrasound (Cichoski et al. 2015). Although the application of ultrasound has been studied for its antimicrobial effect due to cavitation, it is only effective when combined with other treatments such as temperature, pressure, or acids (Piyasena et al. 2003, Turantas et al. 2015). The successful application of ultrasound in biological processes like fermentation, depends on several factors, including acoustic conditions, the type of microorganisms and the food evaluated

(Chisti 2003, Fan et al. 2017, Ojha et al. 2017, Huang et al. 2017, Phull and Mason 2001). There are several studies on the effect of ultrasound on tenderness in meat and meat products (Got et al. 1999, Jayasooriya et al. 2007, McDonell et al. 2014, Stadnik and Dolatowski 2011), as well as on bacterial growth and oxidative reactions during processing and storage in the preparation of products like Italian salami (25 kHz, 128 W, nine min/ 20°C) (de Lima Alves et al. 2017).

The generation of free radicals in the medium and their contribution to lipid oxidation (Alarcon-Rojo et al. 2019, Liao et al. 2007) have been proposed as an important mechanism of action of ultrasound. According to Cunha et al. 2018, the use of natural antioxidants rich in phenolic compounds can improve the oxidative stability of meat products generated with emerging technologies (Cunha et al. 2018). However, more studies are needed to know the optimal conditions of use in meat products without interfering with their sensory parameters. Other authors have reported promising results in the application of ultrasound at a frequency of 25 kHz in meat and meat products, as well as a reduction of the cooling period of chicken meat (Martins Flores et al. 2018), reduction of the bacterial count in poultry meat during cooling (Cichosky et al. 2019) shorter pasteurization time in sausages (Cichosky et al. 2015), higher meat emulsion stability (Cichosky et al. 2019) and reduction of sodium chloride (NaCl) and phosphate in meat emulsions (Basso Pinton et al. 2019, Sena Vaz Leães et al. 2020). Volatile compounds can originate from (1) the activity of endogenous microorganisms (carbohydrate fermentation, amino acid catabolism, and ß-oxidation); (2) lipid autoxidation; (3) ingredients used in the formulations; and (4) storage time and conditions. The volatile compounds formed by the autoxidation of lipids are mainly responsible for the sensory properties of meat products due to the low perception thresholds (Lorenzo et al. 2012). Therefore, oxidative stability is an important parameter for the evaluation of fat quality and depends on both, the chemical composition and quality of the raw material and the processing and storage conditions (Coppin and Pike 2001). To assess oxidative stability or susceptibility to oxidation, the fat is subjected to an accelerated oxidation test under standardized conditions and an appropriate endpoint is chosen to determine signs of oxidative deterioration. Although various procedures are used to accelerate oxidation reactions, including increased temperature, addition of metals, increased oxygen pressure, and exposure to light, heating has proven to be more effective and the most widely used alternative. As reported by da Silva et al., ultrasound assisted technology in the cooking process provided a faster increase in the internal temperature of mortadella and higher temperature homogeneity (da Silva et al. 2020). The treatment did not increase the oxidation of lipids and proteins and provided high microbiological safety to the products. However, for use on an industrial scale, it is necessary to evaluate the impact of ultrasound assisted cooking on other attributes that are also very important for product quality. For instance, the same research group (Cichoski et al. 2021) evaluated the effect of ultrasound assisted cooking (25 kHz, 50 or 100 amplitude percent) on volatile compounds, oxidative stability, and sensory quality of mortadella, and found that ultrasound assisted cooking using 100 amplitude percent is a good alternative to reduce cooking time without affecting product quality.

Finally, the application of ultrasound has been reported to be useful for industrial food processing. In general terms, the tenderness, or the dissolving capacity of water in the meat can significantly affect the properties of the final processed product for consumers. To produce of foods that must be stored in dry form (milk powder) or those that must be programmed (ground meat), water pre-treatment technology is quite crucial. The reconstitution of milk powder is among the leaders of water consumption. Thus, one of the possible solutions for saving energy and improving the quality of the final products is the application of ultrasound. Food sonochemistry (Chandrapala et al. 2012) is a relatively new scientific area that has shown that ultrasound can potentially be applied in the dairy industry (Juliano et al. 2014, Torkamani et al. 2016). The processing technology is based on the idea that at low frequencies (20 kHz) the cavitation effect does not generate excessive free radical formation, but rather a change in the self-organization of water molecules. In a reactor level study including the processing of both, meat, and milk-based products, Krasulya et al., observed that free water molecules have a greater affinity to milk and meat proteins, enhancing their thermo resistivity and controlling microbial growth (Krasulya et al. 2014). Thus, sonochemical reactors have a positive effect on the quality of processed products (ground meat and powdered milk) to extend the shelf life of the final product.

2.4 Ultrasonication of Milk

Ultrasonic processing is a promising food technology to improve the technological and functional properties of milk. Ultrasound has been applied to milk to enhance physicochemical properties, chemical components, functional and sensory characteristics, as well as to control pathogenic microorganisms in the product.

2.4.1 Physicochemical and Functional Properties

The application of HIU in milk has shown positive effects on physicochemical characteristics of milk were presented below.

1. Sample of raw sheep milk; the experimental parameters were ultrasonic probe, 20 kHz, and the effect of ultrasound was HIU did not affect proximate composition, free amino acids, or amino acid profile (Tiwari and Mason 2012).

2. Sample of bovine milk; the experimental parameters were ultrasonic processor, 20 kHz, 1500 W, t = 10 or 15 min, and the effect of ultrasound was hydrogen potential and C* (colourless) were reduced, total solids, physical stability, and L* were increased by HIU (Riesz and Kondo 1992).

3. Sample of whey; the experimental parameters were ultrasonic homogenizer, 0.092, 0.151, and 0.220 W/mL, 20 kHz, t = 10 and 15 min, and the effect of ultrasound was protein rearrangement, aggregate formation, reduction of reactive thiol groups, and changes in secondary structure by HIU (Abadía-García et al. 2016).

4. Sample of buffalo's milk; the experimental parameters were ultrasonic horn. 44, 54 and 66 W, respectively, 20 kHz, 45°C, t = 5–20 min, and the effect of ultrasound was reduction of fat globule and increase in surface area; decreased the size of milk particles. HIU increased free saturated fatty acids (Abesinghe et al. 2020).

5. Sample of whey; the experimental parameters were ultrasonic probe, 550 W, 20 kHz, 10 mm probe, 25–30°C, t = 2.5, five and 7.5 min, respectively, and the effect of ultrasound was modification of the protein structure to make it more susceptible to transglutaminase (Ahmadi et al. 2017).

6. Sample of buffalo's milk; the experimental parameters were ultrasonic homogenizer, 430 and 338 W, respectively, 28 kHz, 20°C, t = 5, 10 and 15 min, and the effect of ultrasound was smaller and more uniform globules of fat as the frequency and duration of the ultrasound increases (Al-Hilphy et al. 2012).

7. Sample of sheep milk; the experimental parameters were ultrasonic equipment, 78 and 104 W respectively, 20 kHz, at 40–69 °C, and the effect of ultrasound was no release of free fatty acids or changes in the protein profile due to ultrasound (Balthazar et al. 2019).

8. Sample of goat milk; the experimental parameters were ultrasonic processor, 100 W, 30 kHz, three, six, and nine min, respectively, and the effect of ultrasound was higher heterogeneity of fat globules, destruction of the fat globule membrane (Karlović et al. 2014).

9. Sample of bovine milk, the experimental parameters were ultrasonic milk fractionation, 283 to625 W, 22.2–57.9°C, and the effect of ultrasound was milk fractionation and formation of a vertically increasing fat concentration gradient, with larger fat globules on the surface (Leong et al. 2016).

10. Sample of whole milk; the experimental parameters were ultrasonic homogenizer, 50 W, 22.5 kHz, 30 min (zero to 500 J/mL), (20–80°C), and the effect of ultrasound was homogenization of fat globules, denaturation of serum proteins, aggregation of fat globules and proteins (Nguyen and Anema 2017).

11. Sample of milk protein, the experimental parameters were 800 W, 13 mm probe, < 50°C; one, three, five, and eight min, respectively, and the effect of ultrasound was improvement in degree of hydrolysis during enzymatic hydrolysis; production of new low molecular weight peptides (Uluko et al. 2015).

12. Sample of raw milk; the experimental parameters were a continuous system, 16 to 20 kHz, t = 14 to 18 min, T = 42 or 54°C, and the effect of ultrasound was sub-micron lipid droplets embedded in protein chains of the curd matrix (van Hekken et al. 2019).

13. Sample of milk and cream; the experimental parameters were a batch sonicator, 77, 104 and 115 W, respectively, 20 kHz, 115 W, t = one and three min, T = 72°C, and the effect of ultrasound was decreased total plasmin activity dependent on

storage time in skim milk.; decrease in fat globule size (Vijayakumar et al. 2015).

14. Sample of casein concentrate; the experimental parameters were cell disruptor, 58 W/L, 20 kHz, t = 0.5, one, two, and five min, respectively, and the effect of ultrasound was increased surface hydrophobicity and reduced particle size: increase in β-sheets and random coils and a reduction in α helix and β-turns (Zhang et al. 2018).

15. Sample of camel milk casein and whey proteins; the experimental parameters were sonicator at 30 kHz and 400 W for 45 min, and the effect of ultrasound was antioxidant activity, aromatic amino acid content, and angiotensin I converting enzyme inhibitory activity increased in whey proteins and caseins (Gammoh et al. 2020).

16. Sample of human milk; the experimental parameters were ultrasound bath at 40 kHz and 100 W, and intensity of 1,591 mW/cm^2, and the effect of ultrasound was higher antioxidant activity in thermosonicated human milk than in raw milk. Similar observations were seen in the retinol content (Parreiras et al. 2020).

17. Sample of chocolate milk beverage; the experimental parameters were ultrasonic probe disruptor, 800 W, at 19 kHz at 0.3, 0.9, 1.8, 2.4 and three kJ/cm^3, respectively, and the effect of ultrasound was the greatest antioxidant activity and ACE inhibitory activity was obtained at the highest energy density (Monteiro et al. 2018).

18. Sample of raw milk beverage; the experimental parameters were 2.4 kW at 16 to 20 kHz. and the effect of ultrasound was the Sonicated milk at 42°C gelled faster and formed firmer curds than the raw milk control (van Hekken et al. 2019).

Improvements have been confirmed, mainly on size of fat globule or fat crystal size, and fat droplets (Ashokkumar 2018). Ultrasonication reduces the particle fat size, and the effect depends on the ultrasound power applied (Bhangu and Ashokkumar 2016). The reduction of droplet size is important in processes such as milk emulsification (Kehinde et al. 2021). Additionally, HIU reduces hydrogen potential of milk and increases total solids. In contrast to pasteurization, the use of HIU allows the preservation of fatty acids in milk (Ashokkumar and Mason 2007).

The time of HIU application is important to get an effect of HIU on physicochemical characteristics of milk. No effect was observed on proximate composition of raw sheep milk when HIU was applied for six min (Tiwari and Mason 2012).

A disadvantage reported from the HIU application in milk, is that ultrasonication modifies milk colour, which could be detrimental for milk appearance (Riesz and Kondo 1992). Colour in thermo sonicated milk changes, having a higher L*, related to a higher homogenization and smaller size of fat globules. The change in colour of milk is apparently promoted by variations in ultrasonication power. Yellowness and lightness are the coordinates more easily modified with sonication powers > 400 W. Cavitation may also promote the release of encapsulated triacylglycerols, cholesterol, and phospholipids in the fat globule, which could change the light dispersion and the milk product (Pérez-Grijalba et al. 2018). Despite variations in the sonication parameters, the size reduction of fat globules and a higher homogenization are the main effects in milk by applying frequencies around 20 kHz. Additionally, denaturation of major proteins in milk by HIU application has also been reported (Abadía-García et al. 2016). Ultrasound modifies the molecular structure of food proteins (Abadía-García et al. 2016) (as mentioned in the topic Ultrasonication Effects on Meat Products).

On the other hand, the application of ultrasound increases the susceptibility to enzymatic hydrolysis, which could generate bioactive peptides (Uluko et al. 2014). Moreover, ultrasonication improves the rheological, functional, and textural properties of whey protein systems (Ahmadi et al. 2017). Some reports show changes in size, physical stability, and encapsulation capacity of casein due to ultrasound (Zhang et al. 2018). The alteration was attributed to the exposure of hydrophobic regions from the interior to the surface of the molecules, as well as the reduction in particle size and changes in secondary structures. Barukčić et al., observed an increase in stability due to the homogenization caused by the effect of ultrasound. (Barukčić et al. 2015) Hence, HIU has the potential to replace conventional pasteurization because whey proteins, being heat labile, do not precipitate.

Thermosonication extends the shelf life of milk due to the inactivation of proteases, as observed after thermosonication of skim milk, in which the total plasmin activity was reduced by up to 94 percent (Vijayakumar et al. 2015).

Structural changes in fat globules of milk by HIU effect modify their integrity, reducing the size and diameter compared to shear homogenization. Abesinghe et al., reported that ultrasonicating buffalo's milk for 15 min significantly reduces the size of the fat globule compared to traditional homogenization (Abesinghe et al. 2020). However, lipolytic rancidity was significantly increased due to the increase in the content of free saturated fatty acids in milk. In goat's milk proteins (Karlović et al. 2014), especially caseins, adsorb to the membrane surface of fat globules, functioning as natural emulsifiers in ultrasound treated milk. Ultrasonication of raw whole cow's milk at two MHz, compared to one MHz, is more effective for manipulating smaller fat cells retained in the later stages of skimming, eliminating 59 percent of fat (Leong et al. 2016). Recent work has demonstrated the potential use of ultrasound processing for the formation of dairy emulsions with tailored textures (Zhang et al. 2021).

Functional compounds are important for their high potential to protect the human body against reactive oxygen species and reduce oxidative stress (Tan et al. 2018). Ultrasonication increases the antioxidant activity in milk (as seen on). Parreiras et al., reported that thermosonicated human milk exhibited higher antioxidant activity than raw and pasteurized human milk (Parreiras et al. 2020). Similarly, ultrasonicated camel milk protein extracts showed higher antioxidant activity than unprocessed extracts (Gammoh et al. 2020).

Milk has peptides with the capacity to inhibit the angiotensin converting enzyme (ACE) and it also has sulphur containing amino acids and vitamins which could provide antioxidant and anticarcinogenic activity and, subsequently, greater benefits to the health of consumers (Khan et al. 2019). The bioactivity of functional compounds of milk is improved with ultrasound processing (Gammoh et al. 2020) since it increases the concentration of these bioactive compounds.

2.4.2 Microorganisms

Ultrasonic processing inactivates microorganisms that cause spoilage; hence it has been considered as a good potential alternative to extend shelf life of milk and milk products (Nguyen and Anema 2017). A couple of references are below:

1. Buffalo milk sample; for the experimental parameters; ultrasonic homogenizer, 430 and 338 W, 28 kHz, at 20°C, t = five, 10, and 15 min, respectively, and the effect of ultrasound was the decrease in counts of aerobic, coliform, and *Staphylococcus* bacteria as the power and time of exposure to ultrasound increased (Al-Hilphy et al. 2012).

2. Milk sample; for the experimental parameters; ultrasonic equipment at 78 and 104 W, respectively, 20 kHz, between 40 and 69°C, t = four, six or eight min, and the effect of ultrasound was the decrease in total aerobic mesophilic bacteria, total coliforms, and *Staphylococcus* spp. (Balthazar et al. 2019).

3. Whey sample; for the experimental parameters; ultrasonic processor, 480 and 600 W, respectively, 20 kHz, at 45 or 55°C, t = 6.5, eight, or 10 min, and the effect of ultrasound was that the best performance was achieved with the conditions of 480 W and 55°C treatment for bacterial inactivation, making it an alternative to pasteurization (Barukčić et al. 2015).

4. Milk sample; the experimental parameters were ultrasonic processor, 200 W, 24 kHz, at 15 to 25°C and t = two, four, eight, and 16 min, respectively, and the effect of ultrasound was that the total viable count and the psychrotrophic count increased significantly during storage at four °C (Chouliara et al. 2010).

5. Milk sample; the experimental parameters were ultrasonic processor, sonifier probe, 150 W (118 W/cm²), 20 kHz, T = 20 and 57°C, respectively, t = one, three, four, or six min, respectively, and the effect of ultrasound was that the ultrasound was effective for reducing total aerobic bacteria in raw milk and *Listeria monocytogenes* (D'Amico et al. 2006).

6. Bovine milk sample; the experimental parameters were ultrasound system at 75 W, 20 kHz, at five °C, t = 15 min, and the effect of ultrasound was the US application was inefficient for microbial control in raw milk (Engin and Karagul Yuceer 2012).

7. Milk sample; the experimental parameters were ultrasonic processor, 500 W, 20 kHz at 10 to 84°C for 0.17 to five min, and the effect of ultrasound was the reduction of thermophilic bacteria and *Bacillus atrophaeus* spores (Ganesan et al. 2015).

8. Milk sample; the experimental parameters were ultrasonic homogenizer and ultrasound generator, 50 W, 850 kHz, t = 10–60 min, and the effect of ultrasound

was the inactivation of *Enterobacter aerogenes* was not achieved (Gao et al. 2014b).

9. Cow milk sample; the experimental parameters were ultrasonic processor, 1,500 W, 20 kHz at 48 to 55°C for 10 or 15 min, and the effect of ultrasound was the decrease in counts of aerobic mesophilic bacteria and *Enterobacteriaceae* during storage (Hernández-Falcón et al. 2018).

10. Raw milk sample; the experimental parameters were ultrasonic equipment, 2,200 W, 20 kHz for 10, 20, 30 and 60 min, respectively, and the effect of ultrasound was the HIU did not reduce the total aerobic bacteria over 50 d at four °C; counts of spore forming bacteria increased (Lim et al. 2019).

11. Milk sample; the experimental parameters were ultrasonic processor, 24 kHz, at 23–60°C during sonication for 5, 10, and 15 min and the effect of ultrasound was that a Gram positive organisms *Staphylococcus aureus* and *Listeria monocytogenes* were more resistant to HIU than *Lactobacillus plantarum* and *Lactobacillus pentosus* (Shamila-Syuhada et al. 2016).

12. Milk sample; the experimental parameters were continuous ultrasound unit with a 2.4 kW dual frequency reactor, and the effect of ultrasound was total aerobic count was reduced and the psychrophilic counts below the detection limit (van Hekken et al. 2019).

The acoustic cavitation causes rupture and shear in the bacterial wall (Gao et al. 2014b). In this regard, it has been observed that *Escherichia coli* (Gram negative) is more sensitive to ultrasonication than *Listeria monocytogenes* (Gera and Doores 2011, Shamila-Syuhada et al. 2016), probably because lactose exerts a protective effect on bacteria and *Escherichia coli* exhibits nonlinear inactivation kinetics while *Listeria monocytogenes* showed linear inactivation kinetics. Ultrasound causes a significant reduction in *Enterobacteriaceae* counts in milk (Juraga et al. 2011) and a significant decrease of *Staphylococcus species* counts in sheep's milk (Balthazar et al. 2019) and buffalo's milk (Al-Hilphy et al. 2012). Specifically, the use of frequencies greater than 20 kHz has produced significant decreases in the counts of aerobic bacteria and coliforms (Al-Hilphy et al. 2012), and of *Staphylococcus aureus, Escherichia coli* and *Listeria. monocytogenes* (Shamila-Syuhada et al. 2016). Recently, it has been proven that ultrasonication of milk concentrates was also effective in reducing the microbial counts in milk powders (Sert et al. 2021). Additionally, van Hekken et al. reported that ultrasonication reduced the total aerobic bacterial count by more than one log cfu/mL, and the number of psychrophiles below the limit of detection (10 cfu/mL) (van Hekken et al. 2019).

Despite several reports of reductions in milk pathogens, other researchers have shown opposite effects of ultrasonication. Very low reductions in the counts of *Escherichia coli* and *Staphylococcus aureus* were observed after ultrasound milk treatments (Engin and Karagul Yuccer 2012). Moreover, there is evidence that both, thermosonication and cold sonication, increase spore forming anaerobic bacteria counts, thus inducing higher rates of deterioration than traditional pasteurization

(Lim et al. 2019). The contrasting effects might be a consequence of differences in other variables such as the temperature.

As seen for other technological properties of food products, the combination of ultrasound with other treatments produces better outcomes also regarding microbial control. For instance, ultrasound combined with the active lactoperoxidase system or the addition of H_2O_2 (from 0.01 to 0.1 percent), is more efficient for reducing microbial counts compared to ultrasound alone (Shamila-Syuhada et al. 2016). Similarly, the effectiveness of combined ultrasonication and heating (72–85°C) on pasteurized milk reduces *Bacillus* spores by one to two log (Ganesan et al. 2015).

2.4.3 Sensory Attributes and Volatile Compounds

HIU minimizes the typical changes in sensory properties induced by pasteurization (Chandrapala and Leong 2015). Nevertheless, this technology is not yet widely accepted in the dairy industry (Zisu and Chandrapala 2015). A rubbery aroma and an off taste that lies between "burnt" and "foreign" have been reported in milk after HIU (200 W) treatment for two min (Chouliara et al. 2010). Some authors have attributed these changes to denaturation of proteins (Nguyen and Anema 2017). Others agree that the undesirable flavour of sonicated milk is due to alterations in physical properties, degradation of components (Riener et al. 2009) and oxidation of polyunsaturated acids PUFA hydroperoxides. This oxidation generates volatile compounds in milk after HIU application (Majid et al. 2015).

The off flavour of sonicated milk seems to increase with sonication time. Jurić et al., ultrasonicated (24 kHz, 200 W) milk for longer than six min and it resulted in a product that tasted like "foreign metal," "burnt," and "rubbery (Jurić et al. 2016)." But no significant differences were found between the intensities of offensive "eggy" and "rubbery" odour attributes after HIU for one to three min of skim milk (Vjayakumar et al. 2015). The rubbery aroma was less intense after reducing the sonication power from 400 to 100 W (Jurić et al. 2016). The rubbery aroma in the ultrasonicated samples may originate from heat induced oxidation of lipids into volatile compounds, rather than from a free radical mechanism (Lim et al. 2019). Lipid oxidation associated to the adverse sensory aroma of sonicated milk, can be controlled by decreasing the sonication time and the temperature in the system (Vijayakumar et al. 2015). Indeed, the application of high power (165 W, 60 s) and low frequency sonication (20 kHz) to milk, dissipates the rubbery aroma very quickly, probably because the cavitation bubbles were larger and less numerous than what would be present at higher frequencies (Lim et al. 2019).

Low frequencies could provide better results in terms of aroma of the treated milk because large bubbles collapse more violently than small bubbles, resulting in fewer free radicals (Juliano et al. 2014). Paniwnyk stated that the highest concentrations of deterioration products were seen at 1,000 kHz and energies above 271 kJ/kg for raw milk and 102 kJ/kg for pasteurized skim milk (Paniwnyk 2017). Milk of other species has also been ultrasonicated, aiming to improve its sensory characteristics and to enhance the bioactive composition, the ultrasound effect on sensory and volatiles of milk was presented below

1. Milk sample; the experimental parameters were US system, comprised of 20 kHz, 75 W for 15 min. The US intensity was 135 J/mL and no major differences were observed in aroma-active compounds and flavor of milk following the HIU treatment. However, some new volatiles were generated by HIU treatment (Engin and Karagul Yuceer 2012).

2. Milk sample; the experimental parameters were ultrasonic processor set at 600 W, and frequency of 20 kHz, and HIU treatment longer than six min resulted in milk tasting like "foreign metal," "burnt," and "rubbery" was the effect of ultrasound reported (Jurić et al. 2016).

3. Milk sample; the experimental parameters were ultrasonic processor at 2,200 W max power and 20 kHz, and a rubbery aroma appeared in the ultrasonicated samples but dissipated over time was the effect of ultrasound reported (Lim et al. 2019).

4. Milk and cream sample; the experimental parameters were a batch sonicator, 2.2 kW at 20 kHz, for one and three min and thermosonication did not significantly increase the intensity of offensive rubbery and eggy odor attributes in skim milk and cream was the effect of ultrasound reported (Vijayakumar et al. 2015).

5. Milk sample; the experimental parameters were ultrasonic processor at 24 kHz and 400 W for 2.5, five, 10, 15 and 20 min at 45°C and ultrasound generated volatiles considered to produce adverse sensory aspects of sonication of milk was the effect of ultrasound reported (Riener et al. 2009).

6. Milk sample; the experimental parameters were ultrasonic processor at 24 kHz for zero, two, four, eight, and 16 min at 15 to 25°C, and volatile compounds increased in concentration with sonication and storage time was the effect of ultrasound reported (Chouliara et al. 2010).

7. Milk sample; the experimental parameters were transducers at one MHz and two MHz applied, 20, 400, 1,000, 1,600, and 2000 kHz; T = four, 20, 45, and 63°C, and ultrasound promoted the derivation of oxidative volatiles above the human sensory threshold was the effect of ultrasound reported (Juliano et al. 2014).

8. Whey sample; the experimental parameters were 20 kHz sonotrode, at 400, 1,000, and 2000 kHz and eight to 390 kJ/kg and no changes in phospholipid composition or in lipid oxidation beyond the detectable odor thresholds for volatile compounds at any tested frequency was the effect of ultrasound reported (Karlović et al. 2014).

9. Camel milk sample; the experimental parameters were ultrasonic processor set at 900 and 20 kHz, and formation of volatile compounds in milk occurred, reductions in some fatty acids was the effect of ultrasound reported (Dhahir et al. 2020).

Regarding the effect of HIU on sensory and volatiles of milk, it is known that HIU induces the formation of volatile compounds due to the oxidation of lipids.

Chouliara et al., applied HIU to milk and found an increase in volatile compounds (pentanal, hexanal, heptanal, and octanal) that were mainly products of lipid oxidation (Chouliara et al. 2010). The flavours associated with these compounds are due to the formation of carbonyls and methional. Dhahir et al., investigated the effect of ultrasound (900 W, 20 kHz) on camel milk (Dhahir et al. 2020). In their study, milk temperature was kept low ($20 \pm 3°C$), therefore any increase in volatile compounds would be attributed to the sonication reactions. The mechanism by which sonication increases the formation of volatile compounds in milk is probably fatty acid oxidation. This is consistent with the decrease in some fatty acids (C18:1 trans, C18:1c9, C20:1n9, and C22:6n3) after such treatment (Dhahir et al. 2020).

The increase in the levels of free fatty acids and oxidation of raw milk after ultrasound application were observed at a frequency above 20 kHz with the presence of undesirable volatile compounds (Chouliara et al. 2010). At frequencies between 10 and 20 kHz, the oxidation of lipids in raw milk and dairy products is below sensory detection levels (Leong et al. 2016) and better preservation of the compounds is obtained than those generated by the conventional process (Monteiro et al. 2018). Additionally, Johansson et al., reported that stable cavitation conditions (beyond one MHz) could avoid exposure to cavitation generated radicals that oxidize fat, consequently eliminating the off flavours (Johansson et al. 2016).

2.5 Ultrasound in Dairy Products

This section addresses the effect of high and low frequency ultrasound treatment on the most common dairy products: yogurt, fermented milks and or dairy drinks, cheeses, and creams. Most of the current studies have focused on the low frequency range (20 to 25 kHz), also known as high power or intensity ultrasound. This is because it generates strong physical, mechanical, chemical, and biochemical effects in liquids, modifying milk components and producing significant effects on the properties of milk and dairy products (Carrillo-Lopez et al. 2020, Paniwnyk 2017). Based on the most recent studies (2017 to date), we have noted that much of the research on low frequency ultrasound has focused on elucidating the effect that ultrasound has on milk components. With this aim, the studied milk components and fractions included sweet whey, whey protein isolate, whey protein concentrates, reconstituted whey powder, isolated proteins (i.e., β-lactoglobulin and α-lactalbumin), anhydrous milk fat, reconstituted micellar casein powder, emulsions and gels, sodium caseinate, micellar casein concentrate and reconstituted lactose solutions, among others. The effect of ultrasound treatment on milk could differ from that of a particular component.

The application of ultrasound in dairy ingredients helps to understand the chemical changes and mechanisms involved in biomolecules (proteins, enzymes, fat, lactose) allowing the development of new products. The effects are dependent on ultrasound parameters and the sample conditions. Reduced fat globule diameter and increased surface area results in other effects such as: increase of free saturated fatty acids, modification of protein structure including whey protein denaturation and secondary structure changes, alpha helix reduction, increase of beta sheets and beta turns, association of fat globules with casein, increase of the degree of enzymatic hydrolysis, proteolysis and lipolysis, among others (Abesinghe et al. 2020, Cheng et al. 2019, Gregersen et al. 2019, Hamdy et al. 2018, Jalilzadeh et al. 2018). However, there are few studies that have addressed the effect of low frequency ultrasound treatment on the quality of commercial or ready to eat dairy products. The most current research on the effect of ultrasonography (low frequency) on the quality of common dairy products have been summarized below. Studies that have used individual dairy ingredients in ultrasound application are not included. The effect of low frequency ultrasound on dairy products is shown below;

Cheeses

1. In pasteurized skimmed milk; the experimental parameters were; 106 and 375 W, three and nine min, 190.4, 570.5, 674.3, and 2,016.9 J, and with 674.3 J/g the firmness of the gel increased but the coagulation time doubled, with 2,016.9 J/g the milk did not coagulate was the ultrasound effect reported (Hammam et al. 2021).

2. In retained ultrafiltered milk, pasteurized, and inoculated; the experimental parameters were: amplitude = 80 percent, F = 20, 60, and 60 kHz, time 20 min; inoculation with *Escherichia coli* O157:H7, *Staphylococcus aureus*, *Penicillium chrysogenum,* and *Clostridium sporogenes, e*valuations after 60 d of maturation, and reduction of counts of all the microorganisms. Improvement of acidity of ripened cheese, acceleration of lipolysis and proteolysis, especially at 60 kHz, increased hardness, and gumminess at 30 d of ripening; improvement of the organoleptic properties with 60 kHz the ultrasound effect reported (Jalilzadeh et al. 2018).

3. In fresh raw milk; the experimental parameters were; amplitude = 50 and 100 percent, T = zero, five and 10 min, evaluations 24 hr post production and increase in yield up to 24.29 percent with 10 min of treatment; decrease of hydrogen potential and increase of yellowness with 10 min of HIU; increase in protein content with five min of HIU; reduction in coliform bacteria counts with 50 percent of amplitude the ultrasound effect reported (Carrillo-Lopez et al. 2020).

Creams and Butters-

1. In fresh cream with 30 percent of fat without heat treatment and not homogenized; the experimental parameters were: F = 20 kHz, power = 100 and 300 W, t = zero, five, 10 and 15 min, T = 50°C, probe = 12 mm and two cm depth, and low potencies improved viscosity and stability, uniform distribution, and particle size reduction, protein denaturation (protein chains opened to cover fat cells), which improved the properties of the whipped cream; color change with 10 min and 300 W, texture improvement with five min and 300 W were the ultrasound effects reported (Amiri et al. 2018b).

2. In pasteurized whipped cream (40 percent of fat) aged at 7°C, 90min; the experimental parameters were: power = 85 W, F = 20 kHz, t = zero, 10, 30, 60, and 90 s, probe: 1.27 cm, amplitude 108 μm, and decrease in solid fat content with 30 to 90 min of HIU due to the increase of temperature (two to six °C); decrease in cream whipping time during butter making and increase of hardness due to crystallization of triacylglycerols, melting of high melting point triacylglycerols with 60 and 90 s, higher crystallization of butter were ultrasound effects reported (Lee and Martini 2019).

Yogurt

1. In milk standardized to three percent of fat and 13 percent of solids non fat; the experimental parameters were: power = 150, 262, 375, 562 and 750 W. F = 20 kHz, t = 10 min, probe = 13 mm diameter, and 2.5 cm depth and yogurts with a low degree of similarity due to the appearance of unpleasant flavours (higher intensity of burnt, spicy and fatty flavour) compared to those produced with homogenized milk; yogurts with a high concentration of ketones, aldehydes, hydrocarbons and dimethyl sulphide was the ultrasound effect reported (Sfakianakis and Tzia 2017).

2. In standardized milk (3.4 percent of protein), inoculated and fermented, hydrogen potential of 5.2; the experimental parameters were: power = 300 W, F = 35 kHz, t = 5 min, T = 42°C, and higher syneresis and softness, increase in the number of visible particles with a less pronounced effect when using starter cultures with high synthesis of exo polycacharides was the ultrasound effect reported (Körzendörfer et al. 2017).

3. In standardized acidified skimmed milk (10 percent of protein, 43.5°C); the experimental parameters were: power = 20 W, F = 20 kHz, sonication until the hydrogen potential decreased from 5.8 to 5.1, and 7 reduction of the maximum torque required to break the gel, reduction of the firmness of the gel, greater softness and fewer aggregates, less cohesive structure and more compact microgels particles (reduced viscosity) were the ultrasound effects reported (Körzendörfer et al. 2019b).

4. In whipped yogurt hydrogen potential 5.0, 4.8 and 4.6; the experimental parameters were; power = 30 W, I = 22.5 W/cm2., F = 20 kHz, amplitude = 10 percent, probe = 13 mm T = 20°C, t = 30 s, and decrease in visual graininess and viscosity with slight effects on particle size was the ultrasound effect reported (Körzendörfer and Hinrichs 2019a).

5. In probiotic goat yogurt, hydrogen potential of 4.4; the experimental parameters were: power = 300 W. I = two W/mL, F = 20 kHz, amplitude = 67 percent, probe = 18 mm, T = 8±2°C, t = three, six, and nine min, and loss in viability of *Lactobacillus bulgaricus* and *Lactobacillus acidophilus* LA-5; reduced post acidification, improvement of viscosity and consistency index, reduction of tyramine content and total biogenic amine with six min of HIU were the ultrasound effects reported (Delgado et al. 2020).

6. In reconstituted milk and fresh buffalo milk; the experimental parameters were: power = 44, 54 and 66 W, F = 20 kHz, t = five, 7.5, 10, 12.5, 15, 17.5, and 20 min, probe three mm diameter and two cm deep, and reduction in the diameter of fat globules and increased surface area, better stability, increased free saturated fatty acids and gel hardness were the ultrasound effects reported (Abesinghe et al. 2020).

7. In reconstituted whole milk powder (13 percent of total solids); the experimental parameters were: power = 50 W, F = 22.5 kHz, T = 22–80°C, t = zero and 30 min and denaturation of whey proteins and aggregation of fat and protein globules with ultrasound without temperature control (>90°C), gels with higher firmness and short gelation times were the ultrasound effects reported (Nguyen and Anema 2017).

Fermented Milk

1. In reconstituted skim milk (3.5 percent of protein, 0.15 fat, 4.5 percent of lactose); the experimental parameters were: I = 20 W/cm², F = 22 ± 1.25 kHz, changing power and length of time: 60, 90 and 120 W/L, (one, three and five min, respectively), post fermentation with kefir, culture for yogurt and mixed culture for kefir and improvement in the accumulation of biologically active compounds and increase in nutritional quality, increased antioxidant capacity, increased accumulation of kefiran polysaccharide (probiotic value), improved consistency and appearance, best treatment: 90 W/L were the ultrasound effects reported (Potoroko et al. 2018).

2. In reconstituted skim milk (0.5 percent of fat, 15 percent of total solids) pasteurized and inoculated; the experimental parameters were: power = 100 W, F = 30 kHz, amplitude = 25 percent, t = five, 10 and 15 min, fermentation after sonication with *Lactobacillus plantarum* AF1 (probiotic) and the ultrasound effects reported were increased cell permeability, increased population of Lactobacillus and β-galactosidase activity, increased amounts of glucose, galactose and lactic acid, increased antioxidant activity (Gholamhosseinpour and Hashemi 2019).

3. In pasteurized reconstituted skim milk inoculated with probiotic strains; the experimental parameters were: power = 116 W, F = 40 kHz, amplitude = 30 percent, t = five, 10, 15 and 20 min, probiotic strains: *Lactobacillus acidophilus* LA-5, *Lactobacillus casei* LC, *Limosilactobacillus reuteri* LR-MM53, *Bifidobacterium bifidum* Bb- 12 and *Bifidobacterium loungm* BB-536 and increased viable cells and total acidity in LA-5, LC and LR -MM53; extracellular release of β-galactosidase by cell disruption in all strains, decreased fermentation time after 10 min US were the ultrasound effect reported (Niamah 2019).

In cheese, low frequency ultrasound treatment has proven to be a technology with high potential to increase gel firmness and hardness of the final product (fresh or ripened cheese). However, the results are variable according to the conditions of the ultrasound treatment. Hammam et al., reported increased firmness

(Hamman et al. 2021) and increased clotting times, while Jalilzadeh et al., observed an increase in hardness after 60 d of maturation (Jalilzadeh et al. 2018). Other reported effects include a decrease in final pH, improvement in sensory attributes due to increased lipolysis and proteolysis and increased yield and protein content (Carrillo-Lopez et al. 2020, Hamman et al. 2021, Jalilzadeh et al. 2018). Ice creams are not a dairy product per se; however they are a commercial product in high demand in the market. Ice cream consists of a frozen multiphase mixture containing fat, air bubbles, foam, and ice crystals dispersed in a serum phase of dissolved proteins, salts, and sugars (Akdeniz and Akalin 2019). The forces generated by high intensity, low frequency ultrasound cause ice crystal fragmentation and increase the nucleation rate and yield. This effect could be used positively to produce high quality ice cream with small and uniform ice crystals.

In creams and butters, low frequency ultrasound treatment has led to a reduction in particle size (fat globules), higher uniformity in size distribution and changes in the properties of triacylglycerols, including protein denaturation, which creates a protective barrier on the blood cells. These effects on the components of milk have led to improvements in the technological and quality properties of the cream, such as improved viscosity, longer stability, and a decrease of the churning time of the cream for the production of butter, as well as an increase in the hardness of the final products (Amiri et al. 2018a, Lee and Martini 2019).

Most of the recently published studies on dairy products are focused on the effect of low frequency ultrasound on the properties of yogurt and fermented milk. In the case of yogurt, ultrasound treatment has been applied to standardized milk before inoculation with lactic acid bacteria and or probiotics, or during and even after the fermentation process. When ultrasound is applied as an alternative technology to milk homogenization before fermentation, off flavours appearance by production of metabolites (ketones and aldehydes) has been observed, when sonication (Probe system, 150 and 750 W) is longer than10 min. In contrast, significant improvements in gel stability and hardness are shown when lower powers (44 to 66 W) and longer treatment (20 to 30 min) are used. Under these conditions, reduction in fat globule diameter, increase in surface area, denaturation of whey proteins, aggregation of fat globules with protein, and increase in free saturated fatty acids have been observed (Abesinghe et al. 2019, Nguyen and Anema 2017, Sfakianakis and Tzia 2017).

In inoculated and fermented milk, low frequency ultrasound produces important improvements in the properties of the gel, such as higher softness (less firmness) and consequently a reduction in the torque required to break the gel and a reduction in viscosity. However, an increase in the number of visible particles has been observed with the use of high power (300 W). Hence, alternatives such as the use of low power (20 W) and the use of starter cultures with high polysaccharide synthesis have led to a reduction in the number of aggregates in the final product (Körzendörfer et al. 2017, Körzendörfer and Hinrichs 2019a). The use of high power (300 W) in yogurt has produced significant decrease in the viability of acidifying and probiotic microorganisms, resulting in reduced post acidification and improvements in viscosity and consistency. Low wattages (30 W) in yogurt seem to produce a significant improvement in the final product by reducing visual graininess and hydrogen potential dependent viscosity of yogurt during ultrasonication (Delgado et al. 2020, Körzendörfer et al. 2017).

Ultrasonication has been applied to milk per se or to inoculated milk, producing effects on fermentation that depend on the ultrasound parameters. For instance, important improvements have been observed in the antioxidant capacity and in the probiotic value of fermented beverages, as well as in their appearance and consistency when ultrasonication is carried out on the milk before inoculation (Potoroko et al. 2018). When sonication is performed on inoculated milk, a significant improvement in the antioxidant capacity of the beverage has been observed, also including a higher number of viable cells and cell disruption that favours the release of enzymes (β-galactosidase), optimization of the fermentation and a higher number of metabolites (lactic acid, glucose, and galactose) (Gholamhosseinpour and Hashemi 2019, Niamah 2019).

High frequency ultrasound is also known as low intensity ultrasound and frequently uses intensities of less than one W/cm^2 and frequencies higher than 100 kHz. It is regularly used for non-destructive analysis techniques in the food industry (Sutariya et al. 2018), such as characterization of food components and measurement of physicochemical properties to ensure quality and to monitor processes such as fermentation. High frequency ultrasound treatment has been reported to induce strong chemical reactions without significant physical changes in liquid systems (Abesinghe et al. 2019, Thi Hong Bui et al. 2020). Generation of large amounts of free radicals that promote the oxidation of milk components (mainly lipids) and that consequently generate a high number of volatile compounds has been reported in milk and dairy products (Thi Hong Bui et al. 2020). Very few recent studies used high frequencies, as shown below:

1. Raw milk sample; the experimental parameters were: power = 700 and 343 W, F = 600 kHz and one MHz, T = five, 25 and 40°C and about the ultrasound effect, the combination of one MHz, five min at 25°C was effective for the rapid and synergistic separation of fat (Leong et al. 2014b).

2. Raw cow milk sample; the experimental parameters were: power = 330, 290 W, F = one and two MHz, t = zero, five, 10, 15 and 20 min and the ultrasound effect was obtained on skimmed milk with low fat concentration (1.7 p/ v percent) with 20 min treatment, fast skimming: 1.6 g fat/min (Leong et al. 2014a,b).

3. Part skim milk and high fat milk sample; the experimental parameters were: flow reactor, one MHz or two MHz of simultaneous sonication at 30°C, 33L/hr and the ultrasound effect was continuous fractionation with higher volumes of product than when the process is performed by batches (Leong et al. 2015).

4. Whole cow milk sample; the experimental parameters were single and double transducer, one and two MHz, on (five to 20 min) or several stages (five min), and the ultrasound effect was formation of gradient of fat concentration and particle size; enriched fat fractions at the top of the container with larger globules and skim milk fractions at the bottom of the container with smaller fat globules. The dual transducer provided higher particle size differentiation; two MHz was more

effective in handling smaller fat globules retained in later stages of skimming (Leong et al. 2016).

5. Reconstituted skim milk suspensions inoculated with Enterobacter aerogenes sample; the experimental parameters were: power = 50 W, F = 850 kHz, T < 20 ℃, t = zero, five, 10, 20, 40, and 60 min and the ultrasound effect were *Enterobacter aerogenes* was not inactivated in milk; high frequencies do not cause physical changes or modifications in the milk components (Gao et al. 2014a).

Other applications include non-destructive characterization of milk quality taking advantage of medium properties such as speed and attenuation of ultrasonic waves (Ouacha et al. 2015). The use of high frequencies was recently explored with ranges up to two MHz, although some studies have shown that frequencies above 1000 kHz generate high concentrations of deterioration products resulting from lipid oxidation (Paniwnyk 2017). High frequencies (> 400 kHz) have potential applications in fat separation/fractionation because a gentle process is generated that facilitates the natural migration of fat globules, resulting in a high creaming rate (Leong et al. 2014a, 2014b).

Some very recent studies have focused on enhancing fractionation and continuous separation, consisting of several fractionation stages and dual transducer configurations been observed, which is a characteristic problem with the use of high frequency batching (Leong et al. 2015, 2016). In this case, a low rate of lipid oxidation and formation of oxidation-derived volatiles, below the threshold of sensory detection have been observed. This is a characteristic problem of the use of high-frequency batching (Leong et al. 2015, 2016, Gao et al. 2014a) being high frequencies in efficient for microbial decontamination in skimmed milk. However, the use of low frequencies produces lethal damage to *Enterobacter aerogenes* in skimmed milk, while high frequencies (800 kHz) do not inactivate this bacterium even when long treatment times (60 min) are used. High frequencies do not modify the properties of milk or its components either (Gao et al. 2014a). Another potential application of high frequency ultrasound is during the fermentation processes, to monitor the change from liquid phase to gel in yogurt. This can be an advantage by measuring the speed of sound and the phase difference of the acoustic wave between the input signals and output measured with an oscilloscope (Ojha et al. 2017). In this case, acoustic monitoring makes it possible to determine the end point of the yoghurt production process.

Finally, the use of high frequency ultrasound constitutes a technology that requires further research due to its growth potential for monitoring the composition and physicochemical properties of the components of milk and dairy products during processing and storage. The main mechanism of this ultrasound modality is acoustic streaming, which consists of movement and mixing within the fluid or liquid without the formation of bubbles, unlike low frequency ultrasound in which acoustic cavitation is induced by the generation, growth and collapse of bubbles that release considerable amounts of energy (Awad et al. 2012, Carrillo-Lopez et al. 2021a).

3. Conclusion

High intensity ultrasonication (HIU) is a potential technology for improvement of physicochemical and functional properties of milk. Few disadvantages are observed in HIU application, such as some colour variations and slight protein oxidation. The off flavours and lipid oxidation developed in sonicated milk seem to diminish with time after ultrasonication. Ultrasound reduces the size of the fat globule, providing an alternative to traditional homogenization. The combination of heat and ultrasound is more appropriate for the control of pathogenic and spoilage bacteria. HIU could generate volatile compounds associated with general rejection of milk, however, this could be minimized at frequencies lower than 50 kHz. Given the evidence, it can be concluded that 20 kHz is an optimum frequency to obtain favourable changes in milk and dairy properties. More research on bioactive compounds and sensory characterization of the sonicated product is needed to determine the ideal parameters for product quality. Future research would allow to scale up this technology to the levels required by the industry.

References

Abadía-García, L., E. Castaño-Tostado, L. Ozimek, S. Romero-Gómez, C. Ozuna and S.L. Amaya-Llano. 2016. Impact of ultrasound pretreatment on whey protein hydrolysis by vegetable proteases. Innovative Food Science and Emerging Technologies 37: 84–90. https://doi.org/10.1016/j.ifset.2016.08.010.

Abesinghe, A.M.N.L., N. Islam, J.K. Vidanarachchi, S. Prakash, K.F.S.T. Silva and M.A. Karim. 2019. Effects of ultrasound on the fermentation profile of fermented milk products incorporated with lactic acid bacteria. International Dairy Journal 90: 1–14. https://doi.org/10.1016/j.idairyj.2018.10.006.

Abesinghe, A.M.N.L., J.K. Vidanarachchi, N. Islam, S. Prakash, K.F.S.T. Silva, B. Bhandari et al. 2020. Effects of ultrasonication on the physicochemical properties of milk fat globules of bubalus bubalis (water buffalo under processing conditions: a comparison with shear-homogenization. Innovative Food Science and Emerging Technologies 59: 102237–102273. https://doi.org/10.1016/j.ifset.2019.102237.

Aguilar, C., J. Serna-Jiménez, E. Benitez, V. Valencia, O. Ochoa and L.I. Sotelo. 2021. Influence of high power ultrasound on natural microflora, pathogen, and lactic acid bacteria in a raw meat emulsion. Ultrasonics Sonochemistry 72: 105415–105422. https://doi.org/10.1016/J.ULTSONCH.2020.105415.

Ahmadi, Z., S.M.A. Razavi, M. Varidi, S. Mohammad, A. Razavi and M. Varidi. 2017. Sequential ultrasound and transglutaminase treatments improve functional, rheological, and textural properties of whey protein concentrate. Innovative Food Science and Emerging Technologies 43: 207–215. https://doi.org/10.1016/j.ifset.2017.08.013.

Akdeniz, V. and A.S. Akalın. 2019. New approach for yoghurt and ice cream production: high-intensity ultrasound. Trends in Food Science and Technology 86: 392–398. https://doi.org/10.1016/j.tifs.2019.02.046.

Al Bsoul, A., J.P. Magnin, N. Commenges-Bernole, N. Gondrexon, J. Willison and C. Petrier. 2010. Effectiveness of ultrasound for the destruction of mycobacterium sp. strain (6PY1). Ultrasonics Sonochemistry 17(1): 106–110. https://doi.org/10.1016/j.ultsonch.2009.04.005.

Alarcón-Rojo, A.D., H. Janacua, J.C. Rodríguez, L. Paniwnyk and T.J.J. Mason. 2015. Power ultrasound in meat processing. Meat Science 107(Supplement C): 86–93. https://doi.org/10.1016/j.meatsci.2015.04.015

Alarcon-Rojo, A.D., L.M. Carrillo-Lopez, R. Reyes-Villagrana, M. Huerta-Jiménez and I.A. Garcia-Galicia. 2019. Ultrasound and meat quality: a review. Ultrasonics Sonochemistry 55: 369–382. https://doi.org/https://doi.org/10.1016/j.ultsonch.2018.09.016.

Almanza-Rubio, J.L., N. Gutiérrez-Méndez, M.Y. Leal-Ramos, D. Sepulveda and I. Salmeron. 2016. Modification of the textural and rheological properties of cream cheese using thermosonicated milk. Journal of Food Engineering 168: 223–230. https://doi.org/10.1016/j.jfoodeng.2015.08.002.

Al-Hilphy, A.R.S., A.K. Niamah and A.B. Al-Temimi. 2012. Effect of ultrasonic treatment on buffalo milk homogenization and numbers of bacteria. International Journal of Food Science and Nutrition Engineering 2(6): 113–118. https://doi.org/10.5923/j.food.20120206.03.

Al-Hilphy, A.R., A.B. Al-Temimi, H.H.M. Al Rubaiy, U. Anand, G. Delgado-Pando and N. Lakhssassi. 2020. Ultrasound applications in poultry meat processing: a systematic review. Journal of Food Science 85(5): 1386–1396. https://doi.org/https://doi.org/10.1111/1750-3841.15135.

Amiri, A., P. Sharifian and N. Soltanizadeh. 2018a. Application of ultrasound treatment for improving the physicochemical, functional and rheological properties of myofibrillar proteins. International Journal of Biological Macromolecules 111: 139–147. https://doi.org/10.1016/j.ijbiomac.2017.12.167.

Amiri, A., A. Mousakhani-Ganjeh, S. Torbati, G. Ghaffarinejhad and R. Esmaeilzadeh Kenari. 2018b. Impact of high-intensity ultrasound duration and intensity on the structural properties of whipped cream. International Dairy Journal 78: 152–158. https://doi.org/10.1016/j.idairyj.2017.12.002.

Ananta, E., D. Voigt, M. Zenker, V. Heinz and D. Knorr. 2005. Cellular injuries upon exposure of escherichia coli and lactobacillus rhamnosus to high-intensity ultrasound. Journal of Applied Microbiology 99(2): 271–278. https://doi.org/10.1111/j.1365-2672.2005.02619.x.

Arvanitoyannis, I.S., K.V. Kotsanopoulos and A.G. Savva. 2017. Use of ultrasounds in the food industry–methods and effects on quality, safety, and organoleptic characteristics of foods: a review. In Critical Reviews in Food Science and Nutrition 57(1): 109–128. https://doi.org/10.1080/10408398.2013.860514.

Ashokkumar, M. and T.J. Mason. 2007. Sonochemistry. pp. 353–372. *In:* John Wiley and Sons [eds.]. Kirk-Othmer Encyclopedia of Chemical Technology https://doi.org/10.1002/0471238961.1915141519211912.a01.

Ashokkumar, M. 2011. The characterization of acoustic cavitation bubbles - an overview. Ultrasonics Sonochemistry 18(4): 864–872. https://doi.org/10.1016/j.ultsonch.2010.11.016.

Ashokkumar, M. 2018. Introductory text to sonochemistry. ChemTexts 4(2): 1–9. https://doi.org/10.1007/s40828-018-0061-4.

Awad, T.S., H.A. Moharram, O.E. Shaltout, D. Asker and M.M. Youssef. 2012. Applications of ultrasound in analysis, processing and quality control of food: a review. Food Research International 48(2): 410–427. https://doi.org/10.1016/j.foodres.2012.05.004.

Balthazar, C.F., A. Santillo, J.T. Guimarães, V. Bevilacqua, M.R. Corbo, M. Caroprese et al. 2019. Ultrasound processing of fresh and frozen semi-skimmed sheep milk and its effects on microbiological and physical-chemical quality. Ultrasonics Sonochemistry 51: 241–248. https://doi.org/10.1016/j.ultsonch.2018.10.017.

Barretto, T.L., M.A.R. Pollonio, J. Telis-Romero and A.C. da Silva Barretto. 2018. Improving sensory acceptance and physicochemical properties by ultrasound application to restructured cooked ham with salt (NaCl) reduction. Meat Science 145: 55–62. https://doi.org/https://doi.org/10.1016/j.meatsci.2018.05.023.

Barukčić, I., K. Lisak Jakopović, Z. Herceg, S. Karlović, R. Božanić and M. Ashokkumar. 2015. Influence of high intensity ultrasound on microbial reduction, physico-chemical characteristics and fermentation of sweet whey. Innovative Food Science and Emerging Technologies 27(2–3): 94–101. https://doi.org/10.1016/j.ifset.2014.10.013.

Basso Pinton, M., L. Pereira Correa, M.M. Xavier Facchi, R.T. Heck, Y.S. Vaz Leães, A.J. Cichoski et al. 2019. Ultrasound: a new approach to reduce phosphate content of meat emulsions. Meat Science 152: 88–95. https://doi.org/10.1016/j.meatsci.2019.02.010.

Basso Pinton, M., B. Alves dos Santos, L. Pereira Correa, Y.S. Vaz Leães, A.J. Cichoski, J.M. Lorenzo et al. 2020. Ultrasound and low-levels of NaCl replacers: a successful combination to produce low-phosphate and low-sodium meat emulsions. Meat Science 170: 108244. https://doi.org/10.1016/j.meatsci.2020.108244.

Berlan, J. and T.J. Mason. 1992. Sonochemistry: from research laboratories to industrial plants. Ultrasonics 30(4): 203–212. https://doi.org/10.1016/0041-624X(92)90078-Z.

Bhangu, S.K. and M. Ashokkumar. 2016. Theory of sonochemistry. Topics in Current Chemistry 374(4): 1–31. https://doi.org/10.1007/s41061-016-0054-y.

Bilek, S.E. and F. Turantaş. 2013. Decontamination efficiency of high power ultrasound in the fruit and vegetable industry, a review. International Journal of Food Microbiology 166(1): 155–162. https://doi.org/10.1016/j.ijfoodmicro.2013.06.028.

Boateng, E.F. and M.M. Nasiru. 2019. Applications of ultrasound in meat processing technology: a review. Food Science and Technology 7(2): 11–15. https://doi.org/10.13189/fst.2019.070201.

Caraveo, O., A.D. Alarcon-Rojo, A. Renteria, E. Santellano and L. Paniwnyk. 2015. Physicochemical and microbiological characteristics of beef treated with high-intensity ultrasound and stored at 4°C. Journal of the Science of Food and Agriculture 95(12): 2487–2493. https://doi.org/10.1002/jsfa.6979.

Carrillo-Lopez, L.M., A.D. Alarcon-Rojo, L. Luna-Rodriguez and R. Reyes-Villagrana. 2017. Modification of food systems by ultrasound. Journal of Food Quality 2017: 1–12. https://doi.org/10.1155/2017/5794931.

Carrillo-Lopez, L.M., M. Huerta-Jimenez, I.A. Garcia-Galicia and A.D. Alarcon-Rojo. 2019. Bacterial control and structural and physicochemical modification of bovine *Longissimus dorsi* by ultrasound. Ultrasonics Sonochemistry 58: 104608–104615. https://doi.org/10.1016/j.ultsonch.2019.05.025.

Carrillo-Lopez, L.M., M.G. Juarez-Morales, I.A. Garcia-Galicia, A.D. Alarcon-Rojo and M. Huerta-Jimenez. 2020. The effect of high-intensity ultrasound on the physicochemical and microbiological properties of mexican panela cheese. Foods 9(3): 1–14. https://doi.org/10.3390/foods9030313.

Carrillo-Lopez, L.M., I.A. Garcia-Galicia, J.M. Tirado-Gallegos, R. Sanchez-Vega, M. Huerta-Jimenez, M. Ashokkumar et al. 2021a. Recent advances in the application of ultrasound in dairy products: effect on functional, physical, chemical, microbiological and sensory properties. Ultrasonics Sonochemistry 73: 105467–501. https://doi.org/10.1016/j.ultsonch.2021.105467.

Carrillo-Lopez, L.M., D. Robledo, V. Martínez, M. Huerta-Jimenez, M. Titulaer, A.D. Alarcon-Rojo et al. 2021b. Post-mortem ultrasound and freezing of rabbit meat: effects on the physicochemical quality and weight loss. Ultrasonics Sonochemistry 79: 105766–10576. https://doi.org/10.1016/j.ultsonch.2021.105766.

Chandrapala, J., C. Oliver, S. Kentish and M. Ashokkumar. 2012. Ultrasonics in food processing - food quality assurance and food safety. Trends in Food Science and Technology 26(2): 88–98. https://doi.org/10.1016/j.tifs.2012.01.010.

Chandrapala, J. and T. Leong. 2015. Ultrasonic processing for dairy applications: recent advances. Food Engineering Reviews 7(2): 143–158. https://doi.org/10.1007/s12393-014-9105-8.

Chang, H. and R. Wong. 2012. Textural and biochemical properties of cobia (rachycentron canadum) sashimi tenderised with the ultrasonic water bath. Food Chemistry 132(3): 1340–1345. https://doi.org/https://doi.org/10.1016/j.foodchem.2011.11.116.

Cheeke, J.D.N. 2000. Fundamentals and applications of ultrasonic waves. CRC Press. Taylor and Francis Group.

Chemat, F., Zill-E-Huma and M.K. Khan. 2011. Applications of ultrasound in food technology: processing, preservation and extraction. Ultrasonics Sonochemistry 18(4): 813–835. https://doi.org/10.1016/j.ultsonch.2010.11.023.

Chen, F., M. Zhang and C.-H. Yang. 2020. Application of ultrasound technology in processing of ready-to-eat fresh food: a review. Ultrasonics Sonochemistry 63:104953. https://doi.org/10.1016/j.ultsonch.2019.104953.

Cheng, Y., P.O. Donkor, X. Ren, J. Wu, K. Agyemang, I. Ayim et al. 2019. Effect of ultrasound pretreatment with mono-frequency and simultaneous dual frequency on the mechanical properties and microstructure of whey protein emulsion gels. Food Hydrocolloids 89: 434–442. https://doi.org/https://doi.org/10.1016/j.foodhyd.2018.11.007.

Chisti, Y. 2003. Sonobioreactors: using ultrasound for enhanced microbial productivity. Trends in Biotechnology 21(2): 89–93. https://doi.org/10.1016/S0167-7799(02)00033-1.

Chouliara, E., K.G. Georgogianni, N. Kanellopoulou and M.G. Kontominas. 2010. Effect of ultrasonication on microbiological, chemical and sensory properties of raw, thermized and pasteurized milk. International Dairy Journal 20(5): 307–313. https://doi.org/10.1016/j.idairyj.2009.12.006.

Cichoski, A.J., C. Rampelotto, M.S. Silva, H.C. de Moura, N.N. Terra, R. Wagner et al. 2015. Ultrasound-assisted post-packaging pasteurization of sausages. Innovative Food Science and Emerging Technologies 30: 132–137. https://doi.org/10.1016/j.ifset.2015.04.011.

Cichoski, A.J., M.S. Silva. Y.S.v Leães, C.C.B. Brasil, C.R. de Menezes, J.S. Barin et al. 2019. Ultrasound: a promising technology to improve the technological quality of meat emulsions. Meat Science 148: 150–155. https://doi.org/10.1016/j.meatsci.2018.10.009.

Cichoski, A.J., J.S. da Silva, Y.S.V. Leães, S.S. Robalo, B.A. dos Santos, S.R. Reis et al. 2021. Effects of ultrasonic-assisted cooking on the volatile compounds, oxidative stability, and sensory quality of mortadella. Ultrasonics Sonochemistry 72: 105443–105453. https://doi.org/10.1016/j.ultsonch.2020.105443.

Coppin, E.A. and O.A. Pike. 2001. Oil stability index correlated with sensory determination of oxidative stabilityin light-exposed soybean oil. JAOCS 78(1): 13–19. DOI:10.1007/S11746-001-0212-4.

Cunha, L.C.M., M.L.G. Monteiro, J.M. Lorenzo, P.E.S. Munekata, V. Muchenje, F.A.L. de Carvalho et al. 2018. Natural antioxidants in processing and storage stability of sheep and goat meat products. Food Research International 111: 379–390. https://doi.org/10.1016/j.foodres.2018.05.041.

Da Silva, J.S., M. Voss, C.R. de Menezes, J.S. Barin, R. Wagner, P.C.B. Campagnol et al. 2020. Is it possible to reduce the cooking time of mortadellas using ultrasound without affecting their oxidative and microbiological quality? Meat Science 159: 107947–107953. https://doi.org/10.1016/j.meatsci.2019.107947.

Dai, C., F. Xiong, R. He, W. Zhang and H. Ma. 2017. Effects of low-intensity ultrasound on the growth, cell membrane permeability and ethanol tolerance of *Saccharomyces cerevisiae*. Ultrasonics Sonochemistry 36: 191–197. https://doi.org/10.1016/j.ultsonch.2016.11.035.

Dang, D.S., C.D. Stafford, M.J. Taylor, J.F. Buhler, K.J. Thornton and S.K. Matarneh. 2022. Ultrasonication of beef improves calpain-1 autolysis and caspase-3 activity by elevating cytosolic calcium and inducing mitochondrial dysfunction. Meat Science 183: 108646–108654. https://doi.org/10.1016/j.meatsci.2021.108646.

de Lima Alves, L., M. Stefanello da Silva, D.R. Martins Flores, D. Rodrigues Athayde, A. Roggia Ruviaro, D. da Silva Brum et al. 2017. Effect of ultrasound on the physicochemical and microbiological characteristics of Italian salami. Food Research International 106: 363–373. https://doi.org/10.1016/j.foodres.2017.12.074.

Delgado, K., C. Vieira, I. Dammak, B. Frasão, A. Brígida, M. Costa et al. 2020. Different ultrasound exposure times influence the physicochemical and microbial quality properties in probiotic goat milk yogurt. Molecules 25(20): 4638. https://doi.org/10.3390/molecules25204638.

Demir, H., S. Çelik and Y.Ç. Sezer. 2021. Effect of ultrasonication and vacuum impregnation pretreatments on the quality of beef marinated in onion juice a natural meat tenderizer. Food Science and Technology International 28(4): 340–352. https://doi.org/10.1177/10820132211012919.

Den Hertog-Meischke, M.J.A., R.J.L. van Laack and F.J.M. Smulders. 1997. The water-holding capacity of fresh meat. Veterinary Quarterly 19(4): 175–181. https://doi.org/10.1080/01652176.1997.9694767.

Dhahir, N., J. Feugang, K. Witrick, S. Park and A. AbuGhazaleh. 2020. Impact of ultrasound processing on some milk-borne microorganisms and the components of camel milk. Emirates Journal of Food and Agriculture 32(4): 245–254. https://doi.org/10.9755/ejfa.2020.v32.i4.2088.

Ding, Q., G. Tian, X. Wang, W. Deng, K. Mao and Y. Sang. 2021. Effect of ultrasonic treatment on the structure and functional properties of mantle proteins from scallops (*Patinopecten yessoensis*). Ultrasonics Sonochemistry 79: 105770. https://doi.org/10.1016/j.ultsonch.2021.105770.

D'Amico, D.J., T.M. Silk, J. Wu, M. Guo, D.J.D. Amico, T.M. Silk et al. 2006. Inactivation of microorganisms in milk and apple cider treated with ultrasound. journal of food protection. 69(3): 556–563. https://doi.org/10.4315/0362-028X-69.3.556

Engin, B. and Y. Karagul Yuceer. 2012. Effects of ultraviolet light and ultrasound on microbial quality and aroma-active components of milk. Journal of the Science of Food and Agriculture 92(6): 1245–1252. https://doi.org/10.1002/jsfa.4689.

Evelyn and F.V.M. Silva. 2015. Use of power ultrasound to enhance the thermal inactivation of clostridium perfringens spores in beef slurry. International Journal of Food Microbiology 206: 17–23. https://doi.org/10.1016/j.ijfoodmicro.2015.04.013.

Fallavena, L.P., L.D. Ferreira Marczak and G.D. Mercali. 2020. Ultrasound application for quality improvement of beef biceps femoris physicochemical characteristics. LWT-Food Science and Technology 118: 108817–108824. https://doi.org/10.1016/j.lwt.2019.108817.

Fan, D., L. Huang, B. Li, J. Huang, J. Zhao, B. Yan et al. Acoustic intensity in ultrasound field and ultrasound-assisted gelling of surimi. LWT- Food Science and Technology 75: 497–504. https://doi.org/10.1016/J.LWT.2016.08.002.

Gammoh, S., M.H. Alu'datt, C.C. Tranchant, D.G. Al-U'datt, M.N. Alhamad, T. Rababah et al. 2020. Modification of the functional and bioactive properties of camel milk casein and whey proteins by ultrasonication and fermentation with lactobacillus delbrueckii subsp. Lactis LWT-Food Science and Technology 129: 109501–109512. https://doi.org/10.1016/j.lwt.2020.109501.

Ganesan, B., S. Martini, J. Solorio and M.K. Walsh. 2015. Determining the effects of high intensity ultrasound on the reduction of microbes in milk and orange juice using response surface methodology. International Journal of Food Science 2015(350719): 1–7. https://doi.org/10.1155/2015/350719.

Gao, S., Y. Hemar, G.D. Lewis and M. Ashokkumar. 2014a. Inactivation of enterobacter aerogenes in reconstituted skim milk by high- and low-frequency ultrasound. Ultrasonics Sonochemistry 21(6): 2099–2106. https://doi.org/10.1016/j.ultsonch.2013.12.008.

Gao, S., G.D. Lewis, M. Ashokkumar and Y. Hemar. 2014b. Inactivation of microorganisms by low-frequency high-power ultrasound: effect of growth phase and capsule properties of the bacteria. Ultrasonics Sonochemistry 21(1): 446–453. https://doi.org/10.1016/j.ultsonch.2013.06.006.

Garcia-Galicia, I.A., V.G. Gonzalez-Vacame, M. Huerta-Jimenez, L.M. Carrillo-Lopez, J.M. Tirado-Gallegos, R.A. Reyes-Villagrana et al. 2020a. Ultrasound versus traditional ageing: physicochemical properties in beef longissimus lumborum. CyTA- Journal of Food 18(1): 675–682. https://doi.org/10.1080/19476337.2020.1834458.

Garcia-Galicia, I.A., M. Huerta-Jimenez, C. Morales-Piñon, S. Diaz-Almanza, L.M. Carrillo-Lopez, R. Reyes-Villagrana et al. 2020b. The impact of ultrasound and vacuum pack on quality properties of beef after modified atmosphere on display. Journal of Food Process Engineering 43(1): 1–10. https://doi.org/10.1111/jfpe.13044.

Garcia-Galicia, I.A., M. Huerta-Jimenez, L.M. Carrillo-Lopez, D. Sanchez-Aldana and A.D. Alarcon-Rojo. 2022. High-intensity ultrasound as a pre-treatment of pork sub-primals for further processing of meat. International Journal of Food Science and Technology 57(1): 466–480. https://doi.org/10.1111/ijfs.15427.

Gera, N. and S. Doores. 2011. Kinetics and mechanism of bacterial inactivation by ultrasound waves and sonoprotective effect of milk components. Journal of Food Science 76(2): 111–119. https://doi.org/10.1111/j.1750-3841.2010.02007.x.

Gholamhosseinpour, A. and S.M.B. Hashemi. 2019. Ultrasound pretreatment of fermented milk containing probiotic lactobacillus plantarum AF1: carbohydrate metabolism and antioxidant activity. Journal of Food Process Engineering 42(1): e12930. https://doi.org/10.1111/jfpe.12930.

Gómez-Salazar, J.A., A. Galván-Navarro, J.M. Lorenzo and M.E. Sosa-Morales. 2021. Ultrasound effect on salt reduction in meat products: a review. Current Opinion in Food Science 38: 71–78. https://doi.org/10.1016/J.COFS.2020.10.030.

Gonzalez-Gonzalez, L., A.D. Alarcon-Rojo, L.M. Carrillo-Lopez, I.A. Garcia-Galicia, M. Huerta-Jimenez and L. Paniwnyk. 2020. Does ultrasound equally improve the quality of beef? an insight into *Longissimus lumborum, infraspinatus* and *cleidooccipitalis*. Meat Science 160: 107963–107973. https://doi.org/10.1016/j.meatsci.2019.107963.

Got, F., J. Culioli, P. Berge, X. Vignon, T. Astruc, J.M.M. Quideau et al. 1999. Effects of high-intensity high-frequency ultrasound on ageing rate, ultrastructure and some physico-chemical properties of beef. Meat Science 51(1): 35–42. https://doi.org/10.1016/S0309-1740(98)00094-1.

Gregersen, S.B., L. Wiking and M. Hammershøj. 2019. Acceleration of acid gel formation by high intensity ultrasound is linked to whey protein denaturation and formation of functional milk fat globule- protein complexes. Journal of Food Engineering 254: 17–24. https://doi.org/10.1016/j.jfoodeng.2019.03.004.

Hamdy, A.M., M.A. Mohran, A.L. Hassan and M.A. Fahmy. 2018. Effects of heat, ultrasound and microwave pretreatments on the antigenicity of whey protein concentrate (β-lactoglobulin). Assiut Journal of Agricultural Sciences 249(4): 75–87. https://doi.org/10.21608/ajas.2018.28370.

Hammam, A.R.A., S.L. Martinez-Monteagudo, L.E. Metzger and K.A. Alsaleem. 2021. Effect of ultrasound intensity on the functional characteristics of rennet-coagulated skim milk. Journal of Food Process Engineering 44(9): e13800-e13806. https://doi.org/10.1111/jfpe.13800.

Heinz, G. and P. Hautzinger. 2007. Meat processing technology for small to medium scale producers. RAP Publication 2007/20. FAO, Bangkok.

Hernández-Falcón, T.A., A. Monter-Arciniega, S. Cruz-Cansino, E. Alanís-García, G.M. Rodríguez-Serrano, A. Castañeda-Ovando et al. 2018. Effect of thermoultrasound on aflatoxin M1 levels, physicochemical and microbiological properties of milk during storage. Ultrasonics - Sonochemistry 48: 396–403. https://doi.org/10.1016/j.ultsonch.2018.06.018.

Huang, G., S. Chen, C. Dai, L. Sun, W. Sun, Y. Tang et al. 2017. Effects of ultrasound on microbial growth and enzyme activity. Ultrasonics Sonochemistry 37: 144–149. https://doi.org/10.1016/j.ultsonch.2016.12.018.

Huff-Lonergan, E. and S.M. Lonergan. 2005. Mechanisms of water-holding capacity of meat: The role of postmortem biochemical and structural changes. Meat Science 71(1): 194–204. https://doi.org/10.1016/j.meatsci.2005.04.022.

Inguglia, E.S., B.K. Tiwari, J.P. Kerry and C.M. Burgess. 2018. Effects of high intensity ultrasound on the inactivation profiles of *Escherichia coli* K12 and listeria innocua with salt and salt replacers. Ultrasonics Sonochemistry 48: 492–498. https://doi.org/10.1016/j.ultsonch.2018.05.007.

Jalilzadeh, A., J. Hesari, S.H. Peighambardoust and I. Javidipour. 2018. The effect of ultrasound treatment on microbial and physicochemical properties of iranian ultrafiltered feta-type cheese. Journal of Dairy Science 101: 1–12. https://doi.org/10.3168/jds.2017-14352.

Jayasooriya, S.D., B.R. Bhandari, P. Torley and B.R. D'Arcy. 2004. Effect of high power ultrasound waves on properties of meat: a review. International Journal of Food Properties 7(2): 301–319. https://doi.org/10.1081/JFP-120030039.

Jayasooriya, S.D., P.J. Torley, B.R. D'Arcy and B.R. Bhandari. 2007. Effect of high power ultrasound and ageing on the physical properties of bovine semitendinosus and longissimus muscles. Meat Science 75(4): 628–639. https://doi.org/10.1016/j.meatsci.2006.09.010.

Johansson, L., T. Singh, T. Leong, R. Mawson, S. McArthur, R. Manasseh et al. 2016. Cavitation and non-cavitation regime for large-scale ultrasonic standing wave particle separation systems - In situ gentle cavitation threshold determination and free radical related oxidation. Ultrasonics Sonochemistry 28: 346–356. https://doi.org/10.1016/j.ultsonch.2015.08.003.

Juliano, P., A.E. Torkamani, T. Leong, V. Kolb, P. Watkins, S. Ajlouni et al. 2014. Ultrasonics sonochemistry lipid oxidation volatiles absent in milk after selected ultrasound processing. Ultrasonics-Sonochemistry 21(6): 2165–2175. https://doi.org/10.1016/j.ultsonch.2014.03.001.

Juraga, E., B.S. Šalamon, Z. Herceg and A.R. Jambrak. 2011. Application of high intensity ultrasound treatment on enterobacteriae count in milk. Mljekarstvo 61(2): 125–134. URI: https://hrcak.srce.hr/69084.

Jurić, A., I. Delaš, T. Vukušić, S. Milošević, A.R. Jambrak and Z. Herceg. 2016. Influence of gas phase plasma and high power ultrasound on fatty acids in goat milk. American Journal of Food Technology 11(4): 125–133. https://doi.org/10.3923/ajft.2016.125.133.

Karlović, S., T. Bosiljkov, M. Brnčić, D. Semenski and F. Dujmić. 2014. Reducing fat globules particle-size in goat milk: ultrasound and high hydrostatic pressures approach. Chemical and Biochemical Engineering Quarterly Journal 28(4): 499–507. https://doi.org/10.15255/CABEQ.2014.19400.

Kehinde, B.A., P. Sharma and S. Kaur. 2021. Recent nano-, micro- and macrotechnological applications of ultrasonication in food-based systems. Critical Reviews in Food Science and Nutrition 61(4): 599–621. Ltd. https://doi.org/10.1080/10408398.2020.1740646.

Khan, I.T., M. Nadeem, M. Imran, R. Ullah, M. Ajmal and M.H. Jaspal. 2019. Antioxidant properties of Milk and dairy products: a comprehensive review of the current knowledge. Lipids in Health and Disease 18(1): 1–13. https://doi.org/10.1186/s12944-019-0969-8.

Kim, T. K., H.I. Yong, S. Jung, H.W. Kim and Y.S. Choi. 2021. Technologies for the production of meat products with a low sodium chloride content and improved quality characteristics—a review. Foods 10(5): 957–967. https://doi.org/10.3390/foods10050957.

Körzendörfer, A., S. Nöbel and J. Hinrichs. 2017. Particle formation induced by sonication during yogurt fermentation–impact of exopolysaccharide-producing starter cultures on physical properties. Food Research International 97: 170–177. https://doi.org/10.1016/j.foodres.2017.04.006.

Körzendörfer, A. and J. Hinrichs. 2019a. Manufacture of high-protein yogurt without generating acid whey–impact of the final pH and the application of power ultrasound on texture properties. International Dairy Journal 99: 104541. https://doi.org/10.1016/j.idairyj.2019.104541.

Körzendörfer, A., J. Schäfer, J. Hinrichs and S. Nöbel. 2019b. Power ultrasound as a tool to improve the processability of protein-enriched fermented milk gels for Greek yogurt manufacture. Journal of Dairy Science 102(9): 7826–7837. https://doi.org/10.3168/jds.2019-16541.

Krasnikova, E.S., N.L. Morgunova, A.v Krasnikov, S.V Akchurin and I.V. Akchurina. 2020. Physical and chemical effects of ultrasonic cavitation on the grain of meat when lamb salting. Journal of Physics: Conference Series 1679(2): 1–4. https://doi.org/10.1088/1742-6596/1679/2/022001.

Krasulya, O., S. Shestakov, V. Bogush, I. Potoroko, P. Cherepanov and B. Krasulya. 2014. Applications of sonochemistry in russian food processing industry. Ultrasonics Sonochemistry 21(6): 2112–2116. https://doi.org/10.1016/j.ultsonch.2014.03.015.

Krasulya, O.N., A.V. Smirnova, E.V. Kazakova and V.I. Bogush. 2020. Assessment of the cavitation effect on colour attributes of chilled pork with different autolysis during brining. IOP Conference Series: Earth and Environmental Science 613: 012062–012070. https://doi.org/10.1088/1755-1315/613/1/012062.

Lee, J. and S. Martini. 2019. Modifying the physical properties of butter using high-intensity ultrasound. Journal of Dairy Science 102(3): 1918–1926. https://doi.org/10.3168/jds.2018-15075.

Leong, T., L. Johansson, P. Juliano, R. Mawson, S. McArthur and R. Manasseh. 2014a. Design parameters for the separation of fat from natural whole milk in an ultrasonic litre-scale vessel. Ultrasonics Sonochemistry 21(4): 1289–1298. https://doi.org/10.1016/j.ultsonch.2014.01.007.

Leong, T., P. Juliano, L. Johansson, R. Mawson, S.L. McArthur and R. Manasseh. 2014b. Temperature effects on the ultrasonic separation of fat from natural whole milk. Ultrasonics Sonochemistry 21(6): 2092–2098. https://doi.org/10.1016/j.ultsonch.2014.02.003.

Leong, T., P. Juliano, L. Johansson, R. Mawson, S. McArthur and R. Manasseh. 2015. Continuous flow ultrasonic skimming of whole milk in a liter-scale vessel. Industrial and Engineering Chemistry Research 54(50): 12671–12681. https://doi.org/10.1021/acs.iecr.5b03142.

Leong, T., L. Johansson, R. Mawson, S.L. McArthur, R. Manasseh and P. Juliano. 2016. Ultrasonically enhanced fractionation of milk fat in a litre-scale prototype vessel. Ultrasonics Sonochemistry 28: 118–129. https://doi.org/10.1016/j.ultsonch.2015.06.023.

Li, K., Z.-L. Kang, Y.-F. Zou, X.-L. Xu and G.-H. Zhou. 2015. Effect of ultrasound treatment on functional properties of reduced-salt chicken breast meat batter. Journal of Food Science and Technology 52(5): 2622–2633. https://doi.org/10.1007/s13197-014-1356-0.

Li, Y., T. Feng, J. Sun, L. Guo, B. Wang, M. Huang et al. 2020. Physicochemical and microstructural attributes of marinated chicken breast influenced by breathing ultrasonic tumbling. Ultrasonics Sonochemistry 64: 105022. https://doi.org/10.1016/j.ultsonch.2020.105022.

Li, Z., J. Wang, B. Zheng and Z. Guo. 2020. Impact of combined ultrasound-microwave treatment on structural and functional properties of golden threadfin bream (nemipterus virgatus) myofibrillar proteins and hydrolysates. Ultrasonics Sonochemistry. 65: 105063. https://doi.org/10.1016/j.ultsonch.2020.105063.

Liao, L.B., W.M. Chen and X.M. Xiao. 2007. The generation and inactivation mechanism of oxidation-reduction potential of electrolyzed oxidizing water. Journal of Food Engineering 78(4): 1326–1332. https://doi.org/10.1016/j.jfoodeng.2006.01.004.

Lim, S.Y., L.C. Benner and S. Clark. 2019. Neither thermosonication nor cold sonication is better than pasteurization for milk shelf life. Journal of Dairy Science 102(5): 1–13. https://doi.org/10.3168/jds.2018-15347.

Lorenzo, J.M., R. Montes, L. Purriños and D. Franco. 2012. Effect of pork fat addition on volatile compounds of foal dry-cured sausage. Meat Science 91: 506–-512. Doi: 10.1016/j.meatsci.2012.03.006.

Majid, I., G.A. Nayik and V. Nanda. 2015. Ultrasonication and food technology: a review. Cogent Food and Agriculture 1(1): 1–11. https://doi.org/10.1080/23311932.2015.1071022.

Martins Flores, D.R., C.C. Bauermann Brasil, P.C. Bastianello Campagnol, E. Jacob-Lopes, L. Queiroz Zepka, R. Wagner et al. 2018. Application of ultrasound in chicken breast during chilling by immersion promotes a fast and uniform cooling. Food Research International 109: 59–64. https://doi.org/10.1016/j.foodres.2018.04.022.

Mason, T.J., L. Paniwnyk, F. Chemat and M. Abert Vian. 2010. Ultrasonic food processing. pp. 387–414 *In*: A. Proctor [ed.]. Alternatives to Conventional Food Processing. The Royal Society of Chemistry Publishing. https://doi.org/10.1039/9781849730976-00387.

McDonnell, C.K., P. Allen, C. Morin and J.G. Lyng. 2014. The effect of ultrasonic salting on protein and water–protein interactions in meat. Food Chemistry 147: 245–251. https://doi.org/10.1016/J. FOODCHEM.2013.09.125.

Mikš-Krajnik, M., L.X. James Feng, W.S. Bang and H.G. Yuk. 2017. Inactivation of listeria monocytogenes and natural microbiota on raw salmon fillets using acidic electrolyzed water, ultraviolet light or/and ultrasounds. Food Control 74: 54–60. https://doi.org/10.1016/j.foodcont.2016.11.033.

Monteiro, S.H.M.C., E.K. Silva, V.O. Alvarenga, J. Moraes, M.Q. Freitas, M.C. Silva et al. 2018. Effects of ultrasound energy density on the non-thermal pasteurization of chocolate milk beverage. Ultrasonics Sonochemistry 42: 1–10. https://doi.org/10.1016/j.ultsonch.2017.11.015.

Morild, R.K., P. Christiansen, A.H. Sørensen, U. Nonboe and S. Aabo. 2011. Inactivation of pathogens on pork by steam-ultrasound treatment. Journal of Food Protection 74(5): 769–775. https://doi.org/10.4315/0362-028X.JFP-10-338.

Nguyen, T.M.P., Y.K. Lee and W. Zhou. 2009. Stimulating fermentative activities of bifidobacteria in milk by highintensity ultrasound. International Dairy Journal 19(6–7): 410–416. https://doi.org/10.1016/j.idairyj.2009.02.004.

Nguyen, T.M.P., Y.K. Lee and W. Zhou. 2012. Effect of high intensity ultrasound on carbohydrate metabolism of bifidobacteria in milk fermentation. Food Chemistry 130(4): 866–874. https://doi.org/10.1016/j.foodchem.2011.07.108.

Nguyen, H.A.A. and S.G. Anema. 2017. Ultrasonication of reconstituted whole milk and its effect on acid gelation. Food Chemistry. 217: 593–601. https://doi.org/10.1016/j.foodchem.2016.08.117.

Niamah, A.K. 2019. Ultrasound treatment (low frequency) effects on probiotic bacteria growth in fermented milk. Future of Food: Journal on Food, Agriculture and Society 7(2): 1–8. https://doi.org/10.17170/kobra-20190709592.

Nowak, K.W., Ewa Ropelewska, A. E.-Din Bekhit and M. Markowski. 2017. Ultrasound aplications in the meat industry. p. 586. *In*: Bekhit, A. E.-D. [eds.]. Advances in Meat Processing CRC Press. Taylor & Francis Group.

Ojha, K.S., T.J. Mason, C.P. O'Donnell, J.P. Kerry and B.K. Tiwari. 2017. Ultrasound technology for food fermentation applications. Ultrasonics Sonochemistry 34: 410–417. https://doi.org/10.1016/j.ultsonch.2016.06.001,

Ouacha, E., B. Faiz, A. Moudden, I. Aboudaoud, H. Banouni, M. Boutaib et al. 2015. Non-destructive characterization of the air influence on the UHT milk quality by ultrasonic technique. International Conference on Electrical and Information Technologies (ICEIT). 2015: 296–299.

Owusu-Ansah, P., X. Yu, R. Osae, C. Zhou, R. Zhang, A.T. Mustapha et al. 2020. Optimization of thermosonication on bacillus cereus from pork: effects on inactivation and physicochemical properties. Journal of Food Process Engineering 43: 2017. https://doi.org/10.1111/jfpe.13401.

Paniwnyk, L. 2017. Applications of ultrasound in processing of liquid foods: a review. Ultrasonics Sonochemistry 38: 794–806. https://doi.org/10.1016/j.ultsonch.2016.12.025.

Parreiras, P.M., J.A. Vieira Nogueira, L. Rodrigues da Cunha, M.C. Passos, N.R. Gomes, G.S. Breguez et al. 2020. Effect of thermosonication on microorganisms, the antioxidant activity and the retinol level of human milk. Food Control 113: 107172. https://doi.org/10.1016/j.foodcont.2020.107172.

Pérez-Grijalba, B., J.C. García-Zebadúa, V.M. Ruíz-Pérez, D.I. Téllez-Medina, S. García-Pinilla, R.I. Guzmán-Gerónimo et al. 2018. Biofunctionality, colorimetric coefficiients and microbiological stability of blackberry (rubus fructicosus var. himalaya) juice under microwave/ultrasound processing. Revista Mexicana de Ingeniería Química. 17(1): 13–28. https://doi.org/10.24275/10.24275/uam/izt/dcbi/revmexingquim/2018v17n1/Perez.

Phull, S.S. and T.J. Mason. 2001. The uses of ultrasound for biological decontamination. pp. 1–24. *In*: Mason, T. J. and A. Tiehm [eds.]. Advances in Sonochemistry (1st ed., Vol. 6). JAI Press.

Piyasena, P., E. Mohareb and R.C. Mckellar. 2003. Inactivation of microbes using ultrasound: a review. International Journal of Food Microbiology 87: 207–216. https://doi.org/10.1016/S0168-1605(03)00075-8.

Potoroko, I., I. Kalinina, V. Botvinnikova, O. Krasulya, R. Fatkullin, U. Bagale et al. 2018. Ultrasound effects based on simulation of milk processing properties. Ultrasonics Sonochemistry 48: 463–472. https://doi.org/10.1016/j.ultsonch.2018.06.019.

Reyes-Villagrana, R.A., M. Huerta-Jimenez, J.L. Salas-Carrazco, L.M. Carrillo-Lopez, A.D. Alarcon-Rojo, R. Sanchez-Vega, et al. 2020. High-intensity ultrasonication of rabbit carcases: a first glance into a small-scale model to improve meat quality traits. Italian Journal of Animal Science 19(1): 544–550. https://doi.org/10.1080/1828051X.2020.1763212.

Riener, J., F. Noci, D.A. Cronin, D.J. Morgan and J.G. Lyng. 2009. Characterisation of volatile compounds generated in milk by high intensity ultrasound. International Dairy Journal 19(4): 269–272. https://doi.org/10.1016/j.idairyj.2008.10.017.

Riesz, P. and T. Kondo. 1992. Free radical formation induced by ultrasound and its biological implication. Free Radical Biology and Medicine 13: 247–270. DOI: 10.1016/0891-5849(92)90021-8.

Rossi, A.P., D.L. Kalschne, A.P. Iglikowski Byler, E.L. de Moraes Flores, O. Donizeti Leite, D. dos Santos et al. 2021. Effect of ultrasound and chlorine dioxide on *salmonella typhimurium* and *Escherichia coli* inactivation in poultry chiller tank water. Ultrasonics Sonochemistry 80: 105815–105823. https://doi.org/10.1016/j.ultsonch.2021.105815.

Sanches, M.A.R., P.M.O. Colombo Silva, R. Darros-Barbosa, J.T. Romero and A.C. Silva-Barretto da S. 2020. Mass transfer in beef: effect of crossbreeding and ultrasound application. Scientia Agricola 78(5): 1–9. https://doi.org/10.1590/1678-992x-2019–0335.

Sanches, M.A.R., P.M.O. Colombo Silva, T.L. Barretto, R. Darros-Barbosa, A.C. Silva-Barretto da S. and J. Telis-Romero. 2021. Technological and diffusion properties in the wet salting of beef assisted by ultrasound. LWT 149: 112036–112044. https://doi.org/https://doi.org/10.1016/j.lwt.2021.112036.

Sena Vaz Leães, Y., M. Basso Pinton, C. Terezinha de Aguiar Rosa, S. Sasso Robalo, R. Wagner, C. Ragagnin de Menezes et al. 2020. Ultrasound and basic electrolyzed water: a green approach to reduce the technological defects caused by NaCl reduction in meat emulsions. Ultrasonics Sonochemistry 61: 104830–104829. https://doi.org/10.1016/j.ultsonch.2019.104830.

Sert, D., E. Mercan, M. Dinkul and S. Aydemir. 2021. Processing of skim milk powder made using sonicated milk concentrates: a study of physicochemical, functional, powder flow and microbiological characteristics. International Dairy Journal 120: 105080–108088. https://doi.org/10.1016/j.idairyj.2021.105080.

Sfakianakis, P. and C. Tzia. 2017. Flavour profiling by gas chromatography–mass spectrometry and sensory analysis of yoghurt derived from ultrasonicated and homogenised milk. International Dairy Journal 75: 120–128. https://doi.org/10.1016/j.idairyj.2017.08.003.

Shamila-Syuhada, A.K., L.O. Chuah, W.A. Wan-nadiah, L.H. Cheng, A.F.M.M. Alkarkhi, M.E. Effarizah et al. 2016. Inactivation of microbiota and selected spoilage and pathogenic bacteria in milk by combinations of ultrasound, hydrogen peroxide, and active lactoperoxidase system. International Dairy Journal 61: 120–125. https://doi.org/10.1016/j.idairyj.2016.05.002.

Shokri, S., S.S. Shekarforoush and S. Hosseinzadeh. 2020. Stimulatory effects of low intensity ultrasound on the growth kinetics and metabolic activity of lactococcus lactis subsp. Lactis. Process Biochemistry 89: 1–8. https://doi.org/10.1016/j.procbio.2019.10.033.

Sienkiewicz, J.J., A. Wesołowski, W. Stankiewicz, R. Kotowski, J. Joanna, S. Andrzej et al. 2017. The influence of ultrasonic treatment on the growth of the strains of salmonella enterica subs. typhimurium. Journal of Food Science and Technology 54(8): 2214–2223. https://doi.org/10.1007/s13197-017-2648-y.

Stadnik, J., Z.J. Dolatowski and H.M. Baranowska. 2008. Effect of ultrasound treatment on water holding properties and microstructure of beef (*M. semimembranosus*) during ageing. LWT - Food Science and Technology 41(10): 2151–2158. https://doi.org/https://doi.org/10.1016/j.lwt.2007.12.003.

Stadnik, J. and Z.J. Dolatowski. 2011. Influence of sonication on warner-bratzler shear force, colour and myoglobin of beef (*M. semimembranosus*). European Food Research and Technology 233(4): 553–559. https://doi.org/10.1007/s00217-011-1550-5.

Suman, S.P. and P. Joseph. 2013. Myoglobin chemistry and meat color. Annual Review of Food Science and Technology 4(1): 79–99. https://doi.org/10.1146/annurev-food-030212-182623.

Suslick, K.S. 1990. Sonochemistry. Science 247(4949): 1439–1445. DOI: 10.1126/ciencia.247.4949.1439

Suslick, K.S. and D.J. Flannigan. 2008. Inside a collapsing bubble: sonoluminescence and the conditions during cavitation. Annual Review of Physical Chemistry 59(1): 659–683. https://doi.org/10.1146/annurev.physchem.59.032607.093739.

Sutariya, S.,V. Sunkesula, K. Kumar and K. Shah. 2018. Emerging applications of ultrasonication and cavitation in dairy industry: a review. In Cogent Food and Agriculture 4(1): 1549187–1549209. https://doi.org/10.1080/23311932.2018.1549187.

Tan, B.L., M.E. Norhaizan, W.P.P. Liew and H.S. Rahman. 2018. Antioxidant and oxidative stress: a mutual interplay in age-related diseases. Frontiers in Pharmacology 9: 1–28. https://doi.org/10.3389/fphar.2018.01162.

Thi Hong Bui, A., D. Cozzolino, B. Zisu and J. Chandrapala. 2020. Effects of high and low frequency ultrasound on the production of volatile compounds in milk and milk products - a review. In Journal of Dairy Research 87(4): 501–512. https://doi.org/10.1017/S0022029920001107.

Tiwari, B.K. and T.J. Mason. 2012. Ultrasound processing of fluid foods. pp. 135–165. In: Cullen P.J. K Brijesh, Tiwari and V.P. Valdramidis [eds.]. Novel Thermal and Non-Thermal Technologies for Fluid Foods. Elsevier Inc. https://doi.org/10.1016/B978-0-12-381470-8.00006-2.

Torkamani, A.E., P. Juliano, P. Fagan, R. Jiménez-flores, S. Ajlouni and T.K. Singh. 2016. Effect of ultrasound-enhanced fat separation on whey powder phospholipid composition and stability. Journal of Dairy Science 99(6): 4169–4177. https://doi.org/10.3168/jds.2015-10422.

Turantaş, F., G.B. Kılıç and B. Kılıç. 2015. Ultrasound in the meat industry: general applications and decontamination efficiency. International Journal of Food Microbiology 198(1): 59–69. https://doi.org/10.1016/j.ijfoodmicro.2014.12.026.

Uluko, H., L. Liu, H. Li, W. Cui, S. Zhang, L. Zhao et al. 2014. Effect of power ultrasound pretreatment on peptidic profiles and angiotensin converting enzyme inhibition of milk protein concentrate hydrolysates. Journal of the Science of Food and Agriculture 94(12): 2420–2428. https://doi.org/10.1002/jsfa.6572.

Uluko, H., S. Zhang, L. Liu, M. Tsakama, J. Lu and J. Lv. 2015. Effects of thermal, microwave, and ultrasound pretreatments on antioxidative capacity of enzymatic milk protein concentrate hydrolysates. Journal of Functional Foods 18(2): 1138–1146. https://doi.org/10.1016/j.jff.2014.11.024.

Valenzuela, C., I.A. Garcia-Galicia, L. Paniwnyk and A.D. Alarcon-Rojo. 2021. Physicochemical characteristics and shelf-life of beef treated with high intensity ultrasound. Journal of Food Processing and Preservation 45(4): e15350-e15360. https://doi.org/https://doi.org/10.1111/jfpp.15350.

van Hekken, D.L., J. Renye Jr., A.J. Bucci and P.M. Tomasula. 2019. Characterization of the physical, microbiological, and chemical properties of sonicated raw bovine milk. Journal of Dairy Science 102(8): 6928–6942. https://doi.org/10.3168/jds.2018-15775.

Vijayakumar, S., D. Grewell, C. Annandarajah, L. Benner and S. Clark. 2015. Quality characteristics and plasmin activity of thermosonicated skim milk and cream. Journal of Dairy Science 98(10): 1–14. https://doi.org/10.3168/jds.2015-9429.

Visy, A.,G. Jónás, D. Szakos, Z. Horváth-Mezőfi, K.I. Hidas, A. Barkó et al. 2021. Evaluation of ultrasound and microbubbles effect on pork meat during brining process. Ultrasonics Sonochemistry 75: 105589-105596. https://doi.org/10.1016/j.ultsonch.2021.105589.

Yu, Z., Y. Su, Y. Zhang, P. Zhu, Z. Mei, X. Zhou et al. 2021. Potential use of ultrasound to promote fermentation, maturation, and properties of fermented foods: a review. In Food Chemistry 357: 129805-129817. https://doi.org/10.1016/j.foodchem.2021.129805.

Zhang, C., Q. Sun, Q. Chen, B. Kong and X. Diao. 2020. Effects of ultrasound-assisted immersion freezing on the muscle quality and physicochemical properties of chicken breast. International Journal of Refrigeration 117: 247–255. https://doi.org/10.1016/j.ijrefrig.2020.05.006.

Zhang, M., X. Xia, Q. Liu, Q. Chen and B. Kong. 2019. Changes in microstructure, quality and water distribution of porcine longissimus muscles subjected to ultrasound-assisted immersion freezing during frozen storage. Meat Science 151: 24–32. https://doi.org/10.1016/j.meatsci.2019.01.002.

Zhang, R., X. Pang, J. Lu, L. Liu, S. Zhang and J. Lv. 2018. Effect of high intensity ultrasound pretreatment on functional and structural properties of micellar casein concentrates. Ultrasonics Sonochemistry 47: 10–16. https://doi.org/https://doi.org/10.1016/j.ultsonch.2018.04.011.

Zhang, R., L. Luo, Z. Yang, M. Ashokkumar and Y. Hemar. 2021. Formation by high power ultrasound of aggregated emulsions stabilised with milk protein concentrate (MPC70). Ultrasonics Sonochemistry 81: 105852–10557. https://doi.org/10.1016/j.ultsonch.2021.105852.

Zisu, B. and J. Chandrapala. 2015. High power ultrasound processing in milk and dairy products. pp. 149–179. In: Datta, N. and T. Peggy M [eds.]. Emerging Dairy Processing Technologies: Opportunities for the Dairy Industry. John Wiley & Sons, Ltd.

Zou, Y., P. Xu, H. Wu, H. Zhang, Z. Sun, C. Sun et al. 2018. Effects of different ultrasound power on physicochemical property and functional performance of chicken actomyosin. International Journal of Biological Macromolecules 113: 640–647. https://doi.org/10.1016/j.ijbiomac.2018.02.39.

CHAPTER 9

Ultrasound Combined with other Technologies to Accelerate Meat Processing

Luis Manuel Carrillo-Lopez,[1,2] *Raheel Suleman,*[3]
Iván Adrian Garcia-Galicia,[1] *Mariana Huerta-Jimenez,*[1,2]
Monserrath Felix-Portillo[1] and *Alma Delia Alarcon-Rojo*[1,*]

1. Introduction

Ultrasound is an emerging technology that has shown significant advantages in food processes. In the case of meat, studies have been focused on important processes for the industry such as mass transfer (marinated), the effect of ultrasound during post-mortem evolution (pre, during, and post rigor application), during maturation (dry and wet and under several types of packaging), during the freezing and storage process, among others. However, the application of ultrasound at the industrial level has been established for a small number of processes. This is due to the lack of standardization of both, ultrasound variables (system, intensity, S amplitude, frequency, time, temperature, etc.) and meat (type of muscle, type and characteristics of the packaging, additives, etc.), which limits the transition from the lab to the industrial level in the short term. In addition, the accompaniment of ultrasound with conventional technologies during meat processing further hinders the standardization of these variables. This is the case in processes such as cold storage, freezing, addition of additives, emulsion formation, drying, curing, among others. The use of emerging technologies in combination with ultrasound could significantly enhance the effect of ultrasound alone or cause synergistic effects. This chapter describes the use of some of these technologies (high pressures, pulsed electric field, irradiation, and cold

---

[1] Facultad de Zootecnia y Ecología, Universidad Autónoma de Chihuahua, Perif. Francisco R. Almada km 1, Chihuahua, Chih., 31453, México.
Emails: igarciag@uach.mx; monserrath.felix@uach.mx; aalarcon@uach.mx.
[2] Consejo Nacional de Ciencia y Tecnología. Av. Insurgentes Sur 1582, Col. Credito Constructor, Del. Benito Juárez, Ciudad de México, C.P. 03940. México.
Emails: mhuertaj@uach.mx; lmcarrillo@uach.mx
[3] Institute of Food Science and Nutrition, Bahauddin Zakariya University, Multan, Pakistan.
Email: raheelsuleman12@yahoo.com.
* Corresponding author: aalarcon@uach.mx

plasma) alone or with ultrasound as an assisted technique during meat processing. The market demand for meat products with enhanced physicochemical, sensory, and nutritional properties attracted the interest of researchers to investigate innovative processing approaches. Non thermal technologies can be used to replace or assist the conventional processing systems present in meat technology to attain the desirable meat properties. Additionally, the innovative technologies can be promising when used as combined strategies to preserve food, reduce processing times, and improve shelf life and quality of meat and meat products.

2. Ultrasound Combined with other Technologies to Accelerate Meat Processing

2.1 High Pressure and Ultrasound for Meat Processing

2.1.1 High Intensity Ultrasound (HIU)

Ultrasound is a kind of acoustic energy that is mechanical, nonionizing, and nonpolluting (Ünver 2016) having a significant potential for usage in high quality food manufacturing operations. It is an innovative technology, an alternative to typical meat aging approaches for meat tenderization, increasingly being employed in food preservation and analysis. Ultrasound modifies the physical, chemical, and functional aspects of food items (Terefe et al. 2016). Low intensity ultrasound has been used to assess the composition of meat, fish, and poultry products through food quality analysis (Knorr et al. 2004) but it has also been reported to be effective in mass transfer (Cárcel et al. 2007), marination, softening, and microorganism inactivation (Ünver 2016). High intensity ultrasound can elicit tenderness due to cavitation effects that destabilize cell structure, release lysosomes and proteases, and cause protein denaturation (Siró et al. 2009). The muscular tissue might get weakened and promotes beef softness (Chang et al. 2012, Stadnik and Dolatowski 2011). As a result, the age period can be decreased while maintaining meat quality metrics (Dolatowski et al. 2007) without impacting meat oxidative stability (Stadnik et al. 2008).

Despite the benefits of high intensity ultrasound (HIU), there are number of elements that influence its usage. The inconsistencies in the results are related to intrinsic (species, age, aging, muscle type) and extrinsic (ultrasonic systems, time, intensity, and frequency) aspects (Carrillo-Lopez et al. 2017). Some studies have indicated the beneficial effects of ultrasound on the preservation of nutritional, organoleptic properties of meat products (Ünver 2016) and microstructural changes to the myofibrils in beef (Stadnik et al. 2008). It may include beneficial tenderizing actions and increasing meat tenderness while significantly reducing aging time (Peña-González et al. 2017). Skeletal muscle tenderness is influenced significantly by myofibrillar proteins, connective tissues (Purslow 2005) and changes in post mortem proteolytic activity (Hildrum et al. 2009). When meat is subjected to HIU, proteolysis accelerates during post-mortem storage, as evidenced by enhanced desmin and troponin-T degradation (Wang et al. 2018).

Several extensive reviews provide specific information on the numerous physical and chemical processes that create ultrasonic effects (Alarcon-Rojo et al. 2015, 2019). Ultrasound has been used in studies for microbial inactivation in meat (Kang et al. 2017b), meat softness (Chang et al. 2015, Warner et al. 2017), and expedited meat processing techniques such as brining and curing (McDonnell et al. 2014a, Ojha et al. 2016). Usage of US in a phosphate reduction strategy, could include improved functionality of ingredients for meat application through pre-treatment with US, improved ingredient distribution within the meat matrix, or the effect of US on meat quality parameters when applied to the manufactured product.

It has been demonstrated that HIU facilitates mass transfer into the meat matrix. C. Ozuna et al., further investigated the effects of ultrasound on pork brining kinetics and discovered that US enhanced the NaCl and moisture effective diffusivities (Ozuna et al. 2013). Similarly, study by C.K. McDonnell et al., demonstrated that by using HIU at pilot scale on pork curing, meat curing time may be decreased by up to 50 percent. There was no significant influence on the quality and sensory qualities of sonicated meat in the same trial (McDonell et al. 2014a). K.S. Ojha et al., also demonstrated that ultrasonic therapy during pork brining might enhance diffusion of a commercially available sodium replacement salt replacer (Ojha et al. 2016). Therefore, there is evidence that the ultrasound may reduce additive requirements, enhance component distribution, and compensate for quality faults induced by phosphate reduction.

Application of ultrasound through a biological structure causes microstructure compressions and depressions, resulting in cavitation, leading to microstructural alterations in meat matrix (Siró et al. 2009). Several tests have been conducted to investigate the influence of ultrasound on the textural qualities of meat (Alarcon-Rojo et al. 2015). The effect of HIU on meat texture is dependent on several processing conditions, as outlined in a comprehensive analysis by (Warner et al. 2017). Ultrasound therapy can cause the generation of free radicals, which can expedite the oxidation of lipids in meat products. According to research, utilizing high intensity ultrasound on meat products promotes lipid and protein oxidation, which may change the textural qualities (Alarcon-Rojo et al. 2019, Chang et al. 2015, Kang et al. 2017a). However, it can be regulated by a variety of parameters such as pressure, temperature, and ultrasonic settings (Pinton et al. 2019). In study made by M.B. Pinton et al., when cooked meat emulsions were treated with ultrasonic power of 25 kHz for nine and 18 min, there is no enhanced lipid oxidation and might also compensate for deficiencies produced by up to 50 percent of phosphate loss, including enhanced cohesiveness and texture scores in sensory analysis (Pinton et al. 2019). Thus, optimizing processing parameters is critical for maintaining quality parameters.

HIU has been demonstrated to be an effective approach for improving the water holding capacity (WHC) and textural features of cooked meat batter made from meat with poor manufacturing capabilities, such as pale, soft, and exudative (PSE like) chicken breast meat (Li et al. 2014). Furthermore, Z. Zhang et al., discovered that adequate HIU treatment might improve MP gel capabilities by changing protein structure via ultrasound induced cavitation effect and micro streaming current (Zhang et al. 2017, Hu et al. 2015), and therefore improve protein interactions in the formulation.

2.1.2 High Pressure Processing (HPP)

High Pressure Processing (HPP) is an important non thermal processing technology. HPP exposes food products to very high hydrostatic pressure from 300 to 600 MPa and mild temperatures (<45°C) which can inactivate microorganisms and enzymes in food products without any effect on product color, flavor and nutritional composition (O´Flynn et al. 2014). HPP can cause conformational changes in proteins leading to protein denaturation, aggregation or gelation which helps to improve the functionality of comminuted meat products. In doing so, HPP also plays a major role in improving the water holding capacity of meat products. Various studies have reported on the effect of HPP on water binding capacity (WBC) of meat products (Zheng et al. 2018). Pressurization of meat products resulted in an improvement in gel forming properties of meat proteins thus enhancing the WHC and textural characteristics of meat product. Results from various studies showed that HPP increased the emulsion stability, chewiness, cohesiveness, hardness, gumminess and decreased cooking and purge loss in meat products (Inguglia et al. 2017). Studies of C.M. Crehan et al., assessed the effect of HPP on frankfurters with various salt levels and reported notable improvements in the juiciness and textural properties (Crehan et al. 2000). Heat set gels formed after HPP treatment in comminuted meat products have improved characteristics with both low and high salt concentrations (Ikeuchi et al. 1992). A. Grossi et al., studied the effect of HPP treatment on salt reduced sausages with carrot fiber and or potato starch as salt replacers (Grossi et al. 2012). Pork sausages with different formulations of salt, carrot fiber and or potato starch were treated with 400, 600, or 800 MPa for five min at five or 40°C. Results of WBC tests proved that the incorporation of HPP and a new functional ingredients improved the water holding capacity of low salt sausages to the same level as high salt sausages. From the experiment it was concluded that HPP at 600 MPa can reduce salt content of hydrocolloid containing pork sausages from 1.8 to 1.2 percent without any negative impact on the WBC, texture, and color. Similar results were obtained when salt reduced hams were treated with 100 MPa (Tamm et al. 2016). O'Flynn et al., investigated use of high pressure processing on phosphate reduced breakfast sausages and its effect on physicochemical and sensory characteristics (O'Flynn et al. 2014). Sausages with zero, 0.25, 0.5 percent of phosphate content were manufactured using the raw minced pork meat which was pre-treated with HPP at 150 or 300 MPa for five min. Analysis found that HPP treated phosphate free sausages had improved emulsion stability compared to the non HPP treated control. However, a slight decrease in the juiciness was observed for the sausages treated with HPP. From the comprehensive results it was concluded that administration of HPP treatment at 150 MPa for five min had a positive effect in reducing the phosphate content in low fat breakfast sausages to 0.25 percent without any negative impact on the functional characteristics. Despite various successful results, evidence from experiments showed that there were some negative effects on the sensory and acceptability characteristics on the meat products. Decreased functional properties in sausages were observed when they are treated with HPP at 300 MPa (O'Flynn et al. 2014). Application of high pressure over 400 MPa reduced the WHC in meat batters thus affecting the sensory characteristics of the meat product.

High pressure disinfection (N 400 MPa) causes breakage of microbe cell membranes and deactivation of enzymes, particularly those involved in their metabolism (Ananta et al. 2001). Furthermore, appropriate HP parameters, particularly moderate HP treatments in the range of 100 to 300 MPa do not cause excessive protein denaturation (Grossi et al. 2016) can modify meat proteins in a way that improves their functionalities, such as textural, rheological, and water binding properties, through pure physical effects (Bajovic et al. 2012). At this pressures range, the hypothesized process is most likely connected with HP induced surface exposure of internal hydrophobic and sulfhydryl groups (Cao et al. 2012), improved hydrogen bonding (Ma et al. 2013), uncoiling of helices, and creation of sheets of myosin (Cao et al. 2012). Aside from myofibrillar proteins, there is a direct impact on other proteins in the system, such as sarcoplasmic proteins (Lee E J et al. 2011). The decrease in cooking loss without the use of conventional additives is a major functional advantage of HPP on gel type beef products (e.g., phosphates, nonmeat proteins, or starches). This corresponds to greater production yields as well as improved moisture sensation of consumer (Ma et al. 2013). Furthermore, improved water holding capacity (WHC), which is often assessed by a centrifugal technique, is thought to result in a high quality beef product based on the correlations between numerous technological elements (Offer et al. 1989). It is generally known that increasing WHC results in greater juiciness (Perry et al. 2001). On the other hand, better WHC of beef patties are produced by using HPP (400 MPa, one min, four °C, four cycles), however the resulting product was rated low by customers for juiciness (Hayes et al. 2014). Further investigation of the correlations between functional metrics (e.g., cooking loss, centrifugal juice loss, etc.,) and sensory qualities (e.g., juiciness) is required. It has been demonstrated that the microstructures of protein gels largely control their functional characteristics (Han et al. 2014, Savadkoohi et al. 2014). Any treatment that alters the microstructure of a gel can have an impact on its functioning. Indeed, quantitative knowledge about microstructure gained by -picture analysis- would be useful in studying changes in gelation qualities. Proteins are predominantly denatured by heat during the thermal gelation process, revealing more buried amino acid groups (e.g., tyrosine, tryptophan, etc.). Concurrently, the system's components interact and aggregate to produce a honeycomb gel structure based on the strength of different contacts (hydrogen bonding, hydrophobic interactions, disulfide bonds, and ionic bonding, for example) between proteins (Liu et al. 2011). As a result, it is acceptable to conclude that protein changes that occur during thermal gelation process have a considerable impact on the components of various chemical interactions in the final gel system. Much of the study on high pressure meat and meat products has been done to document and understand the physiochemical consequences of HPP. So far, only a few researches have been conducted to investigate the benefits of HPP technology in producing a juicy gel type beef product by manipulation of HP parameters. Such investigations are necessary to link juiciness to objective data such as quantitative microstructure analysis or participation of a specific chemical or structural interaction. As a result, following cooking, the water characteristics and juiciness of rabbit sausages pre-treated with HPP were evaluated. Scanning electron microscopy (SEM) was used to analyze images, as well as to determine the chemical interactions of various gels. The goals of this study were to (1) assess the impact of various HPP parameters on the water properties and juiciness of rabbit sausages; (2) identify probable correlations between juiciness and the observed mechanical indicators; and (3) understand the processes behind the obtained results.

2.1.3 High Hydrostatic Pressure (HHP)

High hydrostatic pressure (HHP) applies pressure to a substance to be treated using water medium. HHP can significantly reduce population of harmful Gram-negative bacteria, Gram-positive bacteria, yeast, and mold in a variety of food products, including processed fruits, dairy products, meat and meat products, and aids in food preservation for a longer period. The decrease in microbial load is affected by treatment with pressure and temperature. It is heavily influenced by the sort of food being processed. When food is treated to HHP treatment, it is subjected to high pressure for a brief period. The pressure applied to food during treatment ranges between 200 and 700 MPa (Van Loey et al. 2003). HHP processed food has outstanding nutritional, sensory, and texture quality since it is subjected to treatment conditions for a relatively short length of time (Huang et al. 2020). The HHP therapy is reported to be more effective against eukaryotes, Gram negative bacteria, protozoa, and parasites than yeast and mold, which are inactivated at considerably greater pressures (Rendueles et al. 2011). HHP requires extremely basic and straightforward instrumentation. It comprises of a pressure compartment in which food is stored and water is delivered into chamber; water is then used to pressurize the food (González-Cebrino et al. 2013). As a result, in the absence of high temperatures and chemical additions, HHP treated food has fresh like characteristics. At ambient temperature, pressures of 350–450 MPa are adequate to inactivate Gram negative bacteria, yeast, and mold; however, pressures greater than 1,100 MPa are necessary to inactivate Gram positive bacteria (Daher et al. 2017). The high pressure results in damage to the cell membrane of microbial cells, which changes the permeability of the microbial cell wall and membranes. The coiled protein structure breaks and there is destruction to microbial cell enzymes, which alter the metabolic pathways; finally, the microbial cell dies, leading to a decrease in microbial population in food (Van Loey et al. 2003).

Research was conducted to assess the efficacy of HHP against *Escherichia coli* found in liquid food (orange juice) was held at 80°C before being subjected to HHP treatment at a pressure of 250 MPa for 900 s. For orange juice with pH values of 3.2, 4.5, and 5.8, the HHP treatment decreased microbial burden by 4.88, 4.15, and 4.61 log CFU/ml, respectively. HHP treatment was used to reduce *Salmonella species* burden in beef. The HH pressure of 500 MPa for 60 s was sufficient to inactivate *Salmonella species* in a chicken breast sample while having no influence on the organoleptic and sensory properties of chicken (Cap et al. 2020). It was also shown that dry cured sausage may be stored for more than 60 d with *Listeria monocytogenes* inactivation of 3.2 log CFU/g using a 600 MPa HHP treatment for 480 s. Up to

60 d, there was no oxidative damage to lipids and proteins in meals. Because HHP has no effect on lipid oxidation, it does not contribute to the formation of rancidity in food (Cava et al. 2020). Similar extraction studies have been published for tomato waste (Ninčević Grassino et al. 2020), grape pomace (Cascaes Teles et al. 2021), red microalgae (Suwal et al. 2019), egg yolk (Naderi et al. 2017), and from gooseberry juice. HHP also improves physical and chemical properties of fermented juices while increasing bioactives (Rios-Corripio et al. 2020). Human breast milk may also be preserved with HHP (Malinowska-Pańczyk 2020). It is also useful for improving the technical and functional qualities of milk proteins, allowing them to be used in a wider range of functional and nutraceutical meals (Carullo et al. 2021). HHP has been identified as a potential treatment not only for bacterial inactivation but also for the extraction and enhancement of antioxidant, phenolic, bioactive, and functional components from a variety of sources, implying its application potential in a variety of nutraceutical, pharmaceutical, health, food, and related industries. There are several technological challenges in creating HHP machines that are suitable for high volume food treatment, hence there are few or no HHP treated foods on the market today (Jadhav et al. 2021).

2.1.4 Conclusion

With rising concerns of damaging food quality and flavor using thermal techniques, non-thermal techniques have been introduced into this area with their practical application in benefiting and facilitating meat processing. Three of the non-thermal techniques have been discussed, summarizing their mechanism and results on meat quality after processing. Emerging technique of high intensity ultrasound has revolutionized meat processing through its advantageous aspects from maintaining overall meat tenderness to its microstructure stability after passing through the treatment. Combination of these techniques gives much higher quality results. These collaborative treatments can be considered as solution to deal with food security crisis in impeding years.

2.2 Pulsed Electric Field and Ultrasound for Meat Processing

2.2.1 High Intensity Ultrasound

Ultrasound is acoustic energy and considered a mechanical, non-ionizing, non-polluting type of energy (Chemat et al. 2011). HIU is an emerging technology with a great potential to control, improve, and accelerate processes without damaging the quality of food. HIU (>five W/cm^2 or 10 to 1000 W/cm^2) causes changes in the physical, chemical, or mechanical properties of foods (Ashokkumar and Mason 2007). It generates acoustic cavitation in a liquid medium, developing physical forces that are considered the main mechanism responsible for changes in exposed materials. In meat, HIU has been successfully used to improve processes such as mass transfer, marination, tenderization of meat and inactivation of microorganisms (Alarcon-Rojo et al. 2019). HIU has ability to induce disruption of cell membranes either by physically weakening of muscle structures or by activating biochemical processes that release endogenous enzymes. Moreover, HIU has been considered an alternative to traditional meat ageing methods for improving the quality properties of meat (Alarcon-Rojo et al. 2019).

2.2.2 Pulsed Electric Field (PEF)

PEF is an innovative non thermal processing technology has gained much attention in recent years due to its ability to render food safe for consumption applying short treatment time and minimal heat production.

Additionally, it is an energy efficient and environmentally friendly alternative for food processing (Gómez et al. 2019). The efficacy of PEF is restricted to products with low electrical conductivity and no air bubbles to avoid the dielectric breakdown (Chauhan and Unni 2015). PEF processing consists of placing food between two electrodes and expose it to a pulsed high voltage field (20 to 80 kV/cm) (Barbosa-Cánovas et al. 2000). The short bursts last from several nanoseconds to several microseconds. The intensity can be adjusted based on the geometry and distance of the working electrodes, the voltage delivered, and the conductivity of the material treated (Ricci et al. 2018). The continuous application of these pulses possesses the ability to inactive pathogens and enzymes while maintaining the nutritional and sensory quality and shelf life of foods (Barba et al. 2017, Horita et al. 2018).

Two mechanisms have been proposed to explain the biological effects of PEF, 'electrical breakdown' and 'electroporation' (Barbosa-Cánovas et al. 2000). PEF can modify the structure of the cell membrane of the pathogens resulting in the loss of its functional biological properties (Sale and Hamilton, 1967). The cell transmembrane potential is around 10 mV, if an external electric field is applied, this increases the potential difference across the cell membrane (around one V) and causes pore formation and a reduction in the membrane thickness (Barbosa-Cánovas et al. 2000). The second mechanism of PEF is that of electroporation. When a cell is subjected to a high voltage electric field, the lipid bilayer and proteins of the cell membrane are temporarily destabilized (Barbosa-Cánovas et al. 2000). Changes in the conformation of lipid molecules are induced, existing pores are expanded, and structurally stable hydrophobic pores are formed which can conduct current. Studies on PEF are currently focused on two major applications of this technique including non-thermal microbial inactivation and improvement of mass transfer through cell disruption (Gómez et al. 2019).

2.2.3 Effects of HIU on Meat

HIU increases tenderness and reduces shear force; either, by generating micro structural changes in the tissue or increasing enzyme activity. HIU reduces microbial loads in meat, and modifies the biochemical, functional, and structural properties of the meat proteins resulting in an increase in water holding capacity (Alarcon-Rojo et al. 2019). Detailed effects of HIU on meat properties can be found in the chapter -Sonochemistry in Foods of Animal Origin- of this book. The major beneficial effects of HIU application on meat are the improvement of sensory properties such as colour and flavour, and reduction of microbial counts, and improvement of mass transfer (Carrillo-López et al. 2021b). HIU has also been proposed as an alternative to reduce chemical additives and preservatives, while ensuring safety in the meat industry (Singla and Sit 2021). However, the main advantage of ultrasonicating fresh meat is an increase in tenderness and this effect has been demonstrated in different animal species, including beef (Garcia-Galicia et al. 2020), pork (Visy et al. 2021), rabbit (Carrillo-Lopez et al. 2021a) and chicken (Li et al. 2020, Zhang et al. 2020). Moreover, HIU reduced the freezing time and produced smaller ice crystals, which resulted in less water losses when thawing beef (Valenzuela et al. 2021) and pork (Owusu-Ansah et al. 2020). In addition, HIU significantly reduces the populations of mesophilic, *Psychrophilic, Staphylococcus, Bacillus cereus* and coliforms (Valenzuela et al. 2021). HIU can also be assisted technology for microbial reduction. A recent study shows that when combined chlorine dioxide with HIU the counts of *Salmonella typhimurium* and *Escherichia coli* in poultry chilling water are reduced (Rossi et al. 2021).

2.2.4 *Effects of PEF on Meat*

Interest in using pulsed electric field (PEF) in the meat processing has increased in recent years, as it induces microstructural changes in meat, which can enhance functional properties and quality of meat. The text below was modified from a table presented by Shi et al. 2021.

1. Pre rigor *Longissimus lumborum* (beef) muscle; the PEF parameters were: five, 10 kV per 20, 50, 90 Hz, and the results showed an increase in toughness of meat and degradation of myofibrillar proteins such as myosin heavy chains, troponin-T, and desmin in myofibrils were degraded (Suwandy et al. 2015b).

2. Pre rigor *Longissimus lumborum* (beef) muscle; the PEF parameters were repeated: PEF (1×, 2×, 3×) at 10 kV, 90 Hz, 20 μs 1×, and 2×, and the results showed that PEF had no tenderizing effect 3× PEF reduced the tenderness (TD), protein denaturation, and reduced proteolysis (Bekhit et al. 2016).

3. Pre rigor *Semimembranosus* (beef) muscle; the PEF parameters were: five, 10 kV × 20, 50, 90 Hz, and the results improved tenderness 21.6 percent of reduction in the shear force (SF) (Suwandy et al. 2015b).

4. Pre rigor *Semimembranosus* (beef) muscle; the PEF parameters were: repeat the PEF (1×, 2×, 3×) 10 kV, 90 Hz, 20 μs, and the results showed an improved tenderness 3× PEF produced the lowest SF (Bekhit et al. 2016).

5. *Longissimus thoracis* (beef) muscle; the PEF parameters were: 0.2 to 0.6 kV/cm to one to 50 Hz, and 20 μs, ageing time (one, three d), and the results showed that PEF had no influence on SF (Faridnia et al. 2014).

6. *Longissimus thoracis et lumborum* (beef) muscle; the PEF parameters were: 1.4 kV/cm, 10 Hz, and 20 μs, ageing time (two, 10, 18, 26 d), and the results showed that ageing time had a tendency towards reducing toughness; PEF did not affect the tenderizing process (Arroyo et al. 2015).

7. *Longissimus lumborum* (beef) muscle; PEF parameters were: five, 10 kV × 20, 50, 90 Hz, ageing time (one d) 19.5 percent, and the results showed that reduction in SF regardless of the intensity (Bekhit et al. 2014).

8. *Longissimus lumborum* (beef) muscle; the PEF parameters were: five, 10 kV × 20, 50, 90 Hz, ageing time (three, seven, 14, 21 d) and the results showed that 19.0 percent of reduction in SF regardless of the intensity (Suwandy et al. 2015a).

9. Post rigor *Longissimus Lumborum* (beef) muscle; the PEF parameters were: repeat the PEF (1×, 2×, 3×) 10 kV, 90 Hz, 20 μs, and the results showed a decrease by 2.5 N with every extra PEF (Suwandy et al. 2015c).

10. *Semimembranosus* (beef) muscle; the PEF parameters were five, 10 kV × 20, 50, 90 Hz, ageing time (one, three d), and the results showed a positive correlation between TD, frequency, and 19.1 percent of reduction in the SF (Bekhit et al. 2014).

11. *Semimembranosus* (beef) muscle; the PEF parameters were: five, 10 kV × 20, 50, 90 Hz, ageing time (three, seven, 14, 21 d, and the results showed a positive correlation between TD and frequency, and 19.0 percent of reduction in the SF (Suwandy et al. 2015a).

12. *Semimembranosus* (beef) muscle; the PEF parameters were repeat PEF (1×, 2×, 3×) 10 kV, 90 Hz, 20 μs, and the results showed the PEF had no influence on the TD, and SF (Suwandy et al. 2015c).

13. *Semitendinosus* (beef) muscle; the PEF parameters were: 1.9 kV/cm, 65 Hz, 20 μs, and the results showed the PEF had no influence on the TD, and SF (O'Dowd et al. 2013).

14. Breast meat (turkey) muscle; the PEF parameters were: 7.5, 10, 12.5 kV × 10, 55, 110 Hz fresh, and frozen thawed meat, and the results showed that the ageing time (five d) of PEF had no influence on the TD, and SF in both fresh, and frozen thawed meat (Arroyo et al. 2015).

15. *Biceps femoris* (beef) muscle; the PEF parameters were: 1.7 kV/cm, 50 Hz, 20 μs, ageing time (21 d), and the results showed that tenderness improved, and SF decreased (Faridnia et al. 2016).

16. *Biceps femoris* and *semitendinosus* (beef) muscle; the PEF parameters were: 0.8 to 1.1 kV/cm, 50 Hz, 20 μs fresh, and frozen thawed meat, and the results showed that tenderness improved, and color increased fat oxidation, and saturated fatty acids in frozen thawed (Kantono et al. 2019).

17. Shoulder (pork) muscle; the PEF parameters were: 0.5 to 5 kV/cm, 50 to 1,000 pulses, one to 25 kJ/kg, and the results showed that tenderness of meat increased with ultrasound application (Toepfl 2006).

18. *Longissimus thoracis* et *lumborum* (pork) muscle; the PEF parameters were: 1.2 or 2.3 kV/cm × 100 or 200 Hz × 150 or 300 pulses, and no effect was reported (McDonnell et al. 2014b).

19. Cold boned beef M. *longissimus et lumborum* muscle; the PEF parameters were: EFS 0.23 to 0.68 kV/cm, voltage 2.5, and 10 kV, frequency 200 Hz; pulse width 20 µs, and 2×; and the results showed a low and high PEF treatments can lead to ultrastructural changes in beef LL (Khan et al. 2018).

21. Briskets (deep pectoralis muscle) (beef) muscle; the PEF parameters were: EFS one, and 1.5 kV/cm; range of voltages 15 to 35 kV; specific energy levels of 40 to 50, and 90 to 100 kJ/kg; frequency 50 Hz; pulse width 20 µs, and the results showed that PEF weakens the connective tissue and increase the collagen solubility (Alahakoon et al. 2017).

The efficiency of PEF depends on different factors, such as number and duration of electric pulses and cell properties among others (Martínez et al. 2019). The ability of PEF to induce changes on meat depend on endogenous (species, animal age, muscle type, muscle composition) and exogenous (animal handling, meat pre-treatment) meat characteristics together with PEF treatment, equipment, and application (Alahakoon et al. 2017). PEF uses an electric field to create electroporation in the cell membrane, increasing membrane permeability and giving place to several biochemical reactions that occur during and after PEF treatment. These reactions occur when the electric field strength exceeds 0.5 kV/cm in animal cells, leading to transient increase of membrane permeability (Bekhit et al. 2014). One of the major parameters responsible for the effectiveness of pulsed electric field processing of meat is the electrical conductivity of the meat and also the medium if the meat is immersed in a solution during PEF treatment, which is temperature dependent (Alahakoon et al. 2017). Electroporation caused by PEF seems to be also responsible for proteolysis processes. Bekhit et al., suggested that PEF has the potential to accelerate the release of calcium ions leading to the activation of calcium activated protease calpains (Bekhit et al. 2014). More recently, Bhat et al., demonstrated that PEF increases tenderization during the process of ageing and that this process is of enzymatic nature (Bhat et al. 2019). Early post-mortem activation of calpain-2 was observed in PEF treated samples. An increase in the calpain activity and proteolysis of desmin and troponin-T was observed. However, no significant effect of PEF was found on the shear force of tough muscles from culled dairy animals during the entire ageing period. Another study has shown that PEF pre-treatment of beef briskets, can significantly increase tenderness and collagen solubilization after sous vide processing without having any adverse effects on protein digestibility (Alahakoon et al. 2018). PEF treatment in combination with aging seems to have an additional effect on meat tenderness due to an increased rate of proteolysis (Faridnia et al. 2015). Changes on meat depend on PEF intensity. Khan et al., reported that high PEF (HPEF, 10 kV, 200 Hz and 20 µs) treatment can negatively influence the quality of cold boned beef loins in comparison to the effects observed for low PEF treatments (LPEF, 2.5 kV, 200 Hz and 20 µs). HPEF beef samples had higher shear force, luminosity, and lipid oxidation, as well as lower redness and P, K and Fe, compared to LPEF and control samples (Khan et al. 2018).

PEF has also demonstrated be a novel method to produce healthier reduced sodium meat products. A recent study of Bhat et al., investigated the potential of PEF as a sodium reduction strategy for processed meat (beef jerky) (Bhat et al.

2020). Their results suggest that PEF treatment improved saltiness by influencing the salt diffusion and sodium delivery that led to better perception during chewing with no effect on color, yield, and oxidative and microbial stability. The study of Cropotova et al., reported the combined effect of PEF application (300 and 600 V/cm) and salting in a brine with five and 10 w/w percent NaCl on sea bass samples (Cropotova et al. 2021). They found a significant increase in primary and secondary lipid oxidation products in PEF treated samples yielding yellow pigmentation. PEF has also been applied in meat drying. Ghosh et al., analyzed the effects of low intensity PEF coupled to mechanical press dewatering on meat drying (Ghosh et al. 2020). They observed that the number of pulses had most significant effect on meat dewatering, and they suggested that PEF coupled with mechanical dewatering saves energy in conventional meat drying with air convection.

Vanga et al., stated that apart from the inactivation of pathogens, PEF can also be used for modification of protein secondary structure and functional properties (Vanga et al. 2021). These changes were highly dependent on the intensity of processing, the local food matrix, and properties of the protein present in the given product. In this regard, Dong et al., presented the effects of PEF on the physicochemical properties and conformations of myofibrillar proteins (MPs) extracted from pale, soft, exudative (PSE) like chicken breast meat (Dong et al. 2020). PEF treatment increased the solubility, surface hydrophobicity, and sulfhydryl group content of the MPs. Additionally, the PEF treatment altered the rheological properties of proteins with an increase in α helix structure. In another study, optimized PEF treatment (18 kV/cm) could induce MPs with a relatively small particle size, thus contributing to the production of a more homogeneous gel structure (Dong et al. 2021).

2.2.5 Combined HIU and PEF on Meat

Very few reports present results regarding combined US and PEF treatment of meat systems. From the effects of HIU and PEF on meat, it could be expected that combination of these techniques could be beneficial regarding effectiveness of the treatment, and processed material quality. Literature suggests that the most important applications that the combination of PEF with HIU are the improvement in tenderness and the enhancement of mass transfer. HIU and PEF contribute to myofibrillar degradation by facilitating the release of proteases and both technologies have effect on the microstructure of the meat.

With regards to tenderness, it would be expected that both technologies will tenderize faster the fast twitch glycolytic muscles (type IIb, white) that is much less susceptible to proteolysis than the slow twitch oxidative muscle or fibre (type I, red) (Ouali 1991). This means that aging rate would be higher in white muscles than in red muscles (Muroya et al. 2012). However, the study of Arroyo et al., showed that PEF treatments applied at different times post-mortem up to 26 d had a tendency towards reducing the toughness of beef samples, but that the application of PEF did not affect the tenderization process provided by ageing itself (Arroyo et al. 2015). These authors observed that PEF application of beef muscles exerted no influence on weight loss, color and cook loss (Arroyo et al. 2015). Regarding HIU, the effects on beef color have been very variable. Many authors have concluded that HIU has no impact in this characteristic (Caraveo-Suarez et al. 2022). Although in frozen pork HIU increased the WHC and reduced fluid losses after meat thawing (Garcia-Galicia et al. 2021). Color of PEF treated beef can be affected by the rise of temperature as reported by O'Dowd et al., who observed changes in lightness of beef muscles by the temperature induced by PEF treatment whereas a and b values remain unaffected (O'Dowd et al. 2013). Many of the quality changes in PEF treated meat are due to electroporation of the cell membranes (Alahakoon et al. 2017). In this way, Faridnia et al., reported that the meat structure becomes more porous as the electric field strength is increased (Faridnia et al. 2014), while L.P. O'Dowd et al., observed that PEF induces rupture of the myofibrils of beef *Semitendinosus* along the Z lines which causes a loss of the fibril organization and degradation of the myofibrillar structure (O'Dowd et al. 2013). On the other hand, HIU modifies the fibrillar structure of beef *Longissimus dorsi*, resulting in a weakening of the muscular matrix. Furthermore, muscle cells seem to be disrupted and muscle fibers tend to separate from each other when muscles were treated with 80 min ultrasound, but differences between muscles appear not to be large (Gonzalez-Gonzalez et al. 2020).

Apart from the increased tenderness and improved texture observed in PEF and HIU treated meat, PEF and HIU can be seen as effective salt reduction strategies through pre-treatment or combined synergistic effects. For instance, PEF pre-treatment allows direct reduction of NaCl without any adverse effect on sensory quality, lipid oxidation as well as microbial stability of products (Inguglia et al. 2017). Therefore, PEF could be a novel method to produce healthier meat products with a lower sodium content (Bhat et al. 2020). Similarly, HIU application is attractive to specifically reduce the use of phosphates (Zhang et al. 2021) and sodium in cured pork products, enabling additive reduction claims and contributing to obtain clean label meat products (Al-Hilphy et al. 2020). HIU in combination with sodium bicarbonate assisted curing had a favourable impact on meat tenderization, WHC, and curing efficiency (Xiong et al. 2020). The increase in mass transfer due to HIU treatment results in a faster diffusion of the brine into the pork tissue (Ozuna et al. 2013), with minimal impact on the product quality (Contreras-Lopez et al. 2020). Either by electroporation or by other mass transfer mechanisms, it is obvious that PEF as well as HIU impact positively on meat structure during curing and salting. Regarding the microbial inactivation Gomez-Gomez et al., investigated the impact of individual and combined PEF and high lower ultrasound (HPU) on the inactivation of different microorganisms in meat emulsions (Gomez-Gomez et al. 2021). Their results showed that PEF (152.3 to 176.3 kJ/kg) followed by HPU (three min) have the best synergistic effects to decrease *Aspergillus niger* and *Bacillus pumilus* loads compared to the individual treatments. They concluded that PEF-HPU is a promising technology to inactivate vegetative bacteria or fungal spores in emulsions. However, limited inactivation was achieved for bacterial spores. It has also been reported that the inactivation of larger microbial cells needs less intense field strengths to suffer a similar inactivation than smaller cells. Moreover, cells in their exponential growth phase are more susceptible to PEF treatments than the same cells in lag or stationary phase (Álvarez et al. 2000). When HIU is applied to meat, the cavitation phenomenon makes bacteria vulnerable to sonication treatments due the pressure generated on the surface of bacteria

(Alarcon-Rojo et al. 2019). For instance, a recent study (Aguilar et al. 2021) has reported that the application of HIU pulses with cycles of 7.7 s and 400 W of power for 10 min in a raw meat emulsion, has represented a microbial reduction of 60 percent of the natural microflora (*Lactobacillus delbrueckii* and *Listeria monocytogenes*).

Some negative effects on sensory attributes of meat have been observed after PEF application. PEF processing affects the muscle cell membranes influencing the interaction between fatty acids and cell membrane phospholipids with prooxidant effect in meat (Faridnia et al. 2015). Such interactions can lead to the formation of undesirable compounds able to decompose into secondary products which can cause off flavors and odors in meat and reduce its sensorial and nutritional quality. Ma et al., also reported that the combination of freezing and PEF treatment resulted in higher levels of some aldehydes and alcohols in lamb meat (Ma et al. 2016). However, many studies did not find any significant impact of PEF on lipid oxidation (Arroyo et al. 2015; Suwandy et al. 2015c). This implies there is a need for more research on the effects of PEF on the different quality attributes of meat. Nevertheless, the reports on undesirable changes in sensory attributes cause by HIU on meat are minimum. It has been observed that PEF causes higher alterations of inter cellular structure than HIU. Therefore, the utilization of HIU prior to PEF treatment can enhance the effectiveness of electroporation (Wiktor et al. 2018). Furthermore, Ostermeier et al., reported that the higher diffusion occurring in PEF pre-treated potato tissue will further benefit from improved heat and mass transfer at the product surface due to HIU treatment (Ostermeier et al. 2021). They observed a tendency of higher number of bubbles to be present probably due to the evaporation of water. Since both, HIU and PEF treatments, affect water diffusion and the water evaporation, a possible application of PEF combined with ultrasound could be in the dry meat products processing. However, the negative effects of PEF on meat, such as the tougher effect on beef *Longissimus lumborum* with high intensity PEF treatment (Suwandy et al. 2015a), the increase in proteolysis of low hydrogen potential meat samples (Suwandy et al. 2015b) and the promotion of myoglobin oxidation with HPEF conditions (Alahakoon et al. 2017) are issues that need to be solved before the combination of PEF-HIU is recommended for industry applications and the operations are scaled up at industrial level.

2.2.6 Conclusion

HIU and PEF are promising non thermal technologies with positive effects in food safety and quality or meat and meat products. The main effects of HIU and PEF, as separate technologies, are the result of the physical and chemical effects generated at the microscopic and structural levels of the food matrix. Both methods induce cell membrane permeabilization and enhance the release and activation of endogenous enzymes. As a result, HIU and PEF improves tenderization, marinating, bacterial inhibition and drying of meat caused by the enhancement in mass transfer and electroporation. The synergistic effect of HIU and PEF seems to be greater than when these technologies applied separately in meat. However, much research is needed to optimize this synergic effect to accelerate tenderization and mass transfer in meat processing and avoid the adverse effects on meat attributes caused by protein degradation. Additionally, the combination of HIU with PEF could be used as a new method to develop innovative meat products and processes. Finally, further studies under different operating conditions are needed to verify the potential impact of these two technologies for each processing application.

2.3 Nanotechnology and Ultrasound for Meat Processing

2.3.1 Ultrasound (US) as an Assisted Method of Nanotechnology

High intensity ultrasound can be used in the preparation and or synthesis of nanostructured materials. Regarding irradiation, there is no research that combines this technology with ultrasound for applications in meat or its products. It is known that both irradiation and ultrasound can be effective in decontaminating bacteria, viruses, and yeasts. In fact, the main purpose of using food irradiation is disinfection, microbial inactivation, and extension of shelf life. The application of both technologies could cause positive or synergistic effects in the microbial reduction in sensory and physicochemical quality, which is why it constitutes a virgin field of research in meat and its products. In the degradation of chitosan (Baxter et al. 2005), for example, cleavage at the β ($1{\rightarrow}4$) bonds without significant changes in deacetylation has been possible using long ultrasound treatment times (Li et al. 2008). This mechanism could be used for the synthesis of chitosan nanoparticles with natural antimicrobial and antifungal properties and with potential applications in food preservation, thanks to the short biodegradability time and biocompatibility with human tissues (Gomes et al. 2017). Jia et al., synthesized biocompatible hybrid nanosheets of graphene oxide and chitosan for pharmaceutical applications, using ultrasound as green technology for the synthesis (Jia et al. 2016). The review below summarizes the ultrasound conditions used for the synthesis and/ or dispersion of nanostructures, their applications and or properties (selected studies from the last 10 yr, from 2013 to 2022).

1. 15, and one min, without further specifications of the ultrasonic equipment or operating conditions (n/s) were the US parameters applied; the US objective or synthesized nanomaterials (NM) were: a. disperse GO (graphene oxide) in deionized water for rGO (reduced graphene oxide) synthesis, b. synthesize ZnO-rGO compounds, c. disperse catalyst, and dye or nanocomposites of reduced graphene oxide, and zinc, and the potential applications are green synthesis of the ZnO-rGO nanocomposite; the use in photocatalysis to degrade dyes (rhodamine B) and CO_2 was reported (Li et al. 2013).

2. 15 min, n/s (uniformization), probe, 450 W, 100 min (on four s, off two s) (breakout), one hr, and were the US parameters; the US objective or synthesized NM was achieved following the describe sequence: a. standardize aqueous solution GO, and sodium riboflavin-5'-phosphate dihydrate to prepare rGO, b. break rGO to get nanosized rGO, and then linked nanosized rGO with a drug or nanosized rGO, and the potential applications were green synthesis of rGO using riboflavin-5′-phosphate as reducing agent, and stabilizer or used as nano carrier for administration of a hydrogen potential sensitive drug (Ma et al. 2015).

3. Two hr, bath (dispersion); 120, 240 W, 20 kHz, 30 min, probe (amidation) were the US parameters; the US objective in synthesized NM were: a. dispersing chitosan GO in water containing one percent of acetic acid, b. induction of amidation formation for the synthesis of nanosheets that could be carriers of drugs, green synthesis of hybrid GO biocompatible nanosheets with chitosan or used as a drug carrier due to its stability in acidic, and physiological aqueous solutions; and the potential applications were the green synthesis of hybrid GO biocompatible nanosheets with the use of chitosan as a drug carrier due to its stability in acidic, and physiological aqueous solutions (Jia et al. 2016).

4. Two hr, bath (dispersion); dispersion to characterize NM were the US parameters; the US objective or synthesized NM target proceeded as follows: a. dispersion to prepare GO, b. dispersion of rGO in casein for characterization (Dynamic Light Scattering, and Transmission Electron Microscopy), and potential applications were in green synthesis of rGO nanosheets mediated by casein as a reducing, and stabilizing agent (Maddinedi et al. 2014).

5. 60 min, and probe (dispersion) were the US parameters; the US objective/ synthesized NM was dispersion of pristine graphene/TiO_2-graphene nanoparticle hybrids, and potential applications are degradation of dyes due to their high catalytic efficiency (Leng et al. 2015).

6. Two hr, n/s (dispersion); 750 W, 20 kHz, and probe (degradation) were the US parameters; The NM was synthesized through the following steps: a. dispersion of graphite oxide (GP) and GO in deionized water during the synthesis of GO, and sonocatalysts of ZnO/graphene/TiO_2, and b. the ultrasonic degradation of polluting dyes/ZnO/graphene/TiO_2 nanocomposites, the US activated the catalytic activity of synthesized catalysts obtained by Sonocatalysts, and the potential applications are that the US activated was highly effective in degrading polluting dyes (Nuengmatcha et al. 2016).

7. Zero, two, four, and six min, amplitudes of zero, 50, 75, and 100 percent, 750 W, 13 mm probe, and 25°C were the US parameters; the US objective in synthesized NM were preparation of film forming nanoemulsions of hazelnuts proteins, these films enriched with clove essential oil, and the potential applications are stable nanoemulsions with good homogeneous dispersion of the oil in the protein matrix, antimicrobial, and antioxidant properties, edible films that could be useful as a mechanical, and antimicrobial barrier (Gul et al. 2018).

8. 20 to 47°C, zero to 180s, 30 to 100 μm, and ultrasonic flow cell were the US parameters; the US objective of the NM obtained were formation of nanoemulsions of citronella oil, and sodium alginate, and Tween 80, and the potential applications were production of translucent nanoemulsions with small droplet sizes (four nm) and the high stability of essential oil delivery systems in food products (Salvia-Trujillo et al. 2013).

9. 120 to 600 s, 24 kHz, 105.3 W/cm2, and probe were the US parameters; the US synthesized NM were formation of film forming nano-emulsions from palm oil, and mesquite seed gum; and the potential applications are hydrophobic

emulsions with nanosized, uniformly dispersed droplets and in food packaging due to hydrophobicity, edible films on yellow, and orange fruits, and vegetables (Rodrigues et al. 2016).

10. Zero, 10, 50, and 100 percent of amplitude, n/s were the US parameters; the US goal for the synthesized NM was the preparation of films based on whey protein isolate, and films embedded with TiO_2, and SiO_2 nanoparticles, and potential applications are that sonication enhances the distribution of nanoparticles (NPs) in the matrix of film, and food packaging materials (Kadam et al. 2013).

11. 15 min, n/s (extraction, and biosynthesis) were the US parameters; and the potential applications were that US improves the antidiabetic, and antibacterial properties of ZnO. The NPs compared to chemical synthesis shown an effective antidiabetic, antibacterial, and oxidative properties (Bayrami et al. 2019).

12. 0.5 to 8.5 min, 40 kHz, 130 W, and bath were the US parameters; US objective/ synthesized NM were colorant extraction in juice samples using an adsorbent (nanocomposite) and the potential applications are that US quickly, and efficiently extracts the artificial colorant due to Allura Red's fast sorption-extraction and detection time (Asfaram et al. 2018).

Regarding the synthesis of nano emulsions, ultrasound has a significant effect due to ultrasonic cavitation, which generates localized regions of shear stress and increases in temperature due to bubble collapse. In emulsions, it produces unstable interfacial waves that allow the dispersion of the particulate oil phase by bursting into the aqueous phase and causes the formation of microbubbles that implode and separate the dispersed oil droplets into nano-meter sized droplets (Kentish et al. 2008). Ultrasound treatment can reduce film thickness as a function of sonication amplitude and time, as well as droplet size (Gul et al. 2018, Salvia-Trujillo et al. 2013, Rodrigues et al. 2016). In addition, the mechanical properties can be significantly improved, showing an increase in the breaking strength (by traction or elongation) without changes in stability (Das et al. 2022). This enables their application as edible films in the food industry, since they provide a mechanical barrier against water vapor and oxygen, including antimicrobial and antioxidant properties due to the components in the nano emulsions (e.g., essential oils). D.M. Kadam et al., also reported the benefits of ultrasound treatment to significantly improve the film forming properties of whey protein isolate, as well as the more uniform distribution of SiO_2 and TiO_2 nanoparticles in the films (Kadam et al. 2013).

Ultrasound treatment also has important applications in the dispersion of nanomaterials. In this case, the sonic waves allow stable dispersions to be obtained. Maddinedi et al., synthesized graphene nanosheets using natural milk casein as reducing and stabilizing agent. As part of the preparation process, these researchers used an ultrasonic bath system to disperse the nanosheets (Maddinedi et al. 2014). N. Ma et al., used ultrasound for both the production of reduced graphene oxide and the synthesis of nanometric reduced graphene oxide with potential applications as a nanocarrier for hydrogen potential sensitive drug delivery (Ma et al. 2015) (the review presented at the section Ultrasound (US) as an Assisted Method of Nanotechnology). In this case, the ultrasound treatment allowed obtaining a uniform solution before the reduction process and subsequently facilitated the rupture of the reduced graphene oxide to obtain nanometric sizes. X. Li et al., also used ultrasound to obtain homogeneous dispersions during the synthesis of reduced graphene oxide and nanocomposites of reduced graphene oxide and zinc oxide (Li et al. 2013). However, many applications related to these nanomaterials are oriented towards photocatalysis, where ultrasound treatment is applied as a pre-treatment or for synthesis.

Other applications of ultrasound include the characterization of nanostructured materials. For example, in the measurement of the photocatalytic activity of catalysts (Li et al. 2013). For this, ultrasonication allows the dispersion of the catalyst (nanocomposites) and the solution to photodegrade (rhodamine B). S. Aber et al., applied ultrasound in a bath system (28 kHz, 70 W, three hr) as one of the pre-treatment steps for cleaning the adsorbent surface of catalytically regenerated granular activated carbon with ZnO, resulting in a greater surface area and capacity of absorption (Aber et al. 2019). These studies are not reported at the section Ultrasound (US) as an Assisted Method of Nanotechnology since they have little relevance in food.

Metallic nanoparticles such as zinc oxide have been certified by the United States Food and Drug Administration (FDA) as safe, so they have been widely studied for medical and environmental applications, including food. However, green methods must be used for synthesis, since they are biological materials. In this regards, A. Bayrami et al., synthesized ZnO nanoparticles using plant extract of *Vaccinium arctostaphylos L* (Bayrami et al. 2019). Here, ultrasound treatment facilitated the extraction of metabolites from the plant extract to allow the reduction of metal ions. Sonic waves were also used to facilitate biosynthesis from the extract. ZnO exhibits activities against pathogenic bacteria and is associated with glucose uptake physiology with insulin like properties, however, further studies with *in vivo* models (for example, hyperglycemic rats) are required to ensure its safety. Related to this, Asfaram et al., applied ultrasound for extraction purposes. In this case, an adsorbent (Fe_3O_4@CuS@Ni_2P-CNTs magnetic nanocomposite) was used for the sorption of trace amounts of Allura Red in fruit juice samples, a food additive (synthetic colorant) associated with cancer (Asfaram et al. 2018). Soltani et al., also used ultrasound (37 kHz, 256 W) to increase the removal efficiency (adsorption) of antibiotics (tetracycline), which in combination with ZnO and nanocellulose nanostructures significantly increased the degradation efficiency (Soltani et al. 2019).

Ali Dheyab et al., extensively describe the sonochemical synthesis of nanomaterials based on iron oxide coated iron oxide, gold, and gold nanoparticles, with potential applications in biomedicine, catalysis, biosensing, drug delivery, and phototherapy (Ali Dheyab et al. 2021). R. Monsef et al., also reported the sonochemical synthesis of $PrVO_4$ nanostructures for wastewater remediation purposes (Monsef et al. 2020). Again, the sonochemical fabrication of nanostructures (60 W, 18 kHz) was easy, fast, and cheap. The produced nanostructures showed high efficiency of contaminant dye removal in water. Another field studied is the application of ultrasound in colloidal dispersion and the stability of nanofluids containing different types of nanoparticles (metallic, metal oxide, ceramic, graphite, carbon nanotubes, graphene, etc.). According to Asadi et al., systematic studies are required to estimate the optimal ultrasonication times and powers for nanofluids to have the highest increase in thermal conductivity and the lowest increase in viscosity, which are relevant in heat transfer applications (Asadi et al. 2019). Ultrasound (ultrasonic probe or bath) has also been combined with an electrochemical cell for the synthesis of complex nanomaterials (sonoelectrochemistry). The ultrasonic probe acts as a working electrode to produce nanoparticles (Islam et al. 2019). This field is in early stages with potential applications in areas such as electronics, optics, and electrochemistry.

In foods, the use of ultrasound to facilitate the diffusion of microencapsulated fatty acids in meat has been reported (Ojha et al. 2017), the ultrasound assisted fabrication of nano coatings and nanofilms (Esmaeili et al. 2021, Priyadarshi et al. 2021, Azlin-Hasim et al. 2015), the preparation of nanocarriers loaded with active food ingredients (Koshani and Jafari 2019, Jafari et al. 2022), the synthesis of nano capsules from plant extracts (Rashidaie Abandansarie et al. 2019) and in the production of stable nano emulsions as vehicles for bioactive compounds (Abassi et al. 2019a, Abbasi et al. 2019b) have been studied.

2.3.2 Ultrasound (US) as an Assisted Technology of Nanotechnology in Meat and Meat Products

There are few studies in which ultrasound and nanotechnology are combined for their application in the processing of meat and its products. The most recent studies (from 2015 to date) were summarized below:

1. The US parameters were 30 or 60 min, 25 kHz, n/s; and the finding reduced omega-6 or omega-3 ratio in pork (better fatty acid profile) (Ojha et al. 2017).

2. The US parameters were n/s; the US objective was manufacture of chitosan, and *Lepidium sativum* nano-coatings to increase the shelf life of beef in refrigeration, and the findings were: greater oxidizing activity, delay of lipid oxidation, and improvement of the chemical properties of the meat, and reduction of microbial deterioration, and reduction in the growth of pathogenic bacteria below the permissible limit at the end of the storage period (better fatty acid profile) (Esmaeili et al. 2021).

3. The US parameters were five to 15 min, 50 percent of amplitude, 24 kHz, probe;the US objective was the production of nano-emulsions of linseed oil in water, stabilized by different wall materials for the controlled release of alpha linolenic acid in the gastrointestinal tract, and the findings were that the US produces nano-emulsions stabilized by whey protein, and sodium alginate, with the potential to protect linolenic acid from gastric digestion, so it could be used as a transporter of omega-3 fatty acids (*in vivo* diet in birds) or to enrich the chicken meat (Abbasi et al. 2019a).

4. The US parameters were 15 min, 50 percent of amplitude, 24 kHz, probe; the US objective was the preparation of nano-emulsions with targeted delivery strategy of omega-3 polyunsaturated fatty acids to enrich broiler meat; and the findings or potential applications were controlled release behavior in simulated

gastrointestinal fluids, *in vivo*, flaxseed oil loaded nano-emulsions improved growth performance, lipid profile, and oxidative stability (Abbasi et al. 2019a).

5. The US parameters were 30 min, bath, n/s; the US objective was dispersion of zinc oxide nanoparticles (one to three percent of carboxymethylcellulose) in distilled water, and the findings were films with high antioxidant capacity, UV blocking mechanical properties, and antimicrobial activity against *Escherichia coli*, and *Listeria monocytogenes*, the control of psychotropic bacteria in raw meat after 15 d of storage, and low lipid oxidation (Priyadarshi et al. 2021).

6. The US parameters were five min, 500 W, 40 kHz; the US objective was dispersion of multi wall carbon nanotubes in distilled water, and the findings were the synthesis of a nano bio-composite based on carbon nanotubes, and aspartic acid allowed the sensitive, and selective detection of xanthine in fish meat at different storage times (Yazdanparast et al. 2019).

7. The US parameters were two min, 45°C, 20 kHz, and 95 percent of amplitude, probe; and the US objective was to evaluate the effect of US on the emulsifying, physical, and protein structure properties, and the findings were that the US reduced the size to the nanometer level, and the US reduced droplet size in bovine gelatin emulsions but not in fish gelatin (O′Sullivan et al. 2016).

8. The US parameters were 30 min, and n/s; the US objective was 30 min, n/a., Ag nanoparticles synthesis, and the findings or potential applications were low density polyethylene nanocomposite films with Ag nanoparticles extended the shelf life of chicken breast fillets and improved oxidative stability (Azlin-Hasim et al. 2015).

9. The US parameters were 30 min, 35°C, 35 kHz, and bath; the US objective was the US assisted extraction, and the findings or potential applications were the addition of antioxidants as nano-capsules better inhibited the oxidation process of sunflower oil compared to the use of microcapsules (Jafari et al. 2022).

10. The US parameters were 20 min, 20 kHz, 45°C, and bath; the US objective was the US assisted extraction, and the findings or potential applications where nano encapsulated rosemary extract could be used as a natural preservative in beef, and meat products due to its antioxidant, and antimicrobial activity, and delay in lipid oxidation (better fatty acid profile) (Rashidaie Abandansarie et al. 2019).

As mentioned in the previous section, the main use of ultrasound is nanomaterial synthesis, and dispersion. For these purposes, Ojha et al., used ultrasound (25 kHz, 30 or 60 min) to improve the diffusion of microencapsulated fatty acids in pork (*Semitendinosus*) meat. Pork meat ultrasonicated with nanovesicles of fatty acids exhibited a higher content of omega-3 polyunsaturated fatty acids (PUFA) (eicosapentaenic, docosapentaenoic, and docosahexaenoic acids) (Ojha et al. 2017). Ultrasound significantly improved low diffusion rates in the complex solid matrix of meat.

Recently, there has been a growing interest in the use of natural ingredients, and additives in foods, mainly due to their effects on health. In the case of meat, the reduction of traditional packaging has been sought due to the waste it generates, so that edible coatings are a better option for the food industry because they are biodegradable, they constitute a good barrier against humidity, oxygen, and CO_2, if they are made with safe raw materials such as polysaccharides, lipids, and proteins. On this, Esmaeili et al., investigated the effect of nano coating with chitosan, and cress seed gum during storage of beef under refrigeration temperatures (Esmaeili et al. 2021). In this case, the ultrasound treatment allowed the elaboration-formation of the nanofilms. Priyadarshi et al., used ultrasound in a bath system as one of the steps for the fabrication of a functional film based on carboxymethylcellulose incorporated with grape seed extract, and zinc oxide nanoparticles (Priyadarshi et al. 2021). These researchers demonstrated the application of biofilm as a sustainable material in active packaging applications of high fat meat products such as beef. Some research has been focused on improving the properties of traditional plastic packaging. For example, Azlin-Hasim et al., developed low density polyethylene nanocomposite films containing Ag nanoparticles to wrap chicken breast fillets stored in a modified atmosphere (Azlin-Hasim et al. 2015). Although the results showed improved antimicrobial properties, the safety in the use of these packages must be evaluated due to the danger of migration of AgNPs from the package to the food.

Much research has been focused on the ultrasound assisted preparation of nanocarriers loaded with active food ingredients. Undoubtedly, ultrasound has stood out among several nano synthetic routes due to its safety in use, easy operation, energy efficiency, and potential for scaling at an industrial level (Koshani and Jafari 2019). In this regard, Jafari et al., developed nano, and microcapsules of plant extract as a natural preservative agent to improve the oxidative stability of vegetable oils (Jafari et al. 2022). The capsules have potential for use as food additives, either to prolong shelf life or to improve antioxidant, and antimicrobial capacity in meat, and meat products. According to Rashidaie Abandansarie et al., ultrasound is the best method of extracting phenolic compounds from rosemary compared to solvent, and supercritical fluid extraction (Rashidaie Abandansarie et al. 2019). These researchers produced nano capsules of rosemary extract, using basil seed gum, and soy protein isolate as vehicles, to increase the shelf life of beef. The capsules increased the shelf life of beef up to day 21, and increased antimicrobial properties, and antioxidant capacity.

Nano emulsions are submicron sized emulsions that serve as vehicles for bioactive compounds, whose advantages lie in targeted delivery, and protection against destructive conditions. Ultrasound is capable of increasing the bioavailability of bioactive compounds because it generates small structures that enter the intestinal mucosal layer (Abassi et al. 2019a). F. Abbasi et al., demonstrated that ultrasound produces stable flaxseed oil nano emulsions in water with the ability to protect linolenic acid against gastric digestion, so it can be implemented as a vector for the posterior parts of the chicken gastrointestinal tract *in vivo* (broiler) (Abbasi et al. 2019b). These researchers observed an improvement in the blood lipid profile, and upregulation of hepatic expression of desaturase, and elongase genes. In another study, these same researchers (Abbasi et al. 2019a) demonstrated a better feed conversion rate in birds with dietary treatments (fattening) based on nano emulsions produced with ultrasound. In this case, there was a greater incorporation of linolenic acid, and omega-3 polyunsaturated fatty acids in thigh, and breast meat. In addition, in the *in vitro* study, they observed high resistance of the nano emulsions to gastric digestion and relatively rapid release during intestinal digestion. Regarding the fortification of foods and beverages with vitamin ,Walia et al., formulated oil in water nano emulsions using ultrasound. In this case, vitamin D was encapsulated in fish oil to achieve higher oral bioavailability (Walia et al. 2017). The results showed an efficient delivery in the simulated gastrointestinal tract. The study carried out by J. O'Sullivan et al., reported that ultrasound treatment produces a reduction in the size of animal proteins, including bovine gelatin and fish gelatin, without a reduction in the molecular weight profile of the primary structure (O'Sullivan et al. 2016). Only the ultrasound treated bovine gelatin emulsions had smaller droplet sizes compared to the untreated emulsions.

The use of nanomaterials such as carbon nanotubes in the fabrication of sensors has received attention in recent yr. This is due to the ease of detection due to surface modification. Yazdanparast et al., used ultrasound for scattering purposes to fabricate a nano sensor that allows the detection of xanthine in meat (Yazdanparast et al. 2019). In this case, a multiwall carbon nanotube film and poly (l-aspartic acid) were used for the immobilization of xanthine oxidase. The results showed that the biosensor had a low detection limit (3.5 x 10^{-4} µM) and high selectivity, so it can be successfully applied to determine xanthine in fish as an index of freshness.

Much recent research has focused on the development of novel products based on the use of nanotechnology per se, without the use of ultrasound as an accompanying technology, whose main advantages are improved bioavailability, antimicrobial activity, better sensory properties, and specific delivery of bioactive compounds (Ramachandraiah et al. 2015). However, there is debate and controversy regarding the health risks in terms of accumulation and toxicity, so there are challenges in terms of public acceptance and regulation of these products. The area of greatest progress is meat packaging (intelligent packaging), however, the long-term effects that nanomaterials could have on health and the environment due to migration from the packaging to the food and to the biotic and abiotic environment. This section does not address specific studies on the use of nanotechnology in meat and its products. Basically, it highlights the production of ingredients and supply systems for supplements and nutrients for the purpose of reformulation and improvement in bioavailability (Olmedilla-Alonso et al. 2013). For example, minimization or modification of fat content, reduction of harmful salts (sodium, phosphate, and nitrates), inclusion of prebiotics, probiotics, antioxidants, and other additives such as antimicrobials. Also, here we included studies on improved packaging, active packaging, smart packaging, and edible coatings, being this a growing research area with much more articles published compared to ultrasound and nanotechnology as a combined technology.

2.4 Irradiation and Ultrasound for Meat Processing

To date, there is no wide published evidence of research combining the irradiation and HIU in meat or meat products. Nevertheless, both technologies are probed to

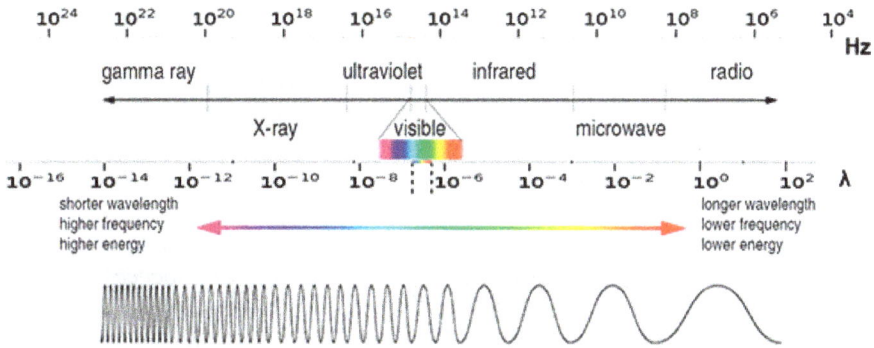

Figure 1. Electromagnetic spectrum, modified from Fan and Niemira (2020).

be useful for decontamination of bacteria, virus, and yeast. This section aimed to present the most recent evidence (2018 to 2022) of different radiation efficiency in microbial reduction of meat and meat products, and its potential combination with HIU. Radiation has been studied and experimentally applied by over a hundred years with various purposes, from disinfection, microbial inactivation, and shelf-life extension (Fan and Niemira, 2020). Nonetheless, its practical or industrial application remains highly limited. Radiation is an energy inside the electromagnetic spectrum (Figure 1). It has been classified into two types, a) nonionizing radiation, and b) ionizing radiation. Nonionizing radiation consists of radioactive energy able to only excite the matter with wavelength > 100 nm, low photon energy < 12.4 eV, and electromagnetic field from one Hz to 3×10^{15} Hz. Some of the technologies considered as nonionizing radiation are microwaves, infrared and ultraviolet (UV) light. Contrarily, ionizing radiation such as X or gamma rays is an energy able to produce charged ions by knocking electrons out of the orbits of the atoms or molecules, consequently, they produce matter ionization. It is an energy released by the atoms in the way of electromagnetic waves or particles ([WHO] 2016). There are three types of ionizing radiation that have been and are allowed to be used in food application, gamma and X rays, and electron beam (Fan and Niemira, 2020).

2.4.1 Ultraviolet Light

UV light radiation has been classified into three groups, based on the wavelength. UV-A, from 315 to 400 nm; UV-B from 280 to 315 nm; and UV-C, with less than 280 nm wavelength. UV-C has been the most used in foods due to its high activity against microorganisms. Nevertheless, UV-C and other groups have demonstrated as well to help in the preservation of the physicochemical properties and to prolong the organoleptic characteristics during shelf life (Silva et al. 2015). UV application has received a huge interest in the food industry because of its capacity to reduce microorganisms and the high safety of its application. In this sense, UV overdose is not possible, there are no residues or by products formation and the quality of most of the foods is not affected (Cutler and Zimmerman 2011). The wavelength more explored for lethal effect on microorganism is 253.7 nm, which is considered the wavelength with more potential of the photons to be absorbed by microorganism DNA (Silva et al. 2015).

In the beginning, the application of UV in foods consisted of the direct exposition of the food to the UV light. Nowadays, other technologies such as gas lamps (i.e., xenon) are also used for the application of UV light to solid and semi solid foods. Those lamps can produce UV pulses (around 100 µs) in the wavelength from 100 to 1,100 nm (Keklik et al. 2012). Pulse application of UV light has been highly efficient for pathogen and contaminants inactivation (*Listeria monocytogenes, Salmonella enterica, Campylobacter* sp. *Escherichia coli, Staphylococcus aureus, molds,* and *yeasts*) in meat and meat products such as beef, goat meat, chicken breast, chicken fillets and tights, and dry cured meat (Cassar et al. 2018, Liu et al. 2019, McLeod et al. 2018, Silva et al. 2015). Lately, McSharry et al., inoculated *Salmonella Typhimurium, Listeria monocytogenes* and *Enterobacteriaceae* from beef broth into diced beef, and later they tried different single and combined UV light wavelengths (255, 265, 285 nm) (McSharry et al. 2022). Populations of all microorganisms were reduced after two min of exposition to UV light emitting diode (UV LED) system. Further, the microbial counts were almost eliminated after 30 or 60 min of exposition to 265 and 285 wavelengths. Additional reduction of microorganism counts was observed when the wavelengths were applied in combination, except in *Salmonella typhimurium*. Less effect was observed with 255 wavelengths. UV LED probed to be an excellent decontamination technology in solid and liquid beef products.

UV light as been reported as an efficient technology to reduce viruses. S.Y. Park et al., applied UV-C (260 nm 60 to 3600 mWs/cm²) to chicken breasts seeking for hepatitis A and murine norovirus reduction (Park and Ha 2015). They achieved a reduction over one log in the titers of both viruses. Nevertheless, they also found negative changes in the physicochemical parameters of the meat. Among the changes were higher lipid oxidation, darker appearance, prominent yellow shades (reduction of lightness and increment in redness and yellowness), and compromised sensory properties in the chicken breast exposed to UV-C.

As mentioned at the beginning of this section, there is no literature yet about the combination of UV light with ultrasound, hence it seems to be a wide field for research. UV radiation inactivates microorganisms by affecting the genetic material (RNA and DNA). After UV photons are absorbed by the genetic material, they have the capacity to create cross linking among pyrimidines close to the DNA chain. Particularly wavelengths between 200 and 280 are described as the germicidal UV. This UV has shown to affect the double bond stability in molecules such as flavins, purines and pyrimides by the formation of dimers (uracils and cytosines) in RNA and DNA (thymine and cytosine) of microbes. In this way, UV forms dimers, interfering with the microbial transcription and replication (Cutler and Zimmerman 2011). However, microorganism have developed various enzymatic pathways to repair the UV damage (Goosen and Moolenaar 2008), hence, time exposition is another important parameter to consider when applying UV light for antimicrobial purposes. On the other hand, ultrasound has shown to mainly affect the bacterial membrane of microorganism by the effect of cavitation. Henceforth, the practical complementarity of the two technologies seems obvious, a technology affecting the surface of the microorganism and the other affecting the core.

UV light have been reported with reduced antimicrobial effect on solid foods (Degala et al. 2018), hence it has been combined with other technologies to boost its effect. For instance, UV-C has been combined with lemongrass oil (0.25, 0.5, and one percent) a natural antimicrobial-antioxidant essential oil to reduce *Escherichia coli* K12 in goat meat. Goat meat was inoculated with *Escherichia coli* K12 and left at 22°C for one hr for bacterial attachment. Later, inoculated meat was sprayed with lemon grass and UV-C (100, 161 and 200 $\mu W/cm^2$) was applied at two, four, six, eight, 10-, and 12 min. *Escherichia coli* was reduced up to 6.7 log CFU in goat meat when combining one percent of lemon grass with 200 μm^{-2} for two min. Most importantly, texture, color, and oxidative stability of goat meat were not affected by the combination of UV-C application and lemon grass addition. UV-C technology has been successfully combined with other compounds and technologies to reduce pathogens on meat. Y. Yeh et al., reduced up to 99 percent *Salmonella species* populations in ground beef by the combination of UV light (254 nm at 800 $\mu W/cm^2$/30 s) with macrophages (S16 and FO1a) (Yeh et al. 2018). Nevertheless, the combination of UV light with lactic and peroxyacetic acid was not successfully used to reduce the *Salmonella* populations on the meat. More recently, Shebs et al., tested the combination of different technologies (UV, macrophages, peroxyacetic acid and acidified sodium chlorite) to reduce the "big six" Shiga toxin producing *Escherichia coli* (STEC) in beef (Shebs et al. 2022). They reported that in both vacuum and aerobic conditions, the combination of UV light and macrophages application led to the highest STEC reduction in meat. UV-C has been used to reduce the fungi population of equipment used in slaughterhouses of Korea. UV-C (15, 30, 90, 150, 300, and 600 mJ/cm^2) combined with high temperatures (60, 65, 70, and 83°C) were applied to steel chips inoculated with the four most frequent fungi (*Aspergillus niger, Penicillium commune, Penicillium oxalicum, and Cladosporium cladosporioides*). UV-C doses of 150 or higher mJ/cm^2 and 60 or more °C reduced up to 6.5 log the fungi incidence. This combination of treatments is declared as highly recommended due to the low price of their application (Lee et al. 2022). In this case, ultrasonication is as well considered one of the most economical technologies for disinfection, which can potentially match with the economic benefits from the UV-C for microbial reduction in meat industry equipment.

2.4.2 Ionizing Radiation

Ionizing radiation has been mostly applied to foods aiming to reduce the microorganisms counts on them. The technology consists of high pressure controlled and direct application of the radiation to foods. Food radiation may be applied by different techniques (energies) such as gamma rays, high energy electrons (e-beam) and x rays. The radiation denatures the molecular links in the chromosomic DNA and or cytoplasm membrane of microorganisms, or cells of insects or other contaminant organisms, causing their death and consequently prolonging the shelf life of the food. This technology has as main disadvantages, the sensorial and visual aspect deterioration in particular foods. As well as face the rejection of some consumers as theyassociate the radiated food as hazardous (Sebranek 2022). Ionizing radiation is generated when an electron is removed from the atom orbit, which lead to ions formation. This radiation affects DNA by denaturalizing the molecular links in the chromosome and the cytoplasmic membrane of bacteria and other microorganisms. Radiation effectiveness for microorganisms' destruction from meat and other foods highly depends on the doses energy absorbed by the organism, which is measured in grays (Gy). Low doses of radiation for pasteurization are in the range from one to 10 kGy, while higher doses (> 40 kGy) are used for procedures such as sterilization and elimination of pathogens like *Clostridium botulinum* from foods (Khan et al. 2017).

2.4.2.1 Gamma Radiation

Gamma radiation is used in more than 26 countries worldwide, and they are the most widely applied for food radiation. Gamma ray is usually generated by using cobalt 60 or cesium 137. This energy has a high capacity of penetration (60 to 80 cm) which makes it ideal for application on bulky or big size packages. Gamma radiation doses that are considered as safe for food radiation are > 10 kGy (Huq et al. 2015), despite, lately radiations up to 30 kGy have been approved and tested in some countries (Begum et al. 2020). Gamma rays have been used in the meat industry since some decades ago. They have been tested for physicochemical, sensorial, and microbial reduction in meat and processed meat from, beef, pork, poultry, lamb, fish, crab, among others (Jayathilakan et al. 2015, Nam et al. 2016). The main advantages of gamma rays (from 7.5 to 15 kGy) in the meat industry are the elevated effectiveness of bacterial reduction (i.e., *Listeria monocytogenes*, *Salmonella*, *Staphylococcus aureus*, *Escherichia coli* O157:H7, *Campylobacter*, and *Yersinia enterocolitica*), nevertheless, the big disadvantage is the lipid oxidation caused by gamma radiation, and the promotion of saturation of fatty acids in meat (Stefanova et al. 2011). Yet, different strategies such as reduction of doses and combination with other technologies have been tested to reduce that negative effect.

In the last three years, there has been a trend in combining gamma and natural antimicrobial-antioxidants on meat such as oregano and cinnamon essential oils, to achieve more efficient ways to control pathogens (*Listeria monocytogenes*) without affection of organoleptic or physicochemical parameters (Huq et al. 2015). R. Akhter et al., combined low gamma ray doses (one kGy) with rosemary extract (0.2 g/ kg of meat), BHT, nisin, and sodium nitrate to be applied in mutton emulsions (Akhter et al. 2021). The combination of gamma ray and rosemary had similar effect than BHT in carbonyl content, thiol losses and TBARS values, suggesting enough protection by the rosemary extract against oxidation induced radiation. Besides, the combination of gamma ray and nisin reduced notably the bacterial populations in the emulsions which suggested that nisin and rosemary extract may be an excellent match combination of natural agents when gamma radiation is applied to meat emulsions, to reduce the lipid oxidation and keeping the bacterial control.

More recently, gamma ray (one kGy) was combined with microencapsulated (sodium alginate) rosemary and mint oils, to treat mutton emulsions and evaluated the effect on cholesterol oxidation, and the fatty acid profile during storing (30 d). Rosemary and mint oils helped to keep the fatty acids profiles similar among radiated and non-radiated emulsions. Nevertheless, rosemary oil promoted a higher PUFA concentration in comparison with mint oil in the emulsions. Both oils reduced the cholesterol content in the meat (Akhter et al. 2022). Gamma radiation (zero, 1.5, three, 4.5, six KGy) has been combined with different concentrations of sodium nitrite (zero, 50 and 150 mg/ kg) aiming to reduce the presence spores of *Clostridium sporogenes*, as well as to evaluate technological and sensorial characteristics in cooked ham. Gamma doses of three kGy were recommended to improve sensory characteristics and ensure the microbiological safety by spore survival reduction of ham elaborated with 50 mg/kg of sodium nitrite, without affecting other technological parameters such as, color or hydrogen potential (Silva et al. 2021).

The potential of gamma radiation combination with other technologies in food is wide, even among radiation energies to reduce the doses of radiations which may impact on the chemistry of the foods. Some examples of those combinations are presented further on this section.

2.4.2.2 X Rays

X rays are produced by a high collision of high energy electrons with a metal target. In food processing, X rays have shown to have a better penetration than electron beam, but lower than gamma rays. It can be considered indeed a new technology for bacterial and other microbial reduction, and it needs to be further researched for use in meat and meat products. Hence, X rays seems a very promising technology for investigation by single or combined application in meat (Nam et al. 2016). X rays (doses range from 0.2 to 0.8 kGy) single application has been tested for bacterial (*Salmonella typhimurium*, *Escherichia coli* O157:H7, and *Listeria monocytogenes*) reduction in inoculated surface of sliced ham. The higher post packaged doses of X ray (0.8 kGy) application resulted in remarkable reductions of the three types of bacteria (up to 7.2 log CFU/g), with no risk of bacterial recovery. This effectiveness to reduce bacterial counts in the ham were confirmed by intracellular enzyme inactivation and DNA damage of the bacteria, with no damage of the bacterial membrane. Remarkably, there was no negative effect of X ray application to color or texture of the ham (Cho and Ha 2019).

The combination of various types of radiations may resulted on physical processes or interactions between the energy and the matter. Ionizing particles in food may produce free radicals after the interaction with water molecules, which can start or potentialize the food molecules to form new radicals or other compounds (Bliznyuk et al. 2022). Hence, the combination of different radiations frequently aims and needs to reduce the doses of each energy, to have a synergy on the positive effects and reduce the negative effects on the physicochemical and sensorial characteristics of the food. X ray (0.76 kGy/hr) has lately been compared with gamma radiation (6.37 kGy/hr) in combination with cinnamon essential oils, aiming to reduce the microbiological counts and to preserve the physicochemical properties of fermented sausages. *Escherichia coli* O157:H7 and *Listeria monocytogenes* cocktails were inoculated to ground beef before the elaboration and fermentation of the sausages. After gamma or X ray irradiation and during storage (20°C for eight w), color, texture, *Escherichia coli*, *L. monocytogenes*, lactic acid bacteria, total mesophilic bacteria, molds, and yeasts were evaluated. The combinations produced high reduction of *Escherichia coli*, *Listeria monocytogenes* and other microorganisms during the storage. Besides,

X rays combinations reduced the redness and the tenderness of the sausages after treatment, in comparison with gamma radiation combinations. Nonetheless, those negative effects were reduced during the storage (Ji et al. 2022).

2.4.2.3 Electron Beam

Electron beam consists of high energy stream of electrons, which is generated by an accelerator. In this way, electrons may be accelerated up to 10 MeV, and most importantly, this can be controlled by an off-on system due to a lack of radioactive source. Electron beam has been probed to have positive effects for bacterial reduction of meat, despite its limited penetration capability. However, it is an easy to control technology in terms of speed, accuracy, and energy efficiency. Further, it is a technology more accepted by consumer, in comparison to gamma or x rays (Nam et al. 2016). Electron beam radiation has been applied to ground beef at different doses from zero to 4.5 kGy to evaluate its effect on color, chemical reactions, and taste. Electron radiation at higher doses (4.5 kGy) produced lipid and protein oxidation, and a decrease in color of the grounded meat. Further, higher drip losses of irradiated beef led to a higher saltiness in the electronic tongue. Also, off odor derived from 4.5 kGy electron radiation was not correlated with dimethyl sulfide produced in the beef. Additionally, ≤ three kGy doses of radiation did produce very limited changes to the sensory quality of the beef (Feng et al. 2019). Electron beam radiation has been tested as well in meat of duck (An et al. 2018), quail (Derakhshan et al. 2018), and other aquatic species (Lv et al. 2018). The main conclusions in those meats are, that electron radiation is an effective technology to reduce bacterial counts and potentially to improve the texture, but it produced lipid oxidation, which it is not necessarily detected in the sensorial analysis. X rays have been combined with electron beam radiation to probe this statement in chilled turkey. Very low doses of X ray (0.5 to 0.75 kGy) and electron beam (0.25 to one kGy) are enough to clearly reduce the total viable bacterial counts. Nevertheless, even slightly higher doses (1.8 ± 0.2, and 1.2 ± 0.1 Gy/s, respectively) did not have negative effects on appearance, color, taste, or smell of the turkey. The presence of alcohols and ketones derived from lipid oxidation are determined by the doses of radiation, tough. Besides, acetone formation may be a marker to determine the exposition of low-fat meat products to radiation (Bliznyuk et al. 2022).

As a conclusion of this section, different radiation technologies are more acceptable as safety processes in foods by the final consumers. This is leading to new potential uses and approaches for industrial application in foods, seeking to preserve the microbial, sensorial, and physicochemical quality of meat and meat products. There is a vast potential field of research on combination of technologies, as it has been growing in the last four years. Yet, no combination of radiation with high intensity ultrasound has been reported to date, which may lead to additional or synergic positive effects, particularly on the microbial reduction. The big challenge for researchers seems to be having the adequate equipment and premises for radiation ultrasonication technology application.

2.5 Cold Plasma and Ultrasound for Meat Processing

2.5.1 Introduction

In recent years, a technique called cold plasma was recognized as a non-thermal emerging technology, which was defined as an ionized gas generated in atmospheric or low-pressure conditions (Fridman 2009). It identified as a sterile and reproducible non thermal technology useful in the food industry for its efficacy in the preservation of organoleptic characteristics in food, useful during the processing of food products, with an effect of antimicrobial activity (Birmingham, 2004, Laroussi 2005, Deng et al. 2006). The cold plasma technique promotes structural changes and is useful in cleaning and disinfecting equipment used in food processing. It is also recognized as a low-cost method (Misra and Jo 2017) for food processing. However, installation cost, safety measures, process specific equipment requirements, and the need for training in laboratory techniques can be a disadvantage at the beginning of its implementation (Ansari et al. 2022).

In recent years, consumers have been demanding convenient and healthy foods. The emerging non thermal technologies such as high pressures, pulsed electric field, ultrasound and nanotechnology methods are gaining the attention of researchers and industry trying to meet the consumer demands. These technologies offer advantages in the preservation of food of animal origin mainly by reducing processing times, improving quality, and extending the shelf life of fresh and processed products. Additionally, the combination of innovative technologies has recently been considered as an interesting tool to offer additional benefits for the meat industry in terms of quality and safety. This section reviews the application of cold plasma technology in the meat industry and the effect of its combination with other emerging technology such as high intensity ultrasound.

2.5.2 Principle of Cold Plasma

Plasma is the fourth and final state of matter. It was produced by exposing air or any gas mixture between electrodes to generate very high electric field strength by using dielectric barrier discharge (DBD), radio frequency, and microwaves power sources. The ionized gases consist of chemically reactive species such as positive and negative ions, radicals, electrons, excited and neutral molecules, ultraviolet photons, and visible light (Lin et al. 2019, Deng et al. 2006). Depending on its thermodynamic state between ions and electrons this technology can be classified into cold and hot plasma. The cold plasma is divided into two classes based on ambient pressure: low plasma pressure and plasma at atmospheric pressure, the latter using at a radio frequency creating ionization by rapid electrical pulses at time intervals and adding gases in the system at different voltages and powers (Ucar et al. 2021). Free ions and electrons in plasma are produced by electrodes that use radio frequency (RF), microwave (MW) or dielectric barrier discharge (DBD). Atmospheric cold plasma (ACP) is an emerging technology in the food industry with a huge antimicrobial potential to improve safety and extend the shelf life of food products. Dielectric barrier discharge is a popular approach for generating ACP. DBD

ACP is a promising technology that can improve food safety with minimal impact on food quality under optimal conditions (Feizollahi et al. 2021). Cold plasma (CP) treatment is an alternative technique for the decontamination of minimally processed foods (Fernandez et al. 2013). The main effect that has been studied of the treatment with CAP is for the decontamination of food surfaces, biofilm forming bacteria (Bourke et al. 2017) and the inactivation of a wide range of Gram positive and Gram-negative microorganisms (Patil et al. 2016). The five physical mechanisms that are recognized in the inactivation of microorganisms are: (1) heat, (2) charged particles, (3) electric field, (4) UV photons and (5) oxygen reactive species (atomic oxygen, metastable oxygen molecules, ozone, and OH) (Deng et al. 2006, Cebrián et al. 2017).

2.5.3   Applications of Cold Plasma (CP) in Meat Processing

Food safety is an issue of concern to government authorities, industry, and consumers. The process of decontamination of meat and meat products arises from the fact that the composition of meat not only makes it highly perishable, but it also favours an extremely high sensitivity in terms of loss of sensory attributes, when subjected to traditional thermal processes (Misra and Jo 2017). Plasma technology is a non-thermal technique applied to food in recent years. From the application of ionized gas, a non-thermal sterilization process is generated. In this sense, the application of this technology in the processing and processing of food, plasma interacts with bacterial cells, inactivating microorganisms, spores, and viruses (Ucar et al. 2021).

The use CP has been applied in packaged foods indirectly, its application has been used in processing conveyor belt systems during spraying or misting, with the goal of preventing contamination in the packaging process (Toyokawa et al. 2017). In this process the plasma does not penetrate the food and only changes are seen on the surface. Another form of application of CP with advantages of sterilization of surfaces of industrial materials such as iron and steel, with effect of the inactivation of pathogenic microorganisms or the process to deterioration, is the effect of the application of ionized ozone gases (excited oxygen and free electrons). These types of oxygen react against organic contaminants, reaction with hydrocarbons of low molecular weight (H_2O, CO_2, and CO) resulting in a sterilization process (Ansari et al. 2022, Laroussi 2005). In studies conducted with the application of plasma, it is reported that the damage to the microbial cell through the formation of reactive species to oxygen, is generated by the production of compounds such as malondialdehyde and this in turn generates damage to the structure of DNA (Laroussi 2002).

In the production chain of meat products there are processes and methods of conservation that includes from cooling, freezing, dehydration, maturation, marinating, cooking processes (grilling), other thermal processes used in the preparation of food (e.g., smoked, fried, roasted, cured). These processes induce changes in structure, texture and sometimes increase the oxidation of lipids, causing changes in the flavour. The disadvantage of thermal processes is the alteration and so undesirable formation of polycyclic aromatic compounds, heterocyclic amines, and some nitrous compounds (Behsnilian et al. 2014). This is unsafe and difficult to adopt during the technological processes of meat products because it stands for a risk to consumer health (Hugas et al. 2002). CP is recognized as an emerging decontamination technology that preserve food and it has been extensively explored for application in raw and processed meat products. In recent years, experiments have shown that CP technology could be a profitable tool for meat safety. The effect of CP on meat products was shown below.

1. Beef patties were the reported meat products; the type of plasma or combination used was cold plasma dielectric barrier discharge; with these processing conditions 50, 60, or 70 kV frequencies of 50 Hz voltage or 60 kV voltage with frequencies of 50, 70, 90, or 110 Hz for 120 s; after treatment, samples were storage at four °C, and produce a change colour value, and metmyoglobin activity and increased TBARS values ($p < 0.05$) (Wang et al. 2021).

2. Fish muscle *(Theragra chalcogramma)* was the reported meat product; the type of plasma or combination used was dielectric barrier discharge, and atmospheric cold plasma; with these processing conditions samples being treated at zero, 20, 30, 40, 50, and 60 kV for 10 min, and producing a greater change in water retention capacity (WHC), textural properties, colour properties ($p < 0.05$) and thermal stability ($p < 0.05$) (Miao et al. 2020).

3. Chicken breast was the reported meat product; the type of plasma or combination used was dielectric barrier discharge, and atmospheric cold plasma; with these processing conditions (100 kV; 233± five W power; zero, one, three, and five min of duration) produced a minimal effect on the colour, hydrogen potential, and WHC of poultry meat. Shelf life extended by six d at four °C (Moutiq et al. 2020).

4. Pacific white shrimp was the reported meat product; the type of plasma or combination used was plasma jet in atmospheric cold plasma, and the processing conditions of 220 V, 282 W, 50 Hz, and 1.9 A; 45, 90, and 150 s produced a lower rate of increase in hydrogen potential, TBARS, free fatty acid, peroxide value, and fluorescent compounds ($p < 0.05$), did not produce undesirable effects on sensory quality, for synthesis during 90 s was the most efficient, and polyphenol oxidase was inhibited by 50 percent during 150 s exposure (Zouelm et al. 2019).

5. Salmon was the reported meat product; the type of plasma or combination used was pulsed discharge plasma, with these processing conditions 16 KHz, 11 kV, one min, no effect to inactivation *Brochothrix thermosphacta, Pseudomonas fragi, Psychrobacter glacincola* was produced (Zhang et al. 2019).

6. Chicken skin was the reported meat product; the type of plasma or combination used was cold plasma jet; with the processing conditions (30 to 180 s, one MHz, two to three kV, and two min) there was no effect to inactivation microorganism of surface *(Campylobacter jejuni)* (Rossow et al. 2018).

7. Egg shells were the reported meat products; the type of plasma or combination used was high voltage atmospheric cold plasma, with these processing conditions: 85 kV, 60 Hz, five to 15 min, and there was no effect of inactivation

microorganism of surface produced (*Salmonella enteritidis, Salmonella enteric serovar Typhimurium*) (Wan et al. 2017).

8. Beef jerky was the reported meat product; the type of plasma or combination used was flexible thin layer plasma system, atmospheric cold plasma; with these processing conditions: 15 kHz for zero (control), 2.5, five, and 10 min treatments, there was a produce of peroxide content, and colour parameters changed. Flavour, off odour, and overall acceptance were negatively affected after 10 min treatment (Yong et al. 2017).

9. Lamb meat was the reported meat product; the type of plasma or combination used was dielectric barrier discharge; with these processing conditions: 80 kV, 50 Hz, five min, there was no effect inactivation *Brochothrix thermosphact produced* (Patange et al. 2017).

10. Egg shells were the product meat reported; the type of plasma or combination used was cold nitrogen plasma; with the processing conditions used (400 W, and 20 min), produced a change inactivation of microorganisms (*Salmonella enteritidis*, and *Salmonella typhimurium)* and shelf life increased up to 14 d (Cui et al. 2016).

11. Beef jerky was the reported meat product; the type of plasma or combination used was radio frequency plasma; with these processing conditions: 200 W, zero to 10 min, there was no effects inactivation on S*taphylococcus aureus* produced (Kim et al. 2014).

12. Pork loin was the reported meat product; the type of plasma or combination used was dielectric barrier discharge, application of DBD on pork loin at 30 KHz for 5 and 10 min resulted in changes in meat pH and inactivation of Listeria monocytogenes (Kim et al. 2013).

13. Fresh pork was the reported meat product; the type of plasma or combination used was microwave powered plasma, with these processing conditions 2.45 GHz, 1.2 kW for 3 min, produced inactivation of *Escherichia coli* O157:H7, and changes in colour meat (Fröhling et al. 2012).

14. Bacon was the reported meat product; the type of plasma or combination used was dielectric barrier discharge, with these processing conditions 13.56 MHz, 125 W, 1.5 min, and there was inactivation Listeria monocytogenes*, Escherichia coli, Salmonella typhimurium*, and prolonged shelf life (Kim et al. 2011).

15. Italian salami was the reported meat product; the type of plasma or combination used was combined, ultrasound assisted extraction (*Sea asparagus*) and discharge plasma; with these processing conditions the extraction was performed for five, and 60 min, in zero to 15 kV, 60 Hz, 2×10^{-2} mbar atmospheric air, 14 W, and did not produce further lipid oxidation or texture deterioration, there was no a negative impact on sensory quality, and inhibition against *Escherichia coli* (Faria et al. 2020).

16. Chicken muscle, rough skin, and smooth skin were the meat reported products; the type of plasma or combination used was combined ultrasound for quality improvement *Escherichia coli*, and *Staphylococcus aureus*, activated water plasma discharge were the procedure conditions; the samples were ultrasonicated at 4, 25, and 40°C for 30, 45, and 60 min (40 Hz, and 220 W), the PAW discharge using Ar gas at a flow rate of three slm for 6.5 min, producing a combination inactivated *Escherichia coli*, and S. aureus at 40°C for 60 min, values of hardness, protein, and lipid were maintained (Royintarat et al. 2020).

Recent studies have suggested that decontamination with nonthermal plasma could be a viable alternative to the traditional methods for the decontamination of heat sensitive materials or food because this technique proves capable of eliminating micro-organisms on surfaces without altering the substrate (Pignata et al. 2017). Chemically reactive species are produced by CP and react with lipopolysaccharide and peptidoglycan present in the bacterial cells causing injury or death (Pignata et al. 2017). In meat, CP has been applied to pork (Kim et al. 2013, Fröhling et al. 2012), chicken (Moutiq et al. 2020), beef jerky (Yong et al. 2017) and in the processing of aquatic products (Miao et al. 2020, Zouelm et al. 2019).

For a decade CP has been studied as a decontamination of animal products. In eggs shells and chicken meat *Salmonella enteritis* is a microorganism commonly found in the surfaces. Cold plasma studies using a Dielectric Barrier Discharge (DBD), jet device at a frequency of two at 10 Hz and a voltage supply of approximately 30 kV, these conditions were found to be sufficient to kill microorganisms such as *Salmonella enteritidis* (Dirks et al. 2012, Lee H.J. et al. 2011). These results differ from the studies conducted by Z. Wan et al., in eggshells under High Voltage Atmospheric Cold Plasma (HVACP) conditions at 85kV, 60 Hz for five and 15 min, reported no effect on the presence of microorganisms such as *Salmonella enteritidis* and *Salmonella enterica serovar Typhimurium* (Wan et al. 2017). In another investigation conducted by Cui et al., applying cold nitrogen plasma at 400 W, for 20 min, in eggshells, a decontamination of the genera of *Salmonella enteritidis* and *Salmonella typhimurium* was observed, and an effect on the increase of storage time (Cui et al. 2016). Similarly, S. Chaplot et al., reported that atmospheric CP reduced 5.3 log CFU/cm^2 *Salmonella typhimurium* in poultry meat (Chaplot et al. 2019). In a more recent study conducted by R. Moutiq et al., applied atmospheric cold plasma with DBD to chicken breast meat (100 kV, 233 W for zero, three, and five min) and reported minimal changes in physical chemical characteristics of meat (Moutiq et al. 2020). They also observed an extension in shelf life of six d at four °C. According to the results presented in eggshells and chicken meat, it is evident that to efficiently decontaminate with the application of cold plasma, it is essential to consider factors such as exposure times, voltage, adequate exposure on the surfaces to be applied, uniformity in treatments (Yong et al. 2014) and the most relevant the plasma source specifications (Misra and Jo 2017). Stratakos and I.R. Grant reported 0.9 and 1.82 log CFU/cm^2 reduction in *Escherichia coli* levels using CP jet (six kV, 20 kHz, 99.5 helium percent and 0.5 oxygen percent) after two and five min, respectively in raw beef (Stratakos and Grant 2019). They also observed that the sub lethally injured cells were unable to recover themselves and eventually died during unfavorable cold storage conditions (four °C). Similarly, Choi et al., found 1.5 log reduction of *Escherichia coli* O157: H7 and more than one log reduction in *Listeria monocytogenes* of pork after CP jet treatments (20 kV for zero to 120 s) (Choi et al. 2017).

Atmospheric Cold Plasma (ACP) has also attracted attention to be used in the processing of aquatic products. Some studies have reported that ACP enhances myosin gelation influenced by reactive oxygen species (ROS). In addition to inactivating endogenous enzymes, improving protein structure, and reducing conformational changes (Nyaisaba et al. 2019). In the study conducted by W. Miao et al., the effect of ACP on the physicochemical and functional properties of myofibrillar proteins was evaluated. They applied different voltage conditions (10, 20, 30, 40, 50 and 60 kV) for 10 min per voltage (Miao et al. 2020). Reporting significant changes in water holding capacity (WHC), texture and colour and thermal stability ($p < 0.05$). These results suggest that atmospheric cold plasma (ACP) treatment could be used as an alternative approach to improve the gelling properties of *Alaska pollock* myofibrillar protein. In the study by Wang et al., it is interesting to note that the impact and frequency of DBP-ACP treatments had an impact on redness reduction (a* value), metmyoglobin accumulation, and increased lipid oxidation reactions in beef patties compared to control samples (Wang et al. 2021). The colour and flavour of the meat are two of the most important attributes as they directly influence consumer acceptance. Colour influences the appearance of meat and consumers' purchasing decisions. Therefore, it is important to generate more studies to minimize the adverse effects of DBD-CP processing in raw meat products, based on the optimal voltages, frequencies, and times in the application of DBD-CP to maintain the colour of fresh meat and extend the shelf life. Rossow et al., applied CP treatments (at one MHz and two to three kV for 30 to 180 s) to chicken skin and breast fillets and reduced *Campylobacter jejuni* from 0.78 to 2.55 log CFU/cm^2 (using argon as feed gas) and 0.65 to 1.42 log CFU/cm^2 using air (Rossow et al. 2018). They reported that ionized argon has a high electron density, thus generating more antimicrobial reactive species. Additionally, increased exposure time significantly led to a higher reduction of *Campylobacter jejuni*. However, CP raised the temperature on the surface of treated samples which could be another reason for higher decontamination. There is few evidence of how the combination of ultrasound and cold plasma can bring advantages in the processing of meat products. V. Gök et al. applied CP treatments (for 180 and 300 s) to reduce *Staphylococcus aureus* (from 5.78 to 0.85 log CFU/cm^2), *Listeria monocytogenes* (5.71 to 0.83 log CFU/cm^2), total aerobic mesophilic bacteria (1.41 log CFU/cm^2) and yeast mold counts (1.66 log CFU/cm^2) of pastirma (dry cured beef product) samples (Gök et al. 2019). A recent study was found that seeks to improve the nutritional value of processed foods with the addition of bioactive and natural phenolic compounds (Faria et al. 2020). Looking for the optimal conditions, as pointed out in the studies carried out by X. Wang et al., (Wang et al. 2021). Cold plasma technology as pre-treatment and ultrasound assisted extraction to maximize the recovery of phenolic compounds is an alternative to use natural additives in processed meat products as an alternative to synthetic additives (Faria et al. 2020). These authors investigated the cold plasma pre-treatment (CPT) on the ultrasound assisted extraction (UAE) of phenolic compounds from sea asparagus *Salicornia neei*. The extract was applied in dry fermented sausage (Italian salami). They found that CPT (discharge power of 14 W for five min) prior to UAE increased the antioxidant activity by 22 and 19 percent measured by the DPPH and ABTS assays, respectively. They observed that despite a slight color change, lipid oxidation and texture parameters were similar to the control at the end of ripening with no impact on its sensory overall acceptability, indicating its potential as a natural alternative to synthetic additives (Faria et al. 2020).

Combined technologies have also been studied as antimicrobial agents. The study of Royintarat et al., shows the synergistic interaction of combined plasma activated water (PAW) and ultrasound to enhance microbial inactivation in meat (Royintarat et al. 2020). They studied the efficacy of ultrasound and PAW on the inactivation of *Escherichia coli* and *Staphylococcus aureus* in chicken muscle, rough skin, and smooth skin. Samples inoculated with bacteria suspension were treated by ultrasound alone and PAW–ultrasound. Combined ultrasound and PAW inactivated up to 1.33 log CFU/ml of *Escherichia coli* K12 and 0.83 log CFU/ml of Staphylococcus aureus at a sample thickness of four mm, at 40°C for 60 min, while PAW alone only reduced *Escherichia coli* K12 by 0.46 log CFU/ml and *Staphylococcus aureus* by 0.33 log CFU/ml under the same condition. The muscle showed a porous structure, which facilitated the penetration of PAW. The color measurements of muscle treated with ultrasound and PAW–ultrasound was dramatically different from the untreated sample, as also perceived by the sensory evaluation panel.

References

Abbasi, F., F. Samadi, S.M. Jafari, S. Ramezanpour and M. Shams-Shargh. 2019a. Production of omega-3 fatty acid-enriched broiler chicken meat by the application of nanoencapsultsed flaxseed oil prepared via ultrasonication. Journal of Functional Foods 57: 373–381. https://doi.org/10.1016/j. jff.2019.04.030.

Abbasi, F., F. Samadi, S.M. Jafari, S. Ramezanpour and M. Shams Shargh. 2019b. Ultrasound-assisted preparation of flaxseed oil nanoemulsions coated with alginate-whey protein for targeted delivery of omega-3 fatty acids into the lower sections of gastrointestinal tract to enrich broiler meat. Ultrasonics Sonochemistry 50: 208–217. https://doi.org/ 10.1016/j.ultsonch.2018.09.014.

Aber, S., R.T. Khajeh and A. Khataee. 2019. Application of immobilized ZnO nanoparticles for the photocatalytic regeneration of ultrasound pretreated-granular activated carbon. Ultrasonics Sonochemistry 58: 104685–104695. https://doi.org/10.1016/j.ultsonch.2019.104685.

Aguilar, C., J. Serna-Jimenez, E. Benitez, V. Valencia, O. Ochoa and L.I. Sotelo. 2021. Influence of high-power ultrasound on natural microflora, pathogen, and lactic acid bacteria in a raw meat emulsion. Ultrasonics Sonochemistry 72(2021): 10541–10548. https://doi.org/10.1016/j. ultsonch.2020.105415.

Akhter, R., F.A. Masoodi, T.A. Wani, S.A. Rather and P.R. Hussain. 2021. Synergistic effect of low dose γ-irradiation, natural antimicrobial, and antioxidant agents on quality of meat emulsions. Radiation Physics and Chemistry 189: 109724–109730. https://doi.org/10.1016/J. RADPHYSCHEM.2021.109724.

Akhter, R., Masoodi, F.A., Wani, T.A. and S.A. Rather. 2022. Impact of microencapsulated natural antioxidants on the lipid profile and cholesterol oxidation of γ-irradiated meat emulsions. LWT. Foord Science and Technology 159: 113155–113160. https://doi.org/10.1016/j.lwt.2022.113155.

Alahakoon, A.U., Faridnia, F., Bremer, P.J., Silcock, P. and I. Oey. 2017. Pulsed electric fields effects on meat tissue quality and functionality. pp. 1–22. *In:* Miklavcic, D. [ed.] Handbook of Electroporation. Vol. 4. Springer International Publishing AG 2016 https://doi.org/10.1007/978-3-319-32886-7_179.

Alahakoon, A.U., I. Oey, P. Bremer and P. Silcock. 2018. Process optimization of pulsed electric fields pre-treatment to reduce the sous vide processing time of beef briskets. International Journal of Food Science and Technology. 54(3): 823–834. https://doi.org/10.1111/ijfs.14002.

Alarcon-Rojo, A.D., H. Janacua, J.C. Rodriguez, L. Paniwnyk and T.J. Mason. 2015. Power ultrasound in meat processing. Meat Science 107: 86–93. https://doi.org/10.1016/j.meatsci.2015.04.015.

Alarcon-Rojo, A.D., L.M. Carrillo-Lopez, R. Reyes-Villagrana, M. Huerta-Jiménez and I.A. Garcia-Galicia. 2019. Ultrasound and meat quality: a review. Ultrasonics Sonochemistry 55: 369–382. https://doi.org/10.1016/j.ultsonch.2018.09.016.

Al-Hilphy, A.R., A.B. Al-Temimi, H.H.M. Rubaiy, U. An, G. Delgado-Pando and N. Lakhssassi. 2020. Ultrasound applications in poultry meat processing: a systematic review. Journal of Food Science 85: 1386–1396. https://doi.org/10.1111/1750-3841.15135.

Ali Dheyab, M., A.A. Aziz and M.S. Jameel. 2021. Recent advances in inorganic nanomaterials synthesis using sonochemistry: a comprehensive review on iron oxide, gold and iron oxide coated gold nanoparticles. Molecules 26(9): 2453–2471. https://doi.org/10.3390/molecules26092453.

Álvarez, I., J. Raso, A. Palop and F.J. Sala. 2000. Influence of different factors on the inactivation of Salmonella senftenberg by pulsed electric fields. International Journal of Food Microbiology 55(1–3): 143–146. https://doi.org/10.1016/S0168-1605(00) 00173-2.

An, K.A., Y. Jo, K. Akram, S.C. Suh and J.H. Kwon. 2018. Assessment of microbial contaminations in commercial frozen duck meats and the application of electron beam irradiation to improve their hygienic quality. Journal of the Science of Food and Agriculture 98(14): 5444–5449. https://doi.org/10.1002/JSFA.9088.

Ananta, E., Heinz, V., O. Schlüter and D. Knorr. 2001. Kinetic studies on high-pressure inactivation of bacillus stearothermophilus spores suspended in food matrices. Innovative Food Science and Emerging Technologies 2(4): 261–272. https://doi.org/10.1016/S1466-8564(01)00046-7.

Ansari, A., K. Parmar and M. Shah. 2022. A comprehensive study on decontamination of food-borne microorganisms by cold plasma. Food Chemistry: Molecular Science 4: 100098–100109. https://doi.org/10.1016/j.fochms.2022.100098.

Arroyo, C., D. Lascorz, L. O'Dowd, F. Noci, J. Arimi and J.G. Lyng. 2015. Effect of pulsed electric field treatment at various stages during conditioning on quality attributes of beef longissimus thoracis et lumborum muscle. Meat Science 99: 52–59. https://doi.org/10.1016/j.meatsci.2014.08.004.

Asadi, A., F. Pourfattah, I. Miklós Szilágyi, M. Afrand, G. Żyła, H. Seon Ahn et al. 2019. Effect of sonication characteristics on stability, thermophysical properties, and heat transfer of nanofluids: acomprehensive review. Ultrasonics Sonochemistry 58: 104701–104716. https://doi.org/10.1016/j.ultsonch.2019.104701.

Asfaram, A., M. Ghaedi, H. Abidi, H. Javadian, M. Zoladl and F. Sadeghfar. 2018. Synthesis of Fe_3O_4@CuS@Ni_2P-CNTs magnetic nanocomposite for sonochemical-assisted sorption and pre-concentration of trace allura red from aqueous samples prior to HPLC-UV detection: CCD-RSM design. Ultrasonics Sonochemistry 44: 240–250. https://doi.org/10.1016/j.ultsonch.2018.02.011.

Ashokkumar, M. and T.J. Mason. 2007. Sonochemistry. pp. 353–372. *In*: John Wiley and Sons [eds.]. Kirk-Othmer Encyclopedia of Chemical Technology. https://doi.org/10.1002/0471238961.191514 1519211912.a01

Azlin-Hasim, S., M.C. Cruz-Romero, M.A. Morris, E. Cummins and J.P. Kerry. 2015. Effects of a combination of antimicrobial silver low density polyethylene nanocomposite films and modified atmosphere packaging on the shelf life of chicken breast fillets. Food Packaging and Shelf Life 4: 26–35. https://doi.org/10.1016/j.fpsl.2015.03.003.

Bajovic, B., Bolumar, T. and V. Heinz. 2012. Quality considerations with high pressure processing of fresh and value added meat products. Meat Science 92(3): 280–289. DOI: 10.1016/j.meatsci.2012.04.024.

Barba, F.J., M. Koubaa, L. do Prado-Silva, V. Orlien and A.D.S. Sant'Ana. 2017. Mild processing applied to the inactivation of the main foodborne bacterial pathogens: a review. Trends in Food Science and Technology 66: 20–35. https://doi.org/10.1016/j. tifs.2017.05.011.

Barbosa-Cánovas, G.V., M.D. Pierson, Q.H. Zhang and D.W. Schaffner. 2000. Pulsed electric fields, journal of food science [Suppl] 2000: 65–81.

Baxter, S., S. Zivanovic and J. Weiss. 2005. Molecular weight and degree of acetylation of high-intensity ultrasonicated chitosan. Food Hydrocolloids 19(5): 821–830. https://doi.org/10.1016/j.foodhyd.2004.11.002.

Bayrami, A., S. Alioghli, S. Rahim Pouran, A. Habibi-Yangjeh, A. Khataee and S. Ramesh. 2019. A facile ultrasonic-aided biosynthesis of ZnO nanoparticles using *Vaccinium arctostaphylos* L. leaf extract and its antidiabetic, antibacterial, and oxidative activity evaluation. Ultrasonics Sonochemistry 55: 57–66. https://doi.org/10.1016/j.ultsonch.2019.03.010.

Begum, T., P.A. Follett, F. Hossain, L. Christopher, S. Salmieri and M. Lacroix. 2020. Microbicidal effectiveness of irradiation from gamma and X-ray sources at different dose rates against the foodborne illness pathogens *Escherichia coli, Salmonella typhimurium* and *Listeria monocytogenes* in rice. LWT 132. https://doi.org/10.1016/j.lwt.2020.109841.

Behsnilian, D., P. Butz, R. Greiner and R. Lautenschlaeger. 2014. Process-induced undesirable compounds: chances of non-thermal approaches. Meat Science 98: 392–403. https://doi.org/10.1016/j.meatsci.2014.06.038.

Bekhit, A.D., R. Van de Ven, V. Suwandy, F. Fahri and D. Hopkins. 2014. Effect of pulsed electric field treatment on cold-boned muscles of different potential tenderness. Food and Bioprocess Technology 7(11): 3136–3146. DOI: 10.1007/s11947-014-1324-8.

Bekhit, A.D., V. Suwandy, A. Carne, R. van de Ven and D.L. Hopkins. 2016. Effect of repeated pulsed electric field treatment on the quality of hot-boned beef loins and topsides. Meat Science 111: 139–146. DOI: 10.1016/j.meatsci.2015.09.001.

Bhat, Z.F., J.D. Morton, S.L. Mason and A.D. Bekhit. 2019. Pulsed electric field operates enzymatically by causing early activation of calpains in beef during ageing. Meat Science 153: 144–151. https://doi.org/10.1016/j.meatsci.2019.03.018.

Bhat, Z.F., J.D. Morton, S.L. Mason and A.D. Bekhit. 2020. The application of pulsed electric field as a sodium reducing strategy for meat products. Food Chemistry 306: 125622. DOI: 10.1016/j.foodchem.2019.125622.

Birmingham, G. 2004. Mechanisms of bacterial spore deactivation using ambient pressure nonthermal discharges. IEEE Transactions on Plasma Science 32(4): 1526–1531. DOI: 10.1109/TPS.2004.832609.

Bliznyuk, U., V. Avdyukhina, P. Borshchegovskaya, T. Bolotnik, V. Ipatova, Z. Nikitina et al. 2022. Effect of electron and X-ray irradiation on microbiological and chemical parameters of chilled turkey. Scientific Reports, 12(1): 750–759. https://doi.org/10.1038/s41598-021-04733-3.

Bourke, P., D. Zuizina, D. Boehm, P.J. Cullen and K. Keener. 2017. Review: the potential of cold plasma for safe and sustainable food production. Article in Press, Trends in Biotechnology 1580: 12. https://doi.org/10.1016/j.tibtech.2017.11.001.

Cao, Y., T. Xia, G. Zhou and X. Xu. 2012. The mechanism of high pressure-induced gels of rabbit myosin. Innovative Food Science and Emerging Technologies 16: 41–46. DOI: 10.1016/j.ifset.2012.04.005.

Cap, M., P.F. Paredes, D. Fernández, M. Mozgovoj, S.R. Vaudagna, and A. Rodriguez. 2020. Effect of high hydrostatic pressure on *salmonella* spp inactivation and meat-quality of frozen chicken breast. LWT Food Science and Technology 118: 108873–108877. DOI: https://doi.org/10.1016/j.lwt.2019.108873.

Caraveo-Suarez, R.O., I.A. Garcia-Galicia, E. Santellano-Estrada, L.M. Carrillo-Lopez, M. Huerta-Jimenez, S. Morales-Rodriguez et al. 2022. Ultrasound as a potential technology to improve the quality of meat produced from a mexican autochthonous bovine breed. Sustainability 14: 3886–3899. https://doi.org/10.3390/ su14073886.

Cárcel, J.A., J. Benedito, J. Bon and A. Mulet. 2007. High intensity ultrasound effects on meat brining. Meat Science 76(4): 611–619. DOI: 10.1016/j.meatsci.2007.01.022.

Carrillo-Lopez, L.M., A.D. Alarcon-Rojo, L. Luna-Rodriguez and R. Reyes-Villagrana. 2017. Modification of food systems by ultrasound. journal of food quality 2017: 5794931. DOI: 10.1155/2017/5794931.

Carrillo-Lopez, L.M., D. Robledo, V. Martínez, M. Huerta-Jimenez, M. Titulaer, A.D. Alarcon-Rojo et al. 2021a. Post-mortem ultrasound and freezing of rabbit meat: Effects on the physicochemical quality and weight loss. Ultrasonics Sonochemistry 79: 105766-105776. https://doi.org/10.1016/j.ultsonch.2021.105766.

Carrillo-Lopez, L.M., I.A. Garcia-Galicia, J.M. Tirado-Gallegos, R. Sanchez-Vega, M. Huerta-Jimenez, M. Ashokkumar et al. 2021b. Recent advances in the application of ultrasound in dairy products: effect on functional, physical, chemical, microbiological and sensory properties. Ultrasonics Sonochemistry 73: 105467–105501. https://doi.org/10.1016/j.ultsonch.2021.105467.

Carullo, D., G.V. Barbosa-Cánovas and G. Ferrari. 2021. Changes of structural and techno-functional properties of high hydrostatic pressure (HHP) treated whey protein isolate over refrigerated storage. LWT-Food Science and Technology 137: 110436–110448. https://doi.org/10.1016/j.lwt.2020.110436.

Cascaes Teles, A.S., D.W. Hidalgo Chávez, M.A. Zarur Coelho, A. Rosenthal, L.M. Fortes, R. Gottschalk et al. 2021. Combination of enzyme-assisted extraction and high hydrostatic pressure for phenolic compounds recovery from grape pomace. Journal of Food Engineering 288: 110128–11035. https://doi.org/10.1016/j.jfoodeng.2020.110128.

Cassar, J.R., E.W Mills, J. Campbell, A. Demirci, J.R. Cassar, E.W. Mills et al. 2018. Pulsed UV light as a microbial reduction intervention for boneless/skinless chicken thigh meat. Meat and Muscle Biology 2(2): 142–142. https://doi.org/10.22175l/RMC2018.126.

Cava, R., J. García-Parra and L. Ladero. 2020. Effect of high hydrostatic pressure processing and storage temperature on food safety, microbial counts, colour and oxidative changes of a traditional dry-cured sausage. LWT-Food Science and Technology 128: 109462–109469. DOI:10.1016/j.lwt.2020.109462.

Cebrián, G., S. Condón and P. Mañas. 2017. Physiology of the inactivation of vegetative bacteria by thermal treatments: mode of action, influence of environmental factors and inactivation kinetics. Foods 6: 107–127. https://doi.org/10.3390/foods6120107.

Chang, H.J., X.L. Xu, G.-H. Zhou, C.B. Li and M. Huang. 2012. Effects of characteristics changes of collagen on meat physicochemical properties of beef semitendinosus muscle during ultrasonic processing. Food and Bioprocess Technology 5(1): 285–297. DOI: 10.1007/s11947-009-0269-9.

Chang, H., Q. Wang, C. Tang and G. Zhou. 2015. Effects of ultrasound treatment on connective tissue collagen and meat quality of beef semitendinosus muscle. Journal of Food Quality 38(4): 256–267. DOI: 10.1111/jfq.12141.

Chaplot, S., B. Yadav, B. Jeon and M.S. Roopesh. 2019. Atmospheric cold plasma and peracetic acid-based hurdle intervention to reduce salmonella on raw poultry meat. Journal of Food Protection 82: 878–888. https://doi.org/10.4315/0362-028X.JFP-18-377.

Chauhan, O.P. and L.E. Unni. 2015. Pulsed electric field (PEF) processing of foods and its combination with electron beam processing. pp. 157–184. *In:* Pillai, S. D. and S. Shayanfar (eds.). Electron beam pasteurization and complementary food processing technologies. (1st ed.). Woodhead Publishing. https://doi.org/10.1533/9781782421085.2. 157.

Chemat, F., Zill-E-Huma and M.K. Khan. 2011. Applications of ultrasound in food technology: processing, preservation and extraction, ultrasonics sonochemistry 18: 813–835. https://doi.org/10.1016/j.ultsonch.2010.11.023.

Cho, G.L. and J.W. Ha. 2019. Application of X-ray for inactivation of foodborne pathogens in ready-to-eat sliced ham and mechanism of the bactericidal action. Food Control 96: 343–350. https://doi.org/10.1016/J.FOODCONT.2018.09.034.

Choi, S., P. Puligundla and C. Mok. 2017. Effect of corona discharge plasma on microbial decontamination of dried squid shres including physico-chemical and sensory evaluation. LWT-Food Science and Technology 75: 323–328. http://dx.doi.org/10.1016/j.lwt.2016.08.063.

Contreras-Lopez, G., A. Carnero-Hernandez, M. Huerta-Jimenez, A.D. Alarcon-Rojo, I. Garcia-Galicia and L.M. Carrillo-López. 2020. High-intensity ultrasound applied on cured pork: Sensory and physicochemical characteristics. Food Science and Nutrition 8: 786–795. DOI: 10.1002/fsn3.1321.

Crehan, C.M., D.J. Troy and D.J. Buckley. 2000. Effects of salt level and high hydrostatic pressure processing on frankfurters formulated with 1.5 and 2.5% salt. Meat Science 55(1): 123–130. DOI: 10.1016/s0309-1740(99)00134-5.

Cropotova, J., S. Tappi, J. Genovese, P. Rocculi, M. Dalla Rosa and T. Rustad. 2021. The combined effect of pulsed electric field treatment and brine salting on changes in the oxidative stability of lipids and proteins and color characteristics of sea bass (*Dicentrarchus labrax*). Heliyon 7: e05947-e05954. https://doi.org/10.1016/j.heliyon.2021.e05947.

Cui, H., C. Ma, C. Li and L. Lin. 2016. Enhancing the antibacterial activity of thyme oil against salmonella on eggshell by plasma-assisted process. Food Control 70: 183–190. https://doi.org/10.1016/j.foodcont.2016.05.056.

Cutler, T.D. and J.J. Zimmerman. 2011. Ultraviolet irradiation and the mechanisms underlying its inactivation of infectious agents. Animal Health Research Reviews/Conference of Research Workers in Animal Diseases 12(1): 15–23. https://doi.org/10.1017/S1466252311000016.

Das, D., S.P. Panesar, C.S. Saini and J.F. Kennedy. 2022. Improvement in properties of edible film through non-thermal treatments and nanocomposite materials: a review. Food Packaging and Shelf Life 32: 100843. https://doi.org/10.1016/j.fpsl.2022.100843.

Daher, D., S. Le Gourrierec and C. Pérez-Lamela. 2017. Effect of high pressure processing on the microbial inactivation in fruit preparations and other vegetable based beverages. Agriculture 7(9): 72. DOI: 10.3390/agriculture7090072.

Degala, H.L., A.K. Mahapatra, A. Demirci and G. Kannan. 2018. Evaluation of non-thermal hurdle technology for ultraviolet-light to inactivate *Escherichia coli* K12 on goat meat surfaces. Food Control, 90: 113–120. https://doi.org/10.1016/J.FOODCONT.2018.02.042.

Deng, X., J. Shi and M.G. Kong. 2006. Physical mechanisms of inactivation of bacillus subtilis spores using cold atmospheric plasmas. IEEE Transactions on Plasma Science 34 (4): 1310–1316. DOI: 10.1109/TPS.2006.877739.

Derakhshan, Z., G. Oliveri Conti, A. Heydari, M.S. Hosseini, F.A. Mohajeri, H. Gheisari et al. 2018. Survey on the effects of electron beam irradiation on chemical quality and sensory properties on quail meat. Food and Chemical Toxicology 112: 416–420. https://doi.org/10.1016/J.FCT.2017.12.015.

Dirks, B.P., D. Dobrynin, G. Fridman, Y. Mukhin, A. Fridman and J.J. Quinlan. 2012. Treatment of raw poultry with nonthermal dielectric barrier discharge plasma to reduce campylobacter jejuni and *salmonella enterica*. Journal of Food Protection 75: 22–28. DOI: 10.4315/0362-028X.JFP-11-153.

Dolatowski, Z.J., J. Stadnik and D. Stasiak. 2007. Applications of ultrasound in food technology. ACTA Scientiarum Polonorum, 63(6): 89–99. ISSN:1644-0730, e-ISSN:1898-9594.

Dong, M., Y. Xu, Y. Zhang, M. Han, P. Wang, X. Xu et al. 2020. Physicochemical and structural properties of myofibrillar proteins isolated from pale, soft, exudative (PSE)-like chicken breast meat: Effects of pulsed electric field (PEF). Innovative Food Science and Emerging Technologies 59: 102277. https://doi.org/10.1016/j.ifset.2019.102277.

Dong, M., H. Tian, Y. Xu, M. Han and X. Xu. 2021. Effects of pulsed electric fields on the conformation and gelation properties of myofibrillar proteins isolated from pale, soft, exudative (PSE)-like chicken breast meat: A molecular dynamics study. Food Chemistry 34: 128306–128316. https://doi.org/10.1016/j.foodchem.2020.128306.

Esmaeili, M., P. Ariaii, L. Rouzbeh and M.Y. Pour. 2021. Comparison of coating and nano-coating of chitosan- Lepidium sativum seed gum composites on quality and shelf life of beef. Journal of Food Measurement and Characterization 15(1): 1–12. https://doi.org/10.1007/s11694-020-00643-6.

Fan, X. and B.A. Niemira. 2020. Gamma ray, electron beam, and X-ray irradiation. *In*: Demirci, A., H. Feng and K. Krishnamurthy [eds.]. Food Safety Engineering (pp. 471–492). Springer International Publishing. https://doi.org/10.1007/978-3-030-42660-6_18.

Faria, G.Y.Y., M.M. Souza, J.R.M. Oliveira, C.S.B. Costa, M.P. Collares and C. Prentice. 2020. Effect of ultrasound-assisted cold plasma pre-treatment to obtain sea asparagus extract and its application in Italian salami. Food Research International 137: 109435–109443. https://doi.org/10.1016/j.foodres.2020.109435.

Faridnia, F., A.D.A. Bekhit, B. Niven and I. Oey. 2014. Impact of pulsed electric fields and post-mortem vacuum ageing on beef longissimus thoracis muscles. International Journal of Food Science and Technology, 49(11): 2339–2347. DOI: 10.1111/ijfs.12532.

Faridnia, F., Q.L. Ma, P.J. Bremer, D.J Burritt, N. Hamid and I. Oey. 2015. Effect of freezing as pre-treatment prior to pulsed electric field processing on quality traits of beef muscles. Innovative Food Science and Emerging Technologies 29: 31–40. https://doi.org/10.1016/j.ifset.2014.09.007.

Faridnia, F., P. Bremer, D.J. Burritt and I. Oey. 2016. Effects of pulsed electric fields on selected quality attributes of beef outside flat (Biceps femoris). *In*: Jarm T. and P. Kramar [eds.]. 1st world congress on electroporation and pulsed electric fields in biology, Medicine and Food and Environmental Technologies (WC 2015), IFMBE Proceedings, vol. 51: 53rd ed. https://doi.org/10.1007/978- 981- 287-817-5_12.

Feizollahi, E., N.N. Misra and M.S. Roopesh. 2021. Factors influencing the antimicrobial efficacy of dielectric barrier discharge (DBD) atmospheric cold plasma (ACP) in food processing applications, critical reviews in food science and nutrition 61(4): 666–689. https://doi.org/10.1080/10408398.20 20.1743967.

Feng, X., C. Jo, K.C. Nam and D.U. Ahn. 2019. Impact of electron-beam irradiation on the quality characteristics of raw ground beef. Innovative Food Science and Emerging Technologies 54: 87–92. https://doi.org/10.1016/J.IFSET.2019.03.010.

Fernandez, A., E. Noriega and A. Thompson. 2013. Inactivation of salmonella enterica serovar typhimurium on fresh produce by cold atmospheric gas plasma technology. Food Microbiology 33(1): 24–29. https://doi.org/10.1016/j.fm.2012.08.007.

Fridman, A. 2009. Introduction to theoretical and applied plasma chemistry. plasma chemistry. pp. 1–11. Cambridge University Press, 2009 Downloaded from Cambridge Books Online. http://dx.doi.org/10.1017/CBO9780511546075.003.

Fröhling, A., J. Durek, U. Schnabel, J. Ehlbeck, J. Bolling and O. Schlüter. 2012. Indirect plasma treatment of fresh pork: decontamination efficiency and effects on quality attributes. Innovative Food Science and Emerging Technologies 16: 381–390. DOI: 10.1016/j.ifset.2012.09.001.

Garcia-Galicia, I.A., V.G. Gonzalez-Vacame, M. Huerta-Jimenez, L.M. Carrillo-Lopez, J.M. Tirado-Gallegos, R.A. Reyes-Villagrana et al. 2020. Ultrasound versus traditional ageing: physicochemical properties in beef longissimus lumborum. CyTA-Journal of Food 18(1): 675–682. https://doi.org/1 0.1080/19476337.2020.1834458.

Garcia-Galicia I.A., M. Huerta-Jimenez, L.M. Carrillo-Lopez, D. Sanchez-Aldana and A.D. Alarcon-Rojo. 2021. High-intensity ultrasound as a pre-treatment of pork sub-primals for further processing of meat. International Journal of Food Science and Technology 57(1): 466–480. https://doi.org/10.1111/ijfs.15427.

Ghosh, S., A. Gillis, K. Levkov, E. Vitkin and A. Golberg. 2020. Saving energy on meat air convection drying with pulsed electric field coupled to mechanical press water removal. Innovative Food Science and Emerging Technologies 66: 102509–102517. https://doi.org/10.1016/j.ifset.2020.102509.

Gök, V., S. Aktop, M. Özkan and O. Tomar. 2019. The effects of atmospheric cold plasma on inactivation of listeria monocytogenes and staphylococcus aureus and some quality characteristics of pastırma—a dry-cured beef product. Innovative Food Science and Emerging Technologies 56: 102188–102195. https://doi.org/10.1016/j.ifset.2019.102188.

Gomes, L.P., V.M.F. Paschoalin and E.M. Del Aguila. 2017. Chitosan nanoparticles: production, physicochemical characteristics and nutraceutical applications. Revista Virtual de Química 9(1): 387–409. https://doi.org/10.21577/1984-6835.20170022.

Gómez, B., P.E.S. Munekata, M. Gavahian, F.J. Barba, F.J. Martí-Quijal, T. Bolumar et al. 2019. Application of pulsed electric fields in meat and fish processing industries: an overview. Food Research International 123: 95–105 https://doi.org/10.1016/j.foodres.2019.04.047.

Gomez-Gomez, A., E. Brito-de la Fuente, C. Gallegos, J.V. Garcia-Perez and J. Benedito. 2021. Combined pulsed electric field and high-power ultrasound treatments for microbial inactivation in oil-in-water emulsions. Food Control, 130: 108348–108357. https://doi.org/10.1016/j.foodcont.2021.108348.

González-Cebrino, F., R. Durán, J. Delgado-Adámez, R. Contador and R. Ramírez. 2013. Changes after high-pressure processing on physicochemical parameters, bioactive compounds, and polyphenol oxidase activity of red flesh and peel plum purée. Innovative Food Science and Emerging Technologies 20: 34–41. DOI: 10.1016/j.ifset.2013.07.008.

Gonzalez-Gonzalez, L., A.D. Alarcon-Rojo, L. Carrillo-Lopez, I. Garcia-Galicia, M. Huerta-Jimenez and L. Paniwnyk. 2020. Does ultrasound equally improve the quality of beef? an insight into longissimus lumborum, infraspinatus and cleidooccipitalis. Meat Science. 160(2020): 107963–107973. https://doi.org/10.1016/j.meatsci.2019.107963.

Goosen, N. and G. F. Moolenar. 2008. Repair of UV damage in bacteria. DNA Repair, 7(3): 353–379. https://doi.org/10.1016/J.DNAREP.2007.09.002.

Grossi, A., J. Søltoft-Jensen, J.C. Knudsen, M. Christensen and V. Orlien. 2012. Reduction of salt in pork sausages by the addition of carrot fibre or potato starch and high pressure treatment. Meat Science 92(4): 481–489. DOI: 10.1016/j.meatsci.2012.05.015.

Grossi, A., K. Olsen, T. Bolumar, Å. Rinnan, L. H. Øgendal and V. Orlien. 2016. The effect of high pressure on the functional properties of pork myofibrillar proteins. Food Chemistry 196: 1005–1015. DOI: 10.1016/j.foodchem.2015.10.062.

Gul, O., F. Turker Saricaoglu, A. Besir, I. Atalar and F. Yazici. 2018. Effect of ultrasound treatment on the properties of nano-emulsion films obtained from hazelnut meal protein and clove essential oil. Ultrasonics Sonochemistry 41: 466–474. https://doi.org/10.1016/j.ultsonch.2017.10.011.

Han, M., P. Wang, X. Xu and G. Zhou. 2014. Low-field NMR study of heat-induced gelation of pork myofibrillar proteins and its relationship with microstructural characteristics. Food Research International 62: 1175–1182. DOI:10.1016/j.foodres.2014.05.062.

Hayes, J.E., C.R. Raines, D.A. DePasquale and C.N. Cutter. 2014. Consumer acceptability of high hydrostatic pressure (HHP)-treated ground beef patties. LWT - Food Science and Technology 56(1): 207–210. DOI: 10.1016/j.lwt.2013.11.014.

Hildrum, K.I., R. Rødbotten, M. Høy, J. Berg, B. Narum and J.P. Wold. 2009. Classification of different bovine muscles according to sensory characteristics and Warner Bratzler shear force. Meat Science 83(2): 302–307. DOI: 10.1016/j.meatsci.2009.05.016.

Horita, C.N., R.C. Baptista, M.Y.R. Caturla, J.M. Lorenzo, F.J. Barba and A.S. Sant'Ana. 2018. Combining reformulation, active packaging and non-thermal post-packa- ging decontamination technologies to increase the microbiological quality and safety of cooked ready-to-eat meat products. Trends in Food Science and Technology 72: 45–61. https://doi.org/10.1016/j.tifs.2017.12.003.

Hu, H., I.W.Y. Cheung, S. Pan and E.C.Y. Li-Chan. 2015. Effect of high intensity ultrasound on physicochemical and functional properties of aggregated soybean β-conglycinin and glycinin. Food Hydrocolloids 45: 102–110. DOI: 10.1016/j.foodhyd.2014.11.004.

Huang, H.W., C.P. Hsu and C.Y. Wang. 2020. Healthy expectations of high hydrostatic pressure treatment in food processing industry. Journal of Food and Drug Analysis 28(1): 1–13. DOI: 10.1016/j.jfda.2019.10.002.

Hugas, M., M. Garriga and J.M. Monfort. 2002. New mild technologies in meat processing: High Pressure as a Model Technology. Meat Science 62(3): 359–371. DOI: 10.1016/s0309-1740(02)00122-5.

Huq, T., K.D. Vu, B. Riedl, J. Bouchard and M. Lacroix. 2015. Synergistic effect of gamma (γ)-irradiation and microencapsulated antimicrobials against listeria monocytogenes on ready-to-eat (RTE) meat. Food Microbiology 46: 507–514. https://doi.org/10.1016/j.fm.2014.09.013.

Ikeuchi, Y., H. Tanji, K. Kim and A. Suzuki. 1992. Dynamic rheological measurements on heat-induced pressurized actomyosin gels. Journal of Agricultural and Food Chemistry 40(10): 1751–1755. DOI: 10.1021/jf00022a005.

Inguglia, E.S., Z. Zhang, B.K. Tiwari, J.P. Kerry and C.M. Burgess. 2017. Salt reduction strategies in processed meat products – a review. Trends in Food Science and Technology 59: 70–78. DOI: 10.1016/j.tifs.2016.10.016.

Islam, M.H., M.T.Y. Paul, O.S. Burheim and B.G. Pollet. 2019. Recent developments in the sonoelectrochemical synthesis of nanomaterials. Ultrasonics Sonochemistry 59: 104711–104718. https://doi.org/10.1016/j.ultsonch.2019.104711.

Jadhav, H.B., U.S. Annapure and R.R. Deshmukh. 2021. Non-thermal technologies for food processing. frontiers in nutrition 8: 657090–65103. DOI: 10.3389/fnut.2021.657090.

Jafari, S.Z., S. Jafarian, M. Hojjati and L. Najafian. 2022. Evaluation of antioxidant activity of nano- and microencapsulated rosemary (*Rosmarinus officinalis* L.) leaves extract in cress (*Lepidium sativum*) and basil (*Ocimum basilicum*) seed gums for enhancing oxidative stability of sunflower oil. Food Science and Nutrition (early view). 10: 2111–2119, https://doi.org/10.1002/fsn3.2827.

Jayathilakan, K., K. Sultana, K. Jalarama Reddy and M.C. Pandey. 2015. Radiation processing of meat and meat products-an overview. International Journal of New Technology and Research (IJNTR), 1(5): 5–12. www.ijntr.org ISSN: 2454–4116.

Ji, J., Z. Allahdad, E. Sarmast, S. Salmieri and M. Lacroix. 2022. Combined effects of microencapsulated essential oils and irradiation from gamma and X-ray sources on microbiological and physicochemical properties of dry fermented sausages during storage. LWT- Food Science and Technology 159: 113180–113188. https://doi.org/10.1016/J.LWT.2022.113180.

Jia, J., Y. Gai, W. Wang and Y. Zhao. 2016. Green synthesis of biocompatiable chitosan–graphene oxide hybrid nanosheet by ultrasonication method. Ultrasonics Sonochemistry 32: 300–306. https://doi.org/10.1016/j.ultsonch.2016.03.027.

Kadam, D.M., M. Thunga, S. Wang, M.R. Kessler, D. Grewell, B. Lamsal et al. 2013. Preparation and characterization of whey protein isolate films reinforced with porous silica coated titania nanoparticles. Journal of Food Engineering 117(1): 133–140. https://doi.org/10.1016/j.jfoodeng.2013.01.046.

Kang, D., X. Gao, Q. Ge, G. Zhou and W. Zhang. 2017a. Effects of ultrasound on the beef structure and water distribution during curing through protein degradation and modification. Ultrasonics Sonochemistry 38: 317–325. DOI: 10.1016/j.ultsonch.2017.03.026.

Kang, D., Y. Jiang, L. Xing, G. Zhou and W. Zhang. 2017b. Inactivation of *Escherichia coli* O157: H7 and bacillus cereus by power ultrasound during the curing processing in brining liquid and beef. Food Research International 102: 717–727. DOI: 10.1016/j.foodres.2017.09.062.

Kantono, K., N. Hamid, I. Oey, S. Wang, Y. Xu, Q. Ma et al. 2019. Physicochemical and sensory properties of beef muscles after pulsed electric field processing. Food Research Inernational, 121: 1–11. DOI: 10.1016/j.foodres.2019.03.020.

Keklik, N.M., K. Krishnamurthy and A. Demirci. 2012. Microbial decontamination of food by ultraviolet (UV) and pulsed UV light. pp. 344–369. Dencim A. and M.O. Ngadi [eds.]. *In*: Microbial Decontamination in the Food Industry: Novel Methods and Applications. Elsevier Ltd. https://doi.or g/10.1533/9780857095756.2.344.

Kentish, S., T.J. Wooster, M. Ashokkumar, S. Balachandran, R. Mawson and L. Simons. 2008. The use of ultrasonics for nanoemulsion preparation. Innovative Food Science and Emerging Technologies 9: 170–175. https://doi.org/10.1016/j.ifset.2007.07.005.

Khan, I., C.N. Tango, S. Miskeen, B.H. Lee and D.H. Oh. 2017. Hurdle technology: a novel approach for enhanced food quality and safety–a review. Food Control. 73: 1426–1444. https://doi.org/10.1016/J. FOODCONT.2016.11.010.

Khan, A.A., M.A. Randhawa, A. Carne, I.A. Mohamed Ahmed, F.Y. Al-Juhaimi, D. Barr et al. 2018. Effect of low and high pulsed electric field processing on macro and micro minerals in beef and chicken. Innovative Food Science and Emerging Technologies 45: 273–279. DOI: doi.org/10.1016/j. ifset.2017.11.012.

Kim, B., H.J. Yun, S. Jung, Y. Jung, H. Jung, W. Choe et al. 2011. Effect of atmospheric pressure plasma on inactivation of pathogens inoculated onto bacon using two different gas compositions. Food Microbiology 28 (1): 9–13. https://doi.org/10.1016/j.fm.2010.07.022.

Kim, H.J., H.I. Yong, S. Park, W. Choe and C. Jo. 2013. Effects of dielectric barrier discharge plasma on pathogen inactivation and the physicochemical and sensory characteristics of pork loin. Current Applied Physics 13(7): 1420–1425. https://doi.org/10.1016/j.cap.2013.04.021.

Kim, J.S., E.J. Lee, E.H. Choi and Y.J. Kim. 2014. Inactivation of *staphylococcus aureus* on the beef jerky by radio-frequency atmospheric pressure plasma discharge treatment. Innovative Food Science and Emerging Technologies 22: 124–130. https:// doi.org/10.1016/j.ifset.2013.12.012.

Knorr, D., M. Zenker, V. Heinz and D.U. Lee. 2004. Applications and potential of ultrasonics in food processing. Trends in Food Science and Technology 15(5): 261–266. DOI:10.1016/j.tifs.2003.12.001.

Koshani, R. and S. Mahdi Jafari. 2019. Ultrasound-assisted preparation of different nanocarriers loaded with food bioactive ingredients. Advances in Colloid and Interface Science 270: 123–146. https:// doi.org/10.1016/j.cis.2019.06.005.

Laroussi, M. 2002. Nonthermal decontamination of biological media by atmospheric-pressure plasmas: review, analysis, and prospects. IEEE Transactions on Plasma Science 30(4): 1409–1415. https:// doi.org/10.1109/TPS.2002.804220.

Laroussi, M. 2005. Low temperature plasma-based sterilization: overview and state-of-the-art. Plasma Processes and Polymers 5(2): 391–400. DOI: 10.1002/ppap.200400078.

Lee, E.-J., Y.-H. Kim, N.-H. Lee, S.I. Hong, K. Yamamoto and Y.J. Kim. 2011. The role of sarcoplasmic protein in hydrostatic pressure-induced myofibrillar protein denaturation. Meat Science 87(3): 219–222. DOI 10.1016/j.meatsci.2010.10.012.

Lee, H.J., H. Jung, W. Choe, J.S. Ham, J.H. Lee and C. Jo. 2011. Inactivation of listeria monocytogenes on agar and processed meat surfaces by atmospheric pressure plasma jets. Food Microbiology 28: 1468–1471. DOI: 10.1016/j.fm.2011.08.002.

Lee, E.-S., J.-H. Kim, S.M. Kang, B.-M. Kim and M.-H. Oh. 2022. Inhibitory effects of ultraviolet-C light and thermal treatment on four fungi isolated from pig slaughterhouses in korea. Journal of Animal Science and Technology 64(2): 343–352. https://doi.org/10.5187/jast.2022.e17.

Leng, Y., Y. Gao, W. Wang and Y. Zhao. 2015. Continuous supercritical solvothermal synthesis of TiO_2–pristine-graphene hybrid as the enhanced photocatalyst. The Journal of Supercritical Fluids 103: 115–121. https://doi.org/10.1016/j.supflu.2015.05.001.

Li, J., J. Cai and L. Fan. 2008. Effect of sonolysis on kinetics and physicochemical properties of treated chitosan. Journal of Applied Polymer Science 109: 2417–2425. https://doi.org/10.1002/app.28339.

Li, K., Z.-L. Kang, Y.-Y. Zhao, X.L Xu and G.H. Zhou. 2014. Use of high-intensity ultrasound to improve functional properties of batter suspensions prepared from PSE-like chicken breast meat. Food and Bioprocess Technology 7(12): 3466–3477. DOI: doi.org/10.1007/s11947-014-1358-y.

Li, X., Q. Wang, Y. Zhao, W. Wu, J. Chen and H. Meng. 2013. Green synthesis and photo-catalytic performances for ZnO-reduced graphene oxide nanocomposites. Journal of Colloid and Interface Science 411: 69–75. https://doi.org/10.1016/j.jcis.2013.08.050.

Li, Y., T. Feng, J. Sun, L. Guo, B. Wang, M. Huang et al. 2020. Physicochemical and microstructural attributes of marinated chicken breast influenced by breathing ultrasonic tumbling. Ultrasonics Sonochemistry 64: 105022–105032. https://doi.org/10.1016/j.ultsonch.2020.105022.

Lin, L., X. Liao and H. Cui. 2019. Cold plasma treated thyme essential oil/silk fibroin nanofibers against *Salmonella typhimurium* in poultry meat. Food Packag. Shelf Life 21: 100337–100344. https://doi.org/10.1016/j.fpsl.2019.100337.

Liu, N., Q. Zhu, X. Zeng, B. Yang, M. Liang, P. Hu et al. 2019. Influences of pulsed light-UV treatment on the storage period of dry-cured meat and shelf life prediction by ASLT method. Journal of Food Science and Technology 56(4): 1744–1756. https://doi.org/10.1007/s13197-019-03603-1.

Liu, R., S.-M. Zhao, B.-J. Xie and S.B. Xiong. 2011. Contribution of protein conformation and intermolecular bonds to fish and pork gelation properties. Food Hydrocolloids 25(5): 898–906. DOI: 10.1016/j.foodhyd.2010.08.016.

Lv, M., K. Mei, H. Zhang, D. Xu and W. Yang. 2018. Effects of electron beam irradiation on the biochemical properties and structure of myofibrillar protein from Tegillarca granosa meat. Food Chemistry 254: 64–69. https://doi.org/10.1016/J.FOODCHEM.2018.01.165.

Ma, F., C. Chen, L. Zheng, C. Zhou, K. Cai and Z. Han. 2013. Effect of high pressure processing on the gel properties of salt-soluble meat protein containing $CaCl_2$ and κ-carrageenan. Meat Science 95(1): 22–26. DOI: 10.1016/j.meatsci.2013.04.025.

Ma, N., B. Zhang, J. Liu, P. Zhang, Z. Li and Y. Luan. 2015. Green fabricated reduced graphene oxide: evaluation of its application as nano-carrier for pH-sensitive drug delivery. International Journal of Pharmaceutics 496(2): 984–992. https://doi.org/10.1016/j.ijpharm.2015.10.081.

Ma, Q., N. Hamid, I. Oey, K. Kantono, F. Faridnia, M. Yoo et al. 2016. Effect of chilled and freezing pre-treatments prior to pulsed electric field processing on volatile profile and sensory attributes of cooked lamb meats. Innovative Food Science and Emerging Technologies. 37(C):359–374. https://doi.org/10.1016/j. ifset.2016.04.009.

Maddinedi, S.B., B. Kumar Mandal, R. Vankayala, P. Kalluru, S. Kumar Tammina and H.A. Kiran Kumar. 2014. Casein mediated green synthesis and decoration of reduced graphene oxide. Spectrochimica Acta Part A: Molecular and Biomolecular Spectroscopy 126: 227–231. https://doi.org/10.1016/j. saa.2014.01.114.

Malinowska-Pańczyk, E. 2020. Can high hydrostatic pressure processing be the best way to preserve human milk?. Trends in Food Science and Technology 101: 133–138. DOI: 10.1016/j.tifs.2020.05.009.

Martínez, J.M., C. Delso, I. Álvarez and J. Raso. 2019. Pulsed electric field permeabilization and extraction of phycoerythrin from porphyridium cruentum. Algal Research 37: 51–56. https://doi.org/10.1016/J.ALGAL.2018.11.005.

McDonnell, C.K., J.G. Lyng, J.M. Arimi and P. Allen. 2014a. The acceleration of pork curing by power ultrasound: a pilot-scale production. Innovative Food Science and Emerging Technologies 26: 191–198. DOI: 10.1016/j.ifset.2014.05.004.

McDonnell, C.K., P. Allen, C. Morin and J.G. Lyng. 2014b. The effect of ultrasonic salting on protein and water–protein interactions in meat. Food Chemistry 147: 245–251. https://doi.org/10.1016/J.FOODCHEM.2013.09.125.

McLeod, A., K. Hovde Liland, J.E. Haugen, O. Sørheim, K.S. Myhrer and A.L. Holck. 2018. Chicken fillets subjected to UV-C and pulsed UV light: reduction of pathogenic and spoilage bacteria, and changes in sensory quality. Journal of Food Safety, 38(1): e12421-e12435. https://doi.org/10.1111/jfs.12421.

McSharry, S., L. Koolman, P. Whyte and D. Bolton. 2022. Inactivation of listeria monocytogenes and *salmonella typhimurium* in beef broth and on diced beef using an ultraviolet light emitting diode (UV-LED) system. LWT.Food Science and Technology 158: 113150–113154. https://doi. org/10.1016/j.lwt.2022.113150.

Miao, W., B.M. Nyaisaba, J.K. Kaddy, M. Chen, S. Hatab and S. Deng. 2020. Effect of cold atmospheric plasma on the physicochemical and functional properties of myofibrillar protein from alaska pollock (theragra chalcogramma). International Journal of Food Science and Technology 55(2): 517–525. https://doi.org/10.1111/ ijfs.14295.

Misra, N.N. and C. Jo. 2017. Applications of cold plasma technology for microbiological safety in meat industry. Trends in Food Science and Technology 64: 74–86. https://doi.org/10.1016/ j.tifs.2017.04.005.

Monsef, R., M. Ghiyasiyan-Arani, O. Amiri and M. Salavati-Niasari. 2020. Sonochemical synthesis, characterization and application of PrVO₄ nanostructures as an effective photocatalyst for discoloration of organic dye contaminants in wastewater. Ultrasonics Sonochemistry 61: 104822–104835. https://doi.org/10.1016/j.ultsonch.2019.104822.

Moutiq, R., N.N. Misra, A. Mendonça and K. Keener. 2020. In-package decontamination of chicken breast using cold plasma technology: microbial, quality and storage studies. Meat Science 159: 107942–1079950. https://doi.org/10.1016/j.meatsci.2019.107942.

Muroya, S., K.E. Neath, I. Nakajima, M. Shibata, K Ojima and K. Chikuni. 2012. Differences in mRNA expressions of calpains, calpastatin isoforms and calpain/cal- pastatin rations among bovine skeletal muscles. Animal Science Journal 83: 252–259. DOI: 10.1111/j.1740-0929.2011.00954.x.

Naderi, N., Y. Pouliot, J.D. House and A. Doyen. 2017. High hydrostatic pressure effect in extraction of 5-methyltetrahydrofolate (5-MTHF) from egg yolk and granule fractions. Innovative Food Science and Emerging Technologies 43:191–200. DOI: 10.1016/j.ifset.2017.08.009.

Nam, K.C., C. Jo and D.U. Ahn. 2016. Irradiation of meat and meat products. pp. 1–431. *In* : Cummins, E.J. and J. G. Lyng [eds.]. Emerging Technologies in Meat Processing: Production, Processing and Technology Wiley Blackwell. https://doi.org/10.1002/9781118350676.

Ninčević Grassino, A., J. Ostojić, V. Miletić, S. Djaković, T. Bosiljkov, Z. Zorić et al. 2020. Application of high hydrostatic pressure and ultrasound-assisted extractions as a novel approach for pectin and polyphenols recovery from tomato peel waste. Innovative Food Science and Emerging Technologies 64: 102424. DOI_ 10.1016/j.ifset.2020.102424.

Nuengmatcha, P., S. Chanthai, R. Mahachai and W.-C. Oh. 2016. Sonocatalytic performance of ZnO/ graphene/TiO₂ nanocomposite for degradation of dye pollutants (methylene blue, texbrite BAC-L, texbrite BBU-L and texbrite NFW-L) under ultrasonic irradiation. Dyes and Pigments 134: 487–497. https://doi.org/10.1016/j.dyepig.2016.08.006.

Nyaisaba, B.M., W. Miao, S. Hatab, A. Siloam, M. Chen and S. Deng. 2019. Effects of cold atmospheric plasma on squid proteases and gel properties of protein concentrate from squid (argentinus ilex) mantle. Food Chemistry 291: 68–76. DOI: 10.1016/j.foodchem.2019.04.012.

O'Dowd, L.P., J.M. Arimi, F. Noci, D.A. Cronin and J.G. Lyng. 2013. An assessment of the effect of pulsed electrical fields on tenderness and selected quality attributes of post rigor beef muscle. Meat Science, 93: 303–309. DOI: 10.1016/j.meatsci.2012.09.010.

O'Flynn, C.C., M.C. Cruz-Romero, D. Troy, A.M. Mullen and J.P. Kerry. 2014. The application of high-pressure treatment in the reduction of salt levels in reduced-phosphate breakfast sausages. Meat Science 96(3): 1266–1274. DOI: 10.1016/j.meatsci.2013.11.010.

Offer, G., P. Knight, R. Jeacocke, R. Almond, T. Cousins, J. Elsey et al. 1989. The structural basis of the water-holding, appearance and toughness of meat and meat products. Food Microstructure 8(1): 151–170. https://digitalcommons.usu.edu/foodmicrostructure/vol8/iss1/17.

Ojha, K.S., D.F. Keenan, A. Bright, J.P. Kerry and B.K. Tiwari. 2016. Ultrasound assisted diffusion of sodium salt replacer and effect on physicochemical properties of pork meat. International Journal of Food Science and Technology 51(1): 37–45. DOI: 10.1111/ijfs.13001.

Ojha, K.S., C.A. Perusello, C.Á. García, J.P. Kerry, D. Pando and B.K. Tiwari. 2017. Ultrasonic-assisted incorporation of nano-encapsulated omega-3 fatty acids to enhance the fatty acid profile of pork meat. Meat Science 132: 99–106. https://doi.org/10.1016/j.meatsci.2017.04.260.

Olmedilla-Alonso, B., F. Jiménez-Colmeneroa and F.J. Sánchez- Muniz. 2013. Development and assessment of healthy properties of meat and meat products designed as functional foods. Meat Science 95: 919–930. https://doi.org/10.1016/j.meatsci.2013.03.030.

Ostermeier, R., K. Hill, A. Dingis, S. Topfl and H. Jager. 2021. Influence of pulsed electric field (PEF) and ultrasound treatment on the frying behavior and quality of potato chips. Innovative Food Science and Emerging Technologies. 67(2021): 102553–102561. https://doi.org/10.1016/j.ifset.2020.102553.

Ouali, A. 1991. Animal biotechnology and the quality of heat production. pp. 85–105. *In:* Fiems, L.O. B.G. Cottyn and D. I. Demeyer [eds.]. Amsterdam: Elsevier, Sci., Pub. B.V.

Owusu-Ansah, P., X. Yu, R. Osae, C. Zhou, R. Zhang, A.T. Mustapha et al. 2020. Optimization of thermosonication on *Bacillus cereus* from pork: effects on inactivation and physicochemical properties. Journal of Food Process Engineering 43: e13401–e13413. https://doi.org/10.1016/j. ifset.2020.102553.

Ozuna, C., A. Puig, J.V. García-Pérez, A. Mulet and J.A. Cárcel. 2013. Influence of high intensity ultrasound application on mass transport, microstructure and textural properties of pork meat (*Longissimus dorsi*) brined at different NaCl concentrations. Journal of Food Engineering 119(1): 84–93. DOI: 10.1016/j.jfoodeng.2013.05.016.

O'Sullivan, J., B. Murray, C. Flynn and I. Norton. 2016. The effect of ultrasound treatment on the structural, physical and emulsifying properties of animal and vegetable proteins. Food Hydrocolloids 53: 141–154. https://doi.org/10.1016/j.foodhyd.2015.02.009.

Park, S.Y. and S.-Do. Ha. 2015. Ultraviolet-C radiation on the fresh chicken breast: inactivation of major foodborne viruses and changes in physicochemical and sensory qualities of product. Food and Bioprocess Technology, 8(4): 895–906. DOI: 10.1007/s11947-014-1452-1.

Patange, A., D. Boehm, C. Bueno-Ferrer, P.J. Cullen and P. Bourke. 2017. Controlling brochothrix thermosphacta as a spoilage risk using in-package atmospheric cold plasma. Food Microbiology 66: 48–54. https://doi.org/ 10.1016/j.fm.2017.04.002.

Patil, S., P. Bourke and P.J. Cullen. 2016. Chapter 6. principles of nonthermal plasma decontamination. pp: 143–177. *In*: [eds.] Misra, N.N., O. Schlüter and P.J. Cullen. Cold Plasma in Food and Agriculture, Academic Press. https://doi.org/10.1016/B978-0-12-801365-6.00006-8.

Peña-González, E.M., A.D. Alarcón-Rojo, A. Rentería, I. García, E. Santellano, A. Quintero et al. 2017. Quality and sensory profile of ultrasound-treated beef. Italian Journal of Food Science 29(3): 463–475. DOI: 10.14674/1120-1770/ijfs.v604.

Perry, D., J.M. Thompson, I.H. Hwang, A. Butchers and A.F. Egan. 2001. Relationship between objective measurements and taste panel assessment of beef quality. Australian Journal of Experimental Agriculture 41(7):981–989. DOI: 10.1071/EA00023.

Pignata, C., D.D'Angelo, E. Fea and G.A. Gilli. 2017. A review on microbiological decontamination of fresh produce with nonthermal plasma. J. Appl. Microbiol. 122: 1438–1455. DOI: 10.1111/ jam.13412.

Pinton, M.B., L.P. Correa, M.M.X. Facchi, R.T. Heck, Y.S.V. Leães, A.J. Cichoski et al. 2019. Ultrasound: a new approach to reduce phosphate content of meat emulsions. Meat Science 152: 88–95. DOI: 10.1016/j.meatsci.2019.02.010.

Priyadarshi, R., S.-M. Kim and J.-W. Rhim. 2021. Carboxymethyl cellulose-based multifunctional film combined with zinc oxide nanoparticles and grape seed extract for the preservation of high-fat meat products. Sustainable Materials and Technologies 29: e00325–e00337. https://doi.org/10.1016/j. susmat.2021.e00325.

Purslow, P.P. 2005. Intramuscular connective tissue and its role in meat quality. Meat Science 70(3): 435–447. DOI: 10.1016/j.meatsci.2004.06.028.

Ramachandraiah, K., S.G. Han and K.B. Chin. 2015. Nanotechnology in meat processing and packaging: potential applications-a review. Asian-Australasian Journal of Animal Science 28(2): 290–302. https://doi.org/10.5713/ajas.14.0607.

Rashidaie Abandansarie, S.S., P. Ariaii and M. Charmchian Langerodi. 2019. Effects of encapsulated rosemary extract on oxidative and microbiological stability of beef meat during refrigerated storage. Food Science and Nutrition 7(12): 3969–3978. https://doi.org/10.1002/fsn3.1258.

Rendueles, E., M.K. Omer, O. Alvseike, C. Alonso-Calleja, R. Capita and M. Prieto. 2011. Microbiological food safety assessment of high hydrostatic pressure processing: a review. LWT-Food Science and Technology 44(5): 1251–1260. DOI: 10.1016/j.lwt.2010.11.001.

Ricci, A., G.P. Parpinello and A. Versari. 2018. Recent advances and applications of pulsed electric fields (PEF) to improve poly- phenol extraction and color release during red winemaking. Beverages, 4(1): 18. https://doi.org/10.3390/beverages4010018.

Rios-Corripio, G., J. Welti-Chanes, V. Rodríguez-Martínez and J.Á. Guerrero-Beltrán. 2020. Influence of high hydrostatic pressure processing on physicochemical characteristics of a fermented pomegranate (*Punica granatum* L.) beverage. Innovative Food Science and Emerging Technologies 59:102249. DOI: 10.1016/j.ifset.2019.102249.

Rodrigues, D.C., A.P. Cunha, E.S. Brito, H.M.C. Azeredo and M.I. Gallão. 2016. Mesquite seed gum and palm fruit oil emulsion edible films: Influence of oil content and sonication. Food Hydrocolloids 56: 227–235. https://doi.org/10.1016/j.foodhyd.2015.12.018.

Rossi, A.P., D.L. Kalschne, A.P. Iglikowski Byler, E.L. de Moraes Flores, O. Donizeti Leite, D. dos Santos et al. 2021. Effect of ultrasound and chlorine dioxide on *Salmonella Typhimurium* and *Escherichia coli* inactivation in poultry chiller tank water. Ultrasonics Sonochemistry, 80: 105815. https://doi.org/10.1016/j.ultsonch.2021.105815.

Rossow, M., M. Ludewig and P.G. Braun. 2018. Effect of cold atmospheric pressure plasma treatment on inactivation of Campylobacter Jejuni on chicken skin and breast fillet. LWT–Food Science and Technology 91: 265–270. https://doi.org/ 10.1016/j.lwt.2018.01.052.

Royintarat, T., E.H. Choi, D. Boonyawan, P. Seesuriyachan and W. Wattanutchariya. 2020. Chemical-free and synergistic interaction of ultrasound combined with plasma-activated water (PAW) to enhance microbial inactivation in chicken meat and skin. Scientific Reports 10(1): 1559–1572. https://doi.org/10.1038/s41598-020-58199-w.

Sale, A. and W. Hamilton. 1967. Effects of high electric fields on microorganisms: i. killing of bacteria and yeasts. Biochimica et Biophysica Acta (BBA)-General Subjects 148: 781–788. DOI: 10.1016/0304-4165(67)90052-9.

Salvia-Trujillo, L., A. Rojas-Graü, R. Soliva-Fortuny and O. Martín-Belloso. 2013. Physicochemical characterization of lemongrass essential oil–alginate nanoemulsions: effect of ultrasound processing parameters. Food and Bioprocess Technology 6: 2439–2446. https://doi.org/10.1007/s11947-012-0881-y.

Savadkoohi, S., H. Hoogenkamp, K. Shamsi and A. Farahnaky. 2014. Color, sensory and textural attributes of beef frankfurter, beef ham and meat-free sausage containing tomato pomace. Meat Science 97(4): 410–418. DOI: 10.1016/j.meatsci.2014.03.017.

Sebranek, J.G. 2022. Irradiation of meat and meat products. In Reference Module in Food Science. Elsevier. https://doi.org/10.1016/B978-0-323-85125-1.00002-8.

Shebs, E.L., F.M. Giotto and A.S. de Mello. 2022. Effects of MS bacteriophages, ultraviolet light, and organic acid applications on beef trim contaminated with STEC O157:H7 and the "Big Six" serotypes after a simulated High Event Period Scenario. Meat Science 188: 108783–108788. https://doi.org/10.1016/j.meatsci.2022.108783.

Shi, H., F. Shahidi, J. Wang, Y. Huang, Y. Zou, W. Xu et al. 2021. Techniques for postmortem tenderisation in meat processing: effectiveness, application and possible mechanisms. Food Production. Processing and Nutrition 3(21): 1.26. https://doi.org/10.1186/s43014-021-00062-0.

Silva, D.R.G., G.B.S. Haddad, A.P. de Moura, P.M. de Souza, A.L.S. Ramos, D.L. Hopkins et al. 2021. Safe cured meat using gamma radiation: effects on spores of clostridium sporogenes and technological and sensorial characteristics of low nitrite cooked ham. LWT-Food Science and Technology 137: 110392–110399. https://doi.org/10.1016/j.lwt.2020.110392.

Silva, H.L.A., M.P. Costa, B.S. Frasao, E.F. Mesquita, S.C.R.P. Mello, C.A. Conte-Junior et al. 2015. Efficacy of ultraviolet-c light to eliminate staphylococcus aureus on precooked shredded bullfrog back meat. Journal of Food Safety, 35(3): 318–323. https://doi.org/https://doi.org/10.1111/jfs.12178.

Singla, M. and N. Sit. 2021. Application of ultrasound in combination with other technologies in food processing: a review. Ultrasonics Sonochemistry, 73: 105506–105518. DOI: 10.1016/j.ultsonch.2021.105506.

Siró, I., C. Vén, C. Balla, G. Jónás, I. Zeke and L. Friedrich. 2009. Application of an ultrasonic assisted curing technique for improving the diffusion of sodium chloride in porcine meat. Journal of Food Engineering 91(2): 353–362. DOI: 10.1016/j.jfoodeng.2008.09.015.

Soltani, R.D.C., M. Mashayekhi, M. Naderi, G. Boczkaj, S. Jorfi and M. Safari. 2019. Sonocatalytic degradation of tetracycline antibiotic using zinc oxide nanostructures loaded on nano-cellulose from waste straw as nanosonocatalyst. Ultrasonics Sonochemistry 55: 117–124. https://doi.org/10.1016/j.ultsonch.2019.03.009.

Stadnik, J., Z.J. Dolatowski and H.M. Baranowska. 2008. Effect of ultrasound treatment on water holding properties and microstructure of beef (*M. semimembranosus*) during ageing. LWT-Food Science and Technology 41(10): 2151–2158. DOI: 10.1016/j.lwt.2007.12.003.

Stadnik, J. and Z.J Dolatowski. 2011. Influence of sonication on warner-bratzler shear force, colour and myoglobin of beef (*M. semimembranosus*). European Food Research and Technology 233(4): 553. DOI: 10.1007/s00217-011-1550-5.

Stefanova, R., S. Toshkov, N.v. Vasilev, N.G. Vassilev and I.N. Marekov. 2011. Effect of gamma-ray irradiation on the fatty acid profile of irradiated beef meat. Food Chemistry 127(2): 461–466. https://doi.org/10.1016/j.foodchem.2010.12.155.

Stratakos, A.C. and I.R. Grant. 2019. Evaluation of the efficacy of multiple physical, biological and natural antimicrobial interventions for control of pathogenic escherichia coli on beef. Food Microbiology 76: 209–218. https://doi.org/10.1016/j.fm.2018.05.011.

Suwal, S., V. Perreault, A. Marciniak, É. Tamigneaux, É. Deslandes, L. Bazinet et al. 2019. Effects of high hydrostatic pressure and polysaccharidases on the extraction of antioxidant compounds from red macroalgae, palmaria palmata and solieria chordalis. Journal of Food Engineering 252: 53–59. DOI: 10.1016/j.jfoodeng.2019.02.014.

Suwandy, V., A. Carne, R. van de Ven, A.D.A. Bekhit and D.L. Hopkins. 2015a. Effect of pulsed electric field on the proteolysis of cold boned beef *M. longissimus lumborum* and *M. semimembranosus*. Meat Science, 100: 222–226. DOI: 0.1016/j.meatsci.2014.10.011.

Suwandy, V., A. Carne, R. van de Ven, A.D.A. Bekhit and D.L. Hopkins. 2015b. Effect of pulsed electric field treatment on hot-boned muscles of different potential tenderness. Meat Science 105: 25–31. DOI: 10.1016/j.meatsci.2015.02.009.

Suwandy, V., A. Carne, R. van de Ven, A.D.A. Bekhit and D.L. Hopkins. 2015c. Effect of repeated pulsed electric field treatment on the quality of cold boned beef loins and topsides. Food and Bioprocess Technology 8(6): 1218– 1228. DOI: 10.1007/s11947-015-1485-0.

Tamm, A., T. Bolumar, B. Bajovic and S. Toepfl. 2016. Salt (NaCl) reduction in cooked ham by a combined approach of high pressure treatment and the salt replacer KCl. Innovative Food Science and Emerging Technologies 36: 294–302. DOI: 10.1016/j.ifset.2016.07.010.

Terefe, N.S., A.L. Sikes and P. Juliano. 2016. Ultrasound for structural modification of food products. pp. 209–230. *In*: Knoerzer, K.P. Juliano and G.B. T.-I. F.P.T. Smithers [eds.]. Woodhead Publishing Series in Food Science, Technology and Nutrition. Woodhead Publishing.

Toepfl, S. 2006. Pulsed electric fields (PEF) for permeabilization of cell membranes in food- and bioprocessing: applications, process and equipment design and cost analysis. PhD thesis, University of Technology, Berlin. https://doi.org/1 0.14279/depositonce-1441.

Toyokawa, Y., Y. Yagyu, T. Misawa and A. Sakudo. 2017. A new roller conveyer system of non-thermal gas plasma as a potential control measure of plant pathogenic bacteria in primary food production. Food Control 72: 62–72. https://doi.org/ 10.1016/j.foodcont.2016.07.031.

Ucar, Y., Z. Ceylan, M. Durmus, O. Tomar and T. Cetinkaya. 2021. Application of cold plasma technology in the food industry and its combination with other emerging technologies. Trends in Food Science and Technology 114: 355–371. DOI: 10.1016/j.tifs.2021.06.004.

Ünver, A. 2016. Applications of ultrasound in food processing. Green Chemistry and Technology Letters 2(3):121–126. DOI: 10.18510/gctl.2016.231.

Valenzuela C., I.A. Garcia-Galicia, L. Paniwnyk and A.D. Alarcon-Rojo. 2021. Physicochemical characteristics and shelf-life of beef treated with high intensity ultrasound. Journal of Food Processing and Preservation, 45(4): 2021. e15350–e15360. https://doi.org/10.1111/jfpp.15350.

Vanga, S.K., J. Wang, S. Jayaram and V. Raghavan. 2021. Effects of pulsed electric fields and ultrasound processing on proteins and enzymes: a review. Processes 9(2021): 722–737. https://doi.org/10.3390/pr9040722.

Van Loey, I.A., C. Smout and M. Hendrickx. 2003. High hydrostatic pressure technology in food preservation. *In:* Food Preservation Techniques. Woodhead Publishing Limited.

Visy, A., G. Jónás, D. Szakos, Z. Horváth-Mezőfi, K.I. Hidas, A. Barkó et al. 2021. Evaluation of ultrasound and microbubbles effect on pork meat during brining process. Ultrasonics Sonochemistry 75: 105589–105596. https://doi.org/10.1016/j.ultsonch.2021.105589.

Walia, N., N. Dasgupta, S. Ranjan, L. Chen and C. Ramalingam. 2017. Fish oil-based vitamin D nanoencapsulation by ultrasonication and bioaccessibility analysis in simulated gastro-intestinal tract. Ultrasonics Sonochemistry 39: 623–635. https://doi.org/10.1016/j.ultsonch.2017.05.021.

Wan, Z., Y. Chen, S.K. Pankaj and K.M. Keener. 2017. High voltage atmospheric cold plasma treatment of refrigerated chicken eggs for control of salmonella enteritidis contamination on eggshell. LWT - Food Science and Technology 76: 124–130. https://doi.org/10.1016/j.lwt.2016.10.051.

Wang, A., D. Kang, W. Zhang, C. Zhang, Y Zou and G. Zhou. 2018. Changes in calpain activity, protein degradation and microstructure of beef M. semitendinosus by the application of ultrasound. Food Chemistry 245: 724–730. DOI: 10.1016/j.foodchem.2017.12.003.

Wang, X., Z. Wang, H. Zhuang, M.M. Nasiru, Y. Yuan, J. Zhang et al. 2021. Changes in colour, myoglobin, and lipid oxidation in beef patties treated by dielectric barrier discharge cold plasma during storage. Meat Science 176: 108456–108464. https://doi.org/10.1016/j.meatsci.2021.108456.

Warner, R.D., C.K. McDonnell, A.E.D. Bekhit, J. Claus, R. Vaskoska, A. Sikes et al. 2017. Systematic review of emerging and innovative technologies for meat tenderisation. Meat Science 132: 72–89. DOI: 10.1016/j.meatsci.2017.04.241.

[WHO]. 2016. World health organization. Ionizing radiation, health effects and protective measures. Ionizing Radiation, Health Effects and Protective Measures. https://www.who.int/es/news-room/fact-sheets/detail/ionizing-radiation-health-effects-and-protective-measures.

Wiktor, A., E. Gondek, E. Jakubczyk, M. Dadan, M. Nowacka, K. Rybak et al. 2018. Acoustic and mechanical properties of carrot tissue treated by pulsed electric field, ultrasound and combination of both. Journal of Food Engineering, 238: 12–21. https://doi.org/10.1016/j.jfoodeng.2018.06.001.

Xiong, G.F. Xiaoyi, P. Dongmei, Q. Jun, X. Xinglian and J. Xuejuan. 2020. Influence of ultrasound-assisted sodium bicarbonate marination on the curing efficiency of chicken breast meat. Ultrasonics Sonochemistry, 60 (2019): 104808–104814. https://doi.org/10.1016/j.ultsonch.2019.104808.

Yazdanparast, S., A. Benvidi, S. Abbasi and M. Rezaeinasab. 2019. Enzyme-based ultrasensitive electrochemical biosensor using poly(l-aspartic acid)/MWCNT bio-nanocomposite for xanthine detection: a meat freshness marker. Microchemical Journal 149: 104000–104009. https://doi.org/10.1016/j.microc.2019.104000.

Yeh, Y., de Moura, F.H. van den Broek, K. and A.S. de Mello. 2018. Effect of ultraviolet light, organic acids, and bacteriophage on Salmonella populations in ground beef. Meat Science 139: 44–48. https://doi.org/10.1016/J.MEATSCI.2018.01.007.

Yong, H.I., H.J. Kim, S. Park, W. Choe, M.W. Oh and C. Jo. 2014. Evaluation of the treatment of both sides of raw chicken breasts with an atmospheric pressure plasma jet for the inactivation of *Escherichia coli*. Foodborne Pathogens and Diseases 11: 652–657. DOI: 10.1089/fpd.2013.1718.

Yong, H.I., H. Lee, S. Park, J. Park, W. Choe, S. Jung et al. 2017. Flexible thin-layer plasma inactivation of bacteria and mold survival in beef jerky packaging and its effects on the meat's physicochemical properties. Meat Science 123: 151–156. https://doi.org/10.1016/j.meatsci.2016.09.016.

Zhang, Z., J. M. Regenstein, P. Zhou and Y. Yang. 2017. Effects of high intensity ultrasound modification on physicochemical property and water in myofibrillar protein gel. Ultrasonics Sonochemistry 34: 960–967. DOI: 10.1016/j.ultsonch.2016.08.008.

Zhang, K., C.A. Perussello, V. Milosavljević, P.J. Cullen, D.-W. Sun and B.K. Tiwari. 2019. Diagnostics of plasma reactive species and induced chemistry of plasma treated foods. Critical Reviews in Food Science and Nutrition 59(5): 812–825. https://doi.org/ 10.1080/10408398.2018.1564731.

Zhang, C., Q. Sun, Q. Chen, B. Kong and X. Diao. 2020. Effects of ultrasound-assisted immersion freezing on the muscle quality and physicochemical properties of chicken breast. International Journal of Refrigeration 117: 247–255. https://doi.org/10.1016/j.ijrefrig.2020.05.006.

Zhang, F., H. Zhao, C. Cao, B. Kong, X. Xia and Q. Liu. 2021. Application of temperature-controlled ultrasound treatment and its potential to reduce phosphate content in frankfurter-type sausages by 50%. Ultrasonics Sonochemistry, 71 (2021): 105379. DOI: 10.1016/j.ultsonch.2020.105379.

Zheng, H., M. Han, M. Yang, X. Xu and G. Zhou. 2018. The effect of pressure-assisted heating on the water holding capacity of chicken batters. Innovative Food Science and Emerging Technologies 45: 280–286. DOI: 10.1016/j.ifset.2017.11.011.

Zouelm, F., K. Abhari, H. Hosseini and M. Khani. 2019. The effects of cold plasma application on quality and chemical spoilage of pacific white shrimp (*Litopenaeus vannamei*) during refrigerated storage. Journal of Aquatic Food Product Technology 28 (6): 624–636. https://doi.org/10.1080/10498850.2019.1627452.

Index

For Product Safety Concerns and Information please contact our EU
representative GPSR@taylorandfrancis.com
Taylor & Francis Verlag GmbH, Kaufingerstraße 24, 80331 München, Germany